Treatise on Concrete

louis i. kahn
exposed concrete and hollow stones
1949-1959

Other books by Roberto Gargiani

Auguste Perret (1993), Electa

Il principio del rivestimento (1994), with G. Fanelli, Laterza

Storia dell'architettura contemporanea (1998), with G. Fanelli, Laterza

Principi e costruzione nell'architettura italiana del Quattrocento (2003), Laterza

Archizoom Associati, 1966-1974 (2007), Electa

Editor of *La colonne. Nouvelle histoire de la construction* (2008), Presses polytechniques et universitaires romandes

Rem Koolhaas/OMA. The Construction of Merveilles (2008), EPFL Press

Superstudio (2010), with B. Lampariello, Laterza

Le Corbusier. Béton Brut and Ineffable Space, 1940-1965, Surface Materials and Psychophysiology of Vision (2011), with A. Rosellini, EPFL Press (German translation, 2014, Edition Detail)

Editor of *L'architrave, le plancher, la plate-forme. Nouvelle histoire de la construction* (2012), Presses polytechniques et universitaires romandes

Concrete, From Archeology to Invention 1700-1769 (2013), EPFL Press

Treatise on Concrete
Directed by Roberto Gargiani

roberto gargiani

louis i. kahn exposed concrete and hollow stones 1949-1959

Translated from the Italian by Stephen Piccolo

EPFL Press

A Swiss academic publisher
distributed by Routledge

EPFL Press

The EPFL Press is the English-language imprint of the Foundation of the Presses polytechniques et universitaires romandes (PPUR). The PPUR publishes mainly works of teaching and research of the Ecole polytechnique fédérale de Lausanne (EPFL), of universities and other institutions of higher education.

Presses polytechniques
et universitaires romandes
EPFL – Rolex Learning Center
Post office box 119
CH-1015 Lausanne, Switzerland
E-mail: ppur@epfl.ch
Phone: +41 (0)21 693 2130
Fax: +41 (0)21 693 4027

www.epflpress.org

ISBN 978-2-940222-76-6 (EPFL Press)
Printed in Spain

Routledge
Taylor & Francis Group

Taylor & Francis Group Ltd
2 Park Square, Milton Park
Abingdon, Oxford, OX14 4RN, UK

Routledge is an imprint of Taylor & Francis Group, An informa business.

Simultaneously published in the USA and Canada by Routledge,
711 Third Avenue, New York 10017

www.routledge.com

Library of Congress
Cataloging-in-Publication Data
A catalog record for this book is available from the Library of Congress.

The graphical layout of the book and the cover design are by Paola Ranzini Pallavicini of Studio Pagina Milan, Italy. The photolithographic services were provided by Walter Bassani, AG-MEDIA, Milan Italy.

This work was translated from the Italian by Stephen Piccolo (www.transiting.eu).

Research for this publication has been financed by the European Research Council grant as part of project ERC2008-AdG, Proposal No. 230394, "The surfaces of cement and reinforced concrete: A history of the formwork and processing of the surface."

EPFL
ÉCOLE POLYTECHNIQUE
FÉDÉRALE DE LAUSANNE

Series Editors for the publication program in Architecture are Professors Jacques Lucan and Luca Ortelli.

The publisher and authors express their thanks to the Ecole polytechnique fédérale de Lausanne (EPFL) for its generous support towards the publication of this book.

The author is grateful to the ENAC faculty of the EPFL, who provided additional support.

Treatise on Concrete

Directed by Roberto Gargiani

The volumes of the *Treatise on Concrete* intend to offer a new and documented vision of the evolution of construction techniques and of the creative potential and formal expressions of concrete, from antiquity to the present time. Each book in the series will be based on specific archival research; the analysis of scientific, technical and iconographic original sources; and the investigation into documents made available by companies, engineers, architects and artists.
The full spectrum of the use of concrete will be covered, including the building of foundations, ports, moles, bridges, vaults, domes, fortresses, bunkers, houses and monuments, as well as the manufacture of artificial stones and sculptures. In addition, the volumes will address the theoretical issues of *béton brut*, Brutalism and the true nature of matter.
All volumes focus on the techniques and materials for the manufacture of formwork, from wood to fabric; on the meaning of the imprint; on the surface processing of concrete by hand or with machine; and on the composition of the mixture.
The series of books aims to provide the scientific community with an unique and updated reference on the historical and contemporary uses of concrete, which will open a new chapter in the knowledge of this crucial material for architectural construction, as well as for masterpieces of engineering and art.

Volumes published:
Concrete, From Archeology to Invention, 1700–1769

Volumes in preparation:
Louis I. Kahn, Towards the Zero Degree of Concrete, 1960-1974
Tadao Ando, Concrete Spaces
Mnemosyne in Concrete
The Concrete of Pier Luigi Nervi
Research on the Roman Cement, 1770-1779
The Origins of Modern Concrete Building in Germany, 1800-1877
Concrete Sculpture, from 1950 to Today: Visions of a Material
Contemporary Ornament in Concrete
Bared Concrete of Natural and Artificial Pozzolanas, 1780-1817
Formless: Beyond Formwork
The Concrete of Perret and Pupils

Table of Contents

chapter one
the discovery of exposed concrete

Monumentality, "away from practice"

Louis Isadore Kahn writes his first important essay in 1944, *Monumentality*, for the anthology compiled by Paul Zucker, *New Architecture and City Planning, A Symposium*, where it appears in the section on *The Problem of a New Monumentality.*[1] The question of the civic meaning of monumentality discussed in Zucker's book is approached above all in the projects and studies of Kahn with Oscar Stonorov for Philadelphia in the early 1940s. Kahn's recently completed constructions, or those still in progress at the time of the symposium, are residential complexes designed with Stonorov and with George Howe, which make use of materials like "cinder blocks," "concrete," "hollow tile" as well as wood.[2]

[1] Louis I. Kahn, *Monumentality*, in, Paul Zucker, ed., *New Architecture and City Planning, A Symposium*, New York, Philosophical Library, 1944, pp. 577-588. For a general overview of Kahn's work see the fundamental essays by David B. Brownlee, David G. De Long, *Louis I. Kahn. In the Realm of Architecture*, New York, Rizzoli, 1991; and Robert McCarter, *Louis I. Kahn*, London, Phaidon, 2005. See also the essay destined to open up a new chapter in American historiography about Kahn, by Sarah Williams Goldhagen, *Louis Kahn's Situated Modernism*, New Haven, London, Yale University, 2001. Also on Kahn's work, see Mateo Kries, Jochen Eisenbrand, Stanislaus von Moos, ed., *Louis Kahn. The Power of Architecture*, Weil am Rhein, Vitra Design Museum, 2012. On Kahn's construction processes and structures see the fundamental essay by Thomas Leslie, *Louis I. Kahn. Building Art, Building Science*, New York, G. Braziller, Inc., 2005; and Réjean Legault, *Louis Kahn and the Life of Materials*, in, Kries, Eisenbrand, Von Moos, *op. cit.*, pp. 219-233. For a profile on the evolutions of Kahn's concrete see also Robert McCarter, *Louis I. Kahn and the Nature of Concrete*, typewritten manuscript, Docomomo United Kingdom, The Concrete Centre, *Masters of Concrete*, Dennis Sharp, ed., 2007.

[2] For the list of Kahn's works, with indications of materials see Louis I. Kahn, letter to Charles H. Brunner Attorney Jr., 5 October 1949, LIK, 030.II.A.61.25.

1. "Tubular forms" for a Civic Center project illustrating *Monumentality* (P. Zucker, ed., *New Architecture and City Planning, A Symposium*, New York, Philosophical Library, 1944, p. 588).

2. D'Arcy W. Thompson, *On Growth and Form*, Cambridge, Cambridge University Press, 1944, p. 981.

3. Study drawings illustrating *Monumentality* (P. Zucker, ed., *New Architecture and City Planning, A Symposium*, New York, Philosophical Library, 1944, p. 582).

4. L. I. Kahn, O. Stonorv, Project for Parasol House type, 1944 (LIK, 030.I.A.156.9.1).

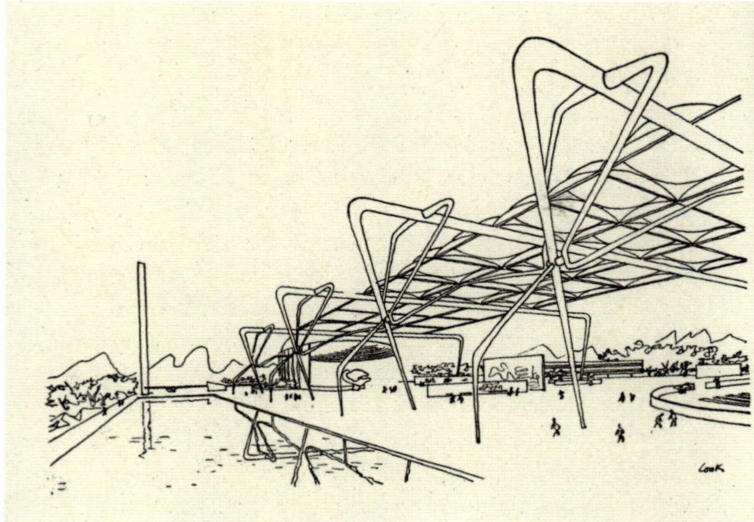

In *Monumentality* Kahn focuses his reasoning on contemporary materials, their qualities and structures. It is the necessary theoretical prelude for what was later to become one of the main formal qualities of his architecture: exposed reinforced concrete. The start of the essay, on gold, "a beautiful material" that "belongs to the sculptor,"[3] introduces the destruction of the principle according to which fine materials proven in historic examples are considered necessary to achieve monumentality in architecture. The most significant passages of his paper have to do with the as yet unexplored expressive potential of contemporary materials, from steel to reinforced concrete, plastic and structural glass, applied and known above all for their physical and technical characteristics. What Kahn has to say about Greek, Roman or Gothic architecture, regarding the use of materials in relation to their strength and the structural systems of those various eras, reveals a vision centered on the evolution of construction, along the lines of Auguste Choisy – the name of the author of the *Histoire de l'Architecture* is mentioned in Kahn's contribution. The objective of the reference to Choisy's theoretical approach is to define the concept of a contemporary monumentality starting with the potential of new materials.

To be able to give form to monumentality through structure alone, Kahn has to disrupt any type of section encoded in construction manuals by the rules of statics and the industrial manufacture of individual members. Against the backdrop of research on a new idea of structure, he underscores the creative potential of welding, because it permits replacement of the simple assembly of I-beams with the continuous structures of "tubular forms," developed case by case to produce the "graceful forms" determined by "stress diagrams."[4] The passage from the I-beam to tubular sections avoids the detail of the assembly of pieces, that "joint" which Kahn considers to be made in a "complex and ugly" way in "common practice."[5] One of the essential criteria of the quality of fabrication of his concrete would come precisely from the obsessive control of the making of joints, in all their various manifestations.

It is worth noting, given the success of the book during the course of the 1940s, that *On Growth and Form*, written by D'Arcy Wentworth

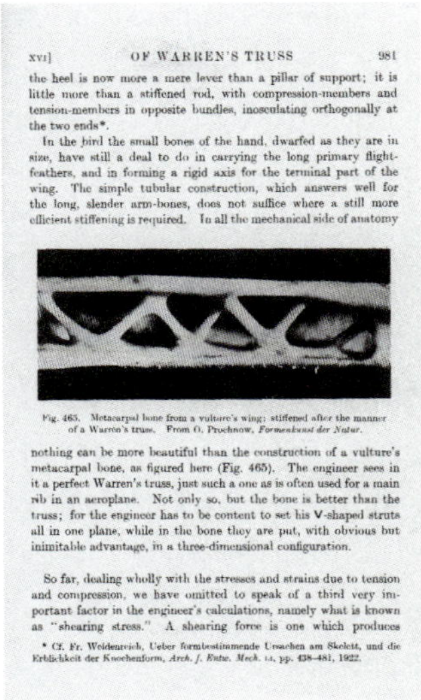

xvi] OF WARREN'S TRUSS 981

the heel is now more a mere lever than a pillar of support; it is little more than a stiffened rod, with compression-members and tension-members in opposite bundles, inosculating orthogonally at the two ends*.

In the bird the small bones of the hand, dwarfed as they are in size, have still a deal to do in carrying the long primary flight-feathers, and in forming a rigid axis for the terminal part of the wing. The simple tubular construction, which answers well for the long, slender arm-bones, does not suffice where a still more efficient stiffening is required. In all the mechanical side of anatomy

Fig. 465. Metacarpal bone from a vulture's wing; stiffened after the manner of a Warren's truss. From O. Prochnow, *Formenkunst der Natur*.

nothing can be more beautiful than the construction of a vulture's metacarpal bone, as figured here (Fig. 465). The engineer sees in it a perfect Warren's truss, just such a one as is often used for a main rib in an aeroplane. Not only so, but the bone is better than the truss; for the engineer has to be content to set his V-shaped struts all in one plane, while in the bone they are put, with obvious but inimitable advantage, in a three-dimensional configuration.

So far, dealing wholly with the stresses and strains due to tension and compression, we have omitted to speak of a third very important factor in the engineer's calculations, namely what is known as "shearing stress." A shearing force is one which produces

* Cf. Fr. Weidenreich, Ueber formbestimmende Ursachen am Skelett, und die Erblichkeit der Knochenform, *Arch. f. Entw. Mech.* LI, pp. 438–481, 1922.

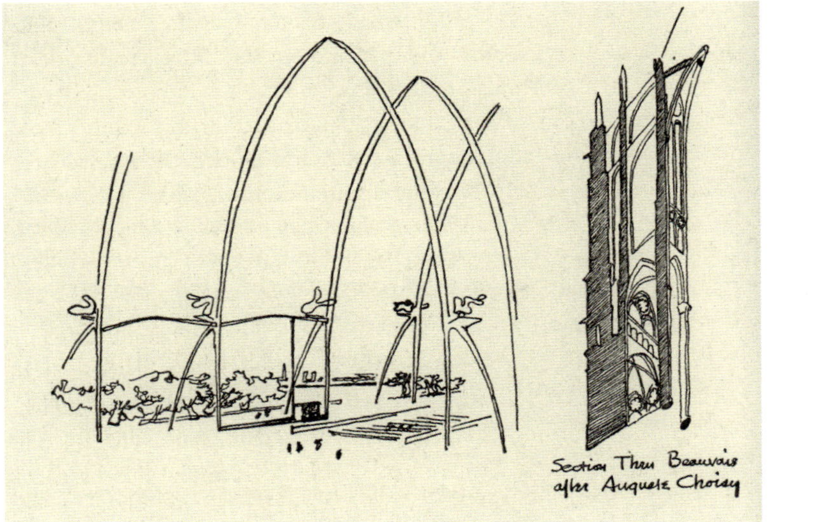

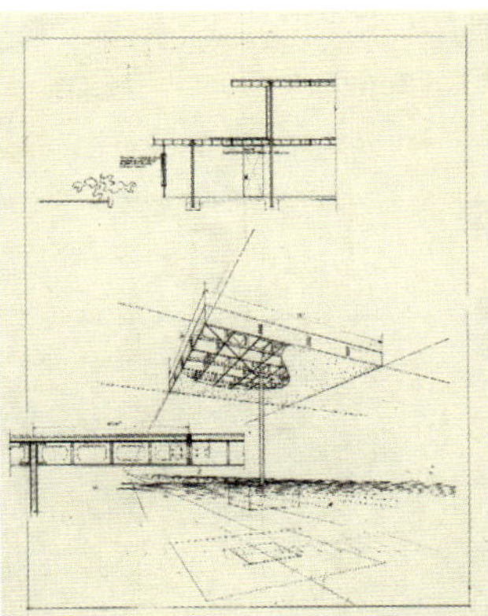

Thompson[6], discusses I-beams and H-beams to underline the qualities of the "hollow tubular pillar" in engineered construction and in living organisms.[7] The sections of *Monumentality* on metal profiles seem to have been written by someone familiar with the pages of *On Growth and Form* devoted to tubular skeletal structures.[8]

The avoidance of any cladding which continuously prompts Kahn to pursue the expressivity of concrete is already outlined in the pages of *Monumentality* in the name of the principle of truth of material, though not yet directly referring to concrete. His statement is incisive: "The giant major skeleton of the structure can assert its right to be seen. It need no longer be clothed for eye appeal."[9] The principle of exposed concrete would be formulated by Le Corbusier only when the worksite of the Unité d'Habitation in Marseille had reached an advanced stage, several years after the publication of *Monumentality*.

In Kahn's reasoning, concrete is but one of the possible materials for the structure. The fact that he identifies the "skeleton" as the form of the contemporary structure explains the way he is looking to the potential of reinforced concrete as the material capable of rivaling steel, also thanks to recent engineering evolutions: "reinforced concrete is emerging from infancy with pre-stressed reinforced concrete, vibration and controlled mixing, promising to aid in its ultimate refinement."[10] This sentence sums up the vision that would radically set apart the research of Kahn on concrete from the various expressions of international brutalism. Those few, significant remarks on concrete are all focused on the intrinsic technical nature of the material, without any reference to the formal quality of design of the

[3] Kahn, *Monumentality* cit., p. 577.
[4] *Ibid.*, p. 580.
[5] *Ibid.*
[6] *Cf.* D'Arcy Wentworth Thompson, *On Growth and Form*, Cambridge, Cambridge University Press, 1917. The edition in 1942, revised by the author himself, was followed by two more, until 1944.
[7] D'Arcy Wentworth Thompson, *On Growth and Form*, Cambridge, Cambridge University Press, 1968, p. 970.
[8] See Kahn, *Monumentality* cit., pp. 579-580, and D'Arcy W. Thompson, *op. cit.*, p. 970.
[9] Kahn, *Monumentality* cit., p. 583.
[10] *Ibid.*, p. 587.

5. L. I. Kahn, O. Stonorv, Memorial Playground, Western Home for Children, Philadelphia, Pennsylvania, 1946-47 (LIK, 030.I.A.265.3).

6. G. Atterbury, precast hollow-cored units (P. Zucker, ed., *New Architecture and City Planning, A Symposium*, New York, Philosophical Library, 1944, p. 179).

surface. It is no coincidence that starting in the 1950s, Kahn begins to use concrete to explore its technical and structural potential, always in collaboration with engineers, who already in *Monumentality* appear as the indispensable counterparts in order to achieve the goal of "gigantic sculptural forms of the skeleton frame."[11] The "ultimate refinement" indicated by Kahn means that the surface of concrete, for him, will never be a place on which to imprint decorative patterns with formwork, or *sculptures moulées*.

To "sketch the faint outlines of a possible structural concept," as Kahn writes, is the purpose of the drawings that illustrate *Monumentality*.[12] The "structural concept" expressed in those drawings is only the one made possible by the "tubular forms" in steel.[13] Though we can find affinities with the metal legs of the furniture designed in the same period by Kahn himself for the Parasol House, the "tubular forms" of the illustrations of *Monumentality* resemble those great prehistoric animal skeletons documented by D'Arcy Wentworth Thompson, and are closer to the skeleton of a stegosaurus[14] with bone details similar to the pelvis of the archaeopteryx and the apatornis[15] than to contemporary engineering structures.

But the drawings of "tubular forms" outline two conflicting visions of structure, probably also as a result of the fact that they were made in different moments. The portals in metal tubes of the dome-covered portico for a civic center belong to the skeletons of organic origin, while the ogival arches with

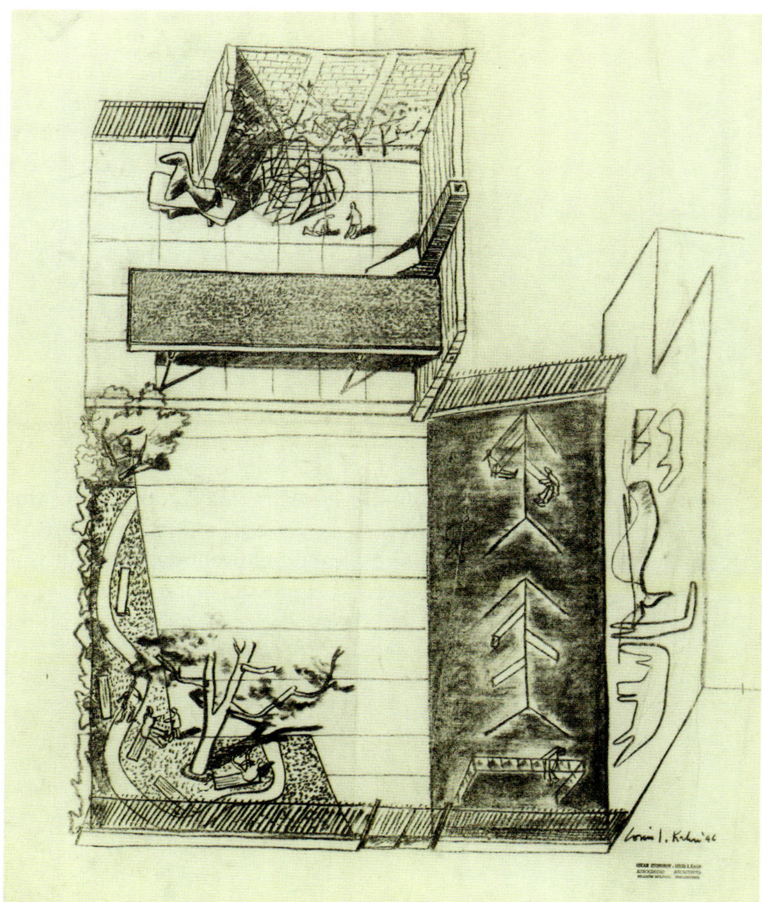

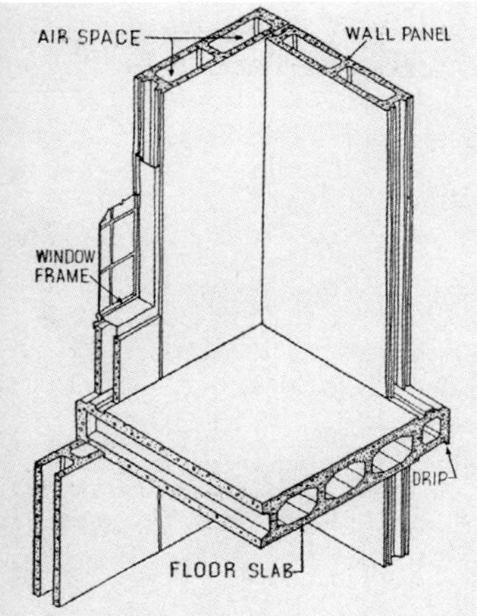

buttresses of a cathedral in steel and glass, though containing certain features of the skeletons documented by D'Arcy Wentworth Thompson, are actually the literal translation in contemporary materials of the masonry of the cathedral of Beauvais. While the portico seems to have been designed against the background of the projects of Kahn and Stonorov for Philadelphia, the steel and glass cathedral is an ideogrammatic scheme with the value of a theoretical manifesto, probably prepared by Kahn specifically for Zucker's anthology and to prompt comparison with Choisy's isometric of the masonry cathedral. Precisely this comparison between the two drawings points to the research Kahn intends to pursue in the years immediately to come: to transpose the figures of traditional masonry construction into contemporary materials.

Though the question of concrete remains marginal in the reasoning on monumentality, in Zucker's anthology Kahn can in any case find information on certain 20th-century experiments with unconventional uses of that material, like the systems of Thomas A. Edison for a "cast-in-place construction," of Grosvenor Atterbury for the pre-cast "hollow-cored units," of Hayes Econocrete for the construction of walls, and of Wallace Neff for the "bubble houses."[16] A few years later, Kahn would have to come to terms with similar issues of economical fabrication of concrete houses.

The success of Kahn's sketches for *Monumentality* is destined to last throughout the 1940s. After Philip Johnson's request to present, in 1945, at the exhibition on War Memorials organized by the Museum of Modern Art in New York, variations of the "steel objects" or "steel cage" Johnson had seen on the pages of *Monumentality*,[17] Kahn's first application of his studies on "tubular forms" is found in the sculptural structures created for the competition project of the Jefferson National Expansion Memorial in St. Louis, Missouri, in 1946-47.[18] Precisely the "tubular forms" also constitute the premise for the sculptural forms Kahn and Stonorov give to the concrete of the Memorial Playground, Western Home for Children, at 715 Christian Street in Philadelphia, designed and constructed in 1946-47 (demolished).[19] The geometry of the "tubular forms" would continue to characterize certain exceptional sculptural figures proposed by Kahn until approximately 1950, confirming their role as the fundamental premise for the discovery of another geometry, better suited to reinforced concrete structures, but still endowed with great sculptural potential.

In the light of the comparison between the steel structure for contemporary monuments and the masonry structure of the cathedral of Beauvais proposed in *Monumentality*, Kahn's decision to spend time studying in Rome, at the American Academy, "away from practice,"[20] takes on a precise connotation as the basis for an architectural vision to come. His first request is sent to New York on 25 April 1947, but is rejected. In the letter explaining his motives, Kahn specifies the reasons for his stay in Rome in

[11] *Ibid.*, p. 585.
[12] *Ibid.*, p. 578.
[13] *Ibid.*, p. 580.
[14] D'Arcy W. Thompson, *op. cit.*, p. 1006.
[15] *Idid.*, p. 1073.
[16] See Harold Sandbank, *Outline summary of prefabrication methods*, in, Zucker, *op. cit.*, pp. 174-189.
[17] Philip C. Johnson, letter to Louis I. Kahn, 2 October 1945, LIK, 030.II.A.61.16; Philip C. Johnson, letter to Louis I. Kahn, 6 October 1945, LIK, 030.II.A.61.16.
[18] *The Louis I. Kahn Archive: Personal Drawings. The Completely Illustrated Catalogue of the Drawings in The Louis I. Kahn Collection, University of Pennsylvania and Pennsylvania Historical and Museum Commission*, New York, Garland, 1987, vol. I, p. 143, 260.17.2.
[19] *Ibid.*, p. 151, 265.1.
[20] Louis I. Kahn, letter to American Academy in Rome, New York, 24 April 1947, LIK, 030.II.A.60.37.

terms of an "environment of the great planning and building works of the past": to develop reflections on "the frames and enclosures of new architectural spaces," in relation to the "continuing evolution of traditional forms," and "to develop material for an illustrated treatise of my architectural and planning ideas."[21]

Vacuum Formed Method for Israel

It might seem like a paradox if the question of monumentality discussed by Kahn is seen as a latent presence in his proposals for economical prefabricated houses formulated at the end of the 1940s. Yet precisely some of these proposals contain the premise for the transformation of "tubular forms" in steel into structures in reinforced concrete conceived and built in the early 1950s.

Starting in 1949, Kahn is involved together with other architects, engineers, philanthropists and economists in the Israel Housing Survey Committee created by Philip Klutznick and called on by the Palestine Economic Corporation to make proposals for the immediate construction of about 50,000 housing units for immigrants to Israel.[22] Before going to Israel, Kahn acquires technical information on the latest methods of fabrication and prefabrication of low-cost housing. In this context, concrete, shaped in different forms and deployed using different techniques, stands out as the main material.

Kahn turns to the International Basic Economy Corporation (IBEC) headed by Nelson A. Rockefeller, with offices at 30 Rockefeller Plaza in New York. IBEC is divided into two departments, the IBEC Housing Corporation directed by George A. Dudley and the IBEC Technical Services Corporation, run by Wallace K. Harrison; the former focuses on the improvement of housing quality in developing nations, especially Latin America, and has technicians who "have developed a method of low-cost mass construction of concrete houses"[23] based on formwork for quick assembly and stripping thanks to a colossal wooden structure that moves along tracks at the foot of the buildings.[24] For the construction of walls with the IBEC Method, after vibration of the poured concrete, "a patented vacuum process is applied which extracts the water from the concrete and accelerates its setting."[25]

The IBEC Housing Corporation sends Kahn documentation on the Marshall Manor complex of 204 housing units, in Norfolk, Virginia, being

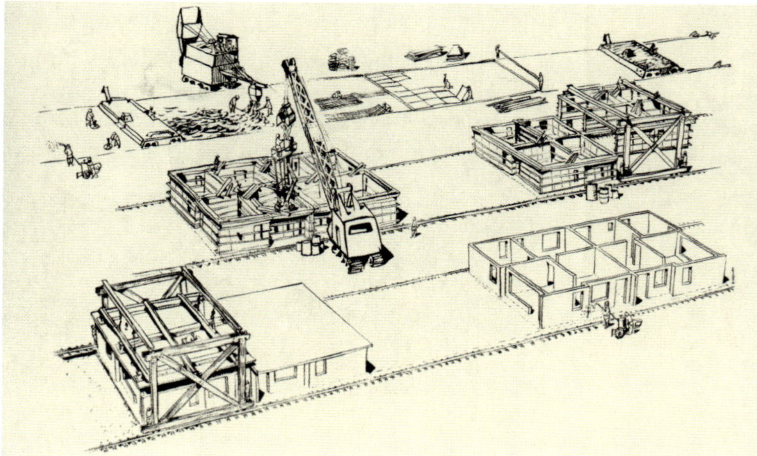

7-8. G. A. Dudley, in collaboration with T.D. FitzGibbon, complex of 204 houses, Marshall Manor, Norfolk, Virginia (Ibec Housing Corporation, "The Ibec method of low-cost house construction," 10 March 1949, LIK, 030.II.A.35.32).

9-12. Construction of prefabricated panels for the realisation of a house (booklet by Lockheed Aircraft Service Company, LIK, 030.II.A.36.33).

FILLING

SCREEDING

VIBRATING

FINISHING

[21] *Ibid.*
[22] See Susan G. Solomon, *Louis I. Kahn's Trenton Jewish Community Center*, New York, Princeton Architectural Press, 2000, pp. 13-28.
[23] Ibec Housing Corporation, "The Ibec method of low-cost house construction," 10 March 1949, LIK, 030.II.A.35.32.
[24] See the photograph "Ibec Method of Low-Cost House Construction," LIK, 030.II.A.35.32. "In simples terms," we read in the information received by Kahn, "the Ibec Method is a concrete 'road' with the efficiency of standard American road-building techniques, pouring complete concrete houses on each alternate slab and later lifting the intermediate slab to form the roof. […]. The material is conventional poured concrete. The special wall forms produce smooth wall surfaces which are ready for painting: no plaster is required. Structure, shelter, and insulation are achieved in a single operation."
[25] Ibec Housing Corporation, "The Ibec method of low-cost house construction," 10 March 1949, LIK, 030.II.A.35.32.

built with the IBEC Method on commission by the local Henry Clay Hofheimer Concrete Corporation – the design is by Dudley, in collaboration with T.D. FitzGibbon, an architect of the IBEC Housing Corporation, and Hofheimer himself.[26]

Other information obtained by Kahn on prefabricated construction includes the booklet of the Lockheed Aircraft Service Company, part of the aeronautical giant, which illustrates the phases of the construction of houses with walls and roofs in reinforced concrete poured on site, using the so-called "tilt-up" method after having poured the platform of the house, used as the work area (we find documentation of houses in Austin, Texas, Manila in the Philippines, and in Hawaii).[27]

From the director of the Housing & Cement Products Bureau of the Portland Cement Association of Chicago, Carl A. Menzel, also consulted for the question of the houses to be built in Israel, Kahn receives a series of booklets on various machinery for the fabrication of concrete, as well as the book *The Reinforced Concrete House*, which among the publications of the association contains, according to Menzel, the answers to certain technical queries – "we understand you are also interested in forms for concrete walls."[28] A series of examples documented by drawings and photographs in the book illustrate the various systems of construction of walls and the finishing of concrete surface. Regarding construction systems for walls, three fundamental types are indicated: "solid wall," "hollow double wall" and "ribbed wall." The formal quality of the concrete relies on ways of making the formwork, with plywood or with planed or rugged boards, depending on the level of rusticity desired for the imprints to be obtained on the surfaces of the walls. "Reinforced concrete walls," the Portland Cement Association reports, "may be smooth, rough or deeply textured, depending upon requirements of the architectural styles. These are accomplished by proper selection of form faces. Smooth forms such as metal, plywood or smooth-finish lumber are used in producing smooth, even surfaces. Rough-surfaced forms produce rough textures. Interesting effects are often produced, for example, by the impression of the grain and joint-lines of form boards."[29] The range of surfaces shown in *The Reinforced Concrete House* covers the "smooth plywood texture," indicated as "extremely popular for all types of concrete walls";[30] the "smooth board-marked texture";[31] the "emphasized joint-line texture";[32] the "grain-marked texture," also obtained with boards;[33] the "grain and joint-marked texture," where the boards are

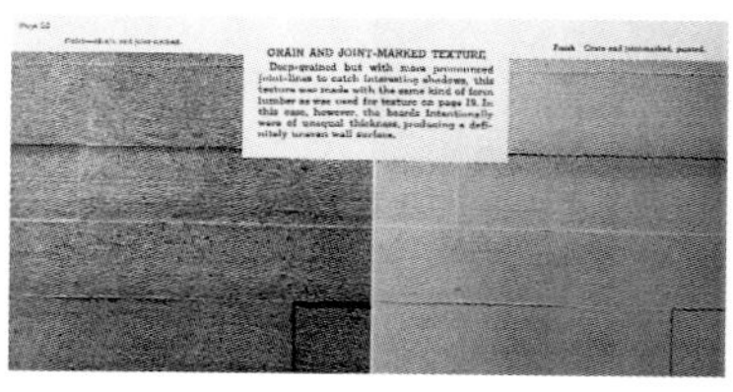

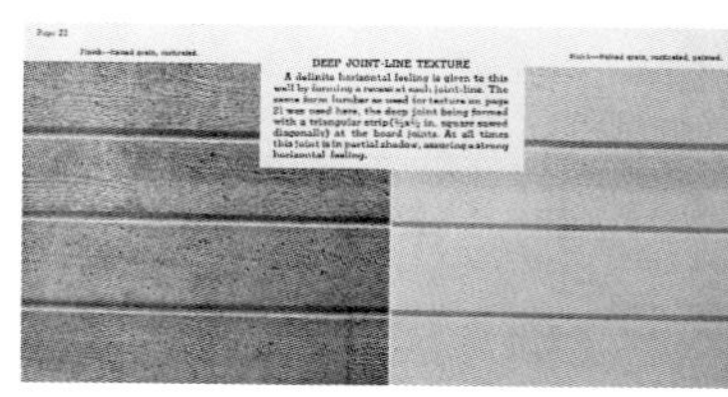

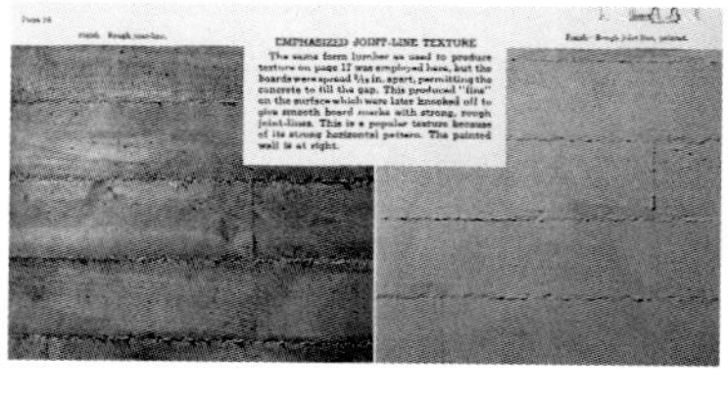

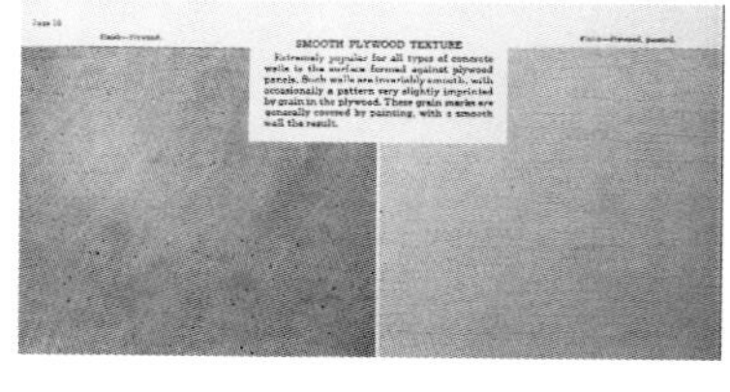

13-16. Details of concrete surfaces (Portland Cement Association, *Reinforced Concrete House*, Chicago, n.d., [1938 circa], pp. 20, 22, 18, 16).

arranged to create chiaroscuro effects;[34] the "emphasized grain-marked texture";[35] and the "deep joint-line texture," obtained with "a triangular strip" applied to the boards.[36] In all the solutions, a variation is proposed with the application of a layer of paint on the concrete that does not, in any case, erase the imprints of the forms.

One of Kahn's discoveries during the research for the Israel Housing Survey Committee, destined to lead to some of his proposals for Israel, has to do with the potential of the Precast Vacuum Concrete Method – the one also used by the IBEC Housing Corporation for the housing complex built in Norfolk, to speed up the setting of the concrete. The inventor of Vacuum Concrete, Karl Pauli Billner, a Swede who moved to the United States in approximately 1900, founded Vacuum Concrete Inc. precisely in Philadelphia in 1935, specializing in the "process to extract excess water from newly poured concrete."[37] Kahn contacts Billner directly and receives the booklet that illustrates the invention and the various applications of Vacuum Concrete, *The Story of ... the Precast Vacuum Concrete Method ... Told in Pictures*, published in 1947.[38] Already negotiating a contract that calls for the construction of one thousand "two-story units" in Egypt, Billner tells Kahn he is willing to make the staff and expertise of Vacuum Concrete Inc. available, to open true centers specializing in the prefabrication of reinforced concrete parts in Israel, with the outlook of transforming Israel into "the center of Prefabrication in the Near East."[39] Also through Billner, Kahn learns about the system created by Petros Doanites, an

17-18. Applications of Vacuum Concrete Method (*The story of ... the precast vacuum concrete method ... told in pictures*, booklet, May 1947, LIK, 030.II.A.61.23).

[26] "After the setting of the floor slab," is the explanation of the Ibec Housing Corporation sent to Kahn "a mechanized wall form is put in position by a crane. This form has been developed as a single piece of mechanized equipment. The concrete is poured to the full height of the walls, which are eight inches thick. The use of concrete aggregate, Solite, provides insulation and fire protection, with the permanence of conventional concrete. The roof slab is lifted into position by a patented vacuum lifting mechanism" (Ibec Housing Corporation, for the "Morning Papers," 3 April 1949, LIK, 030.II.A.35.32; followed by assembly diagram). For the shipment of all the documents regarding Norfolk see George A. Dudley, letter to Louis I. Kahn, 4 April 1949, LIK, 030.II.A.35.32.

[27] Lockheed Aircraft Service Company, booklet, LIK, 030.II.A.36.33.

[28] Carl A. Menzer, letter to Louis I. Kahn, 10 February 1949, LIK, 030.II.A.61.23.

[29] Portland Cement Association, *Reinforced Concrete House*, Chicago, n.d., [1938 circa], p. 12.

[30] *Ibid.*, p. 16.

[31] *Ibid.*, p. 17.

[32] *Ibid.*, p. 18.

[33] *Ibid.*, p. 19.

[34] *Ibid.*, p. 20.

[35] *Ibid.*, p. 21.

[36] *Ibid.*, p. 22.

[37] *Patented July 7, 1936, 2,046,867. United States Patent Office. Method of and apparatus for treating concrete. Karl P. Billner, New York, N.Y., assignor to Vacuum Concrete Corporation, a corporation of Delaware.*

[38] *The story of ... the precast vacuum concrete method ... told in pictures*, May 1947, LIK, 030.II.A.61.23. For the sending of the booklet to Kahn, see George V. Carlson, letter to Louis I. Kahn, 1 March 1949, LIK, 030.II.A.61.23. The booklet explains the technical characteristics of Vacuum Concrete, based on the "utilization of atmospheric pressure," which permits elimination of excess water after pouring, and of the Vacuum Concrete Lifter, the device that makes it possible "to remove precast walls and roofs from the molds much sooner than if ordinary mechanical lifting means were employed" - "It eliminates bolts, special reinforcement and cracks." The photographs in the booklet show residences built by companies using Vacuum Concrete and the Vacuum Concrete Lifter. One of these firms is the George A. Fuller Company, destined to build one of the most important works in concrete by Kahn.

[39] Louis I. Kahn, "Outline Report," n.d., [March 1949], LIK, 030.II.A.61.10.

19. Y. Ben Sira, housing for immigrants and ex-militaries, photograph of the construction site (Y. Ben Sira, "Statement on housing in Israel," 14 January 1949, LIK, 030.II.A.35.34).

architect from Athens, for "an effective precast system for multi-story construction."[40]

Another concrete construction method brought by Kahn to the attention of the Israel Housing Survey Committee is the one invented by Atterbury, who is building the Amsterdam Houses in Manhattan in 1949[41]. Kahn has already seen Atterbury's prefabrication system in Zucker's anthology; now he learns of his "boiling methods" thanks to information supplied by Theodore Crane, professor of Architectural Engineering at Yale University. "Crane was consulted on new concrete method - 'Atterbury system of boiled concrete has economic potentialities as good as Vacuum,'" Kahn reports.[42]

All the information gathered by Kahn for the construction of houses in Israel documents the vast range of current solutions for the making of hollow elements in concrete of different sizes, from the block to the panel. Alongside the theoretical principles outlined in *Monumentality*, this type of information influences the evolution of Kahn's perspective on concrete. The hollow blocks, "hollow tiles" or "hollow-cored units" are more than mere construction examples of prefabricated concrete; they lay the groundwork for the later, fundamental definition of Kahn for this contemporary form of concrete, which recurs in his works: "hollow stones."

Kahn prepares a preliminary report on concrete construction methods, the "Outline Report," and sends it to the Israel Housing Survey Committee. One of the central chapters for techniques has to do with the "prefabrication and mass production methods" and the "advanced design techniques of concrete construction," where he lists the various systems, from that of Billner to that of Atterbury.[43] Once in Israel, one of the tasks of the commission, according to Kahn in the "Outline Report," will be to gather technical information on local construction systems: "Be informed on sources and types of materials available, construction methods, production and manufacturing plants, construction organizations, skilled and unskilled manpower available, and communication and power systems."[44]

When Kahn reaches Israel, together with other members of the committee, at the end of April (he remains until the end of May), he learns about the various systems of construction of low-cost housing, also thanks to meetings with local architects and engineers.[45]

From Yaacov Ben Sira, municipal engineer of Tel Aviv, he receives the photographic documentation of the housing complex for immigrants and military veterans, designed by Ben Sira himself, showing the construction of three-story houses with brick walls erected with the connections open, where the pillars and beams of the reinforced concrete skeleton are inserted later, using rudimentary wooden formwork (the reinforcements are put in place during the phase of construction of the masonry).[46] The engineer H. Neumann, professor at the Hebrew Technical College of Haifa, explains the different construction systems to Kahn, from the most conventional and widespread system using brick walls and reinforced concrete slabs poured in shuttering, to the masonry of "hollow-blocks," all the way to the "big prefabricated building elements."[47]

The information and the worksite visits enable Kahn to get an idea of the degree of backwardness of the construction systems – "construction methods are very primitive."[48] But one of his remarks reveals a certain fascination with that sort of construction seen prior to the application of the cladding, confirming the concept developed in *Monumentality* about the truth of materials: "the housing construction, before the covering of cement and plaster is put over it, is fantastically crude."[49] After all, the very beauty of the Israeli landscape admired during the visits is conveyed by Kahn in similar terms - "rugged beauty of the Landscape."[50]

The discussions with the architects Arieh Sharon and Robert Barnett – the latter a member of the Jewish Agency – on the Agency's temporary lodgings lead Kahn to develop the idea of transforming the wooden barracks into a first phase for the construction of definitive masonry housing. The search for alternative solutions to the temporary housing is one of Sharon's concerns, as Kahn himself reports: "Sharon explained the scheme he developed for combining the construction of permanent shelter (instead of temporary shelter) with a program of employment of the settler, namely to have the settler work on the construction of his house while living in a highly reduced minimum shed in situ."[51]

To transform the wooden shelters of the Jewish Agency into the first phase of a definitive shelter, Kahn proposes changing the arrangement of the frame and the timber: "the order be reversed by exposing the studs to the outside and sheathing the interior. By this method, the interior finish is provided first, and an insulated exterior wall can be added later. This wall may be of silicar brick, panels, sheets, applied when money is available or built by the settler, as the case may be. I also suggested that instead of the wood sheathing [...], that materials such as homosote, asbestos, plaster board, plywood, etc. might possibly be shipped from the States."[52]

[40] *Ibid.* Thanks to Olga Moatsou for helping to decipher the name of the Greek architect.
[41] On Atterbury see Margaret Crawford, *Building the Workingman's Paradise. The Design of American Company Towns*, London, New York, Verso, 1995, pp. 100-128.
[42] Kahn, "Outline Report" cit.
[43] *Ibid. Cf.* Louis I. Kahn, letter to Julius Simon, President Palestine Economic Corporation of New York, 21 March 1949, LIK, 030.II.A.61.10.
[44] Kahn, "Outline Report" cit.
[45] The trip to Israel also offers a chance for a visit to Rome, as well as France and Greece - "You indicated that you wish to spend some time in Rome on your return" (Esther Trebach, of the Palestine Economic Corporation, letter to Louis I. Kahn, 23 February 1949, LIK, 030.II.A.61.10).
[46] Yaacov Ben Sira, Municipal Engineer, Tel Aviv, "Statement on housing in Israel," 14 January 1949, LIK, 030.II.A.35.34.
[47] H. Neumann, engineer of the Hebrew Technical College in Haifa, letter to Louis I. Kahn, 12 May 1949, LIK, 030.II.A.35.35.
[48] Louis I. Kahn, "Report," 13 May 1949, LIK, 030.II.A.35.14.
[49] *Ibid.*
[50] Louis. I. Kahn, "Dates: May 11, 1949," LIK, 030.II.A.35.14.
[51] *Ibid.*
[52] *Ibid.*

20. "Parabolic concrete segments butted to form shell of a house," Vacuum Concrete Inc., 3 June 1949 (L. I. Kahn, "Preliminary report on housing in Israel," 7 June 1949, LIK, 030.II.A.35.33).

21. E. Saarinen, United States Jefferson, National Expansion Memorial, St Louis, Missouri, 1947-65 (E.-L. Pelkonen, D. Albrecht, eds., *Eero Saarinen, Shaping the Future*, New Haven, Yale University Press, 2006, p. XVI).

22. G. A. Fuller Company, transportable shack in corrugated sheet and with semi-circular section, called Quonset (L. R. Crandall, *The George A. Fuller Company, War and Peace, 1940-1947*, New York, George A. Fuller Company, 1947).

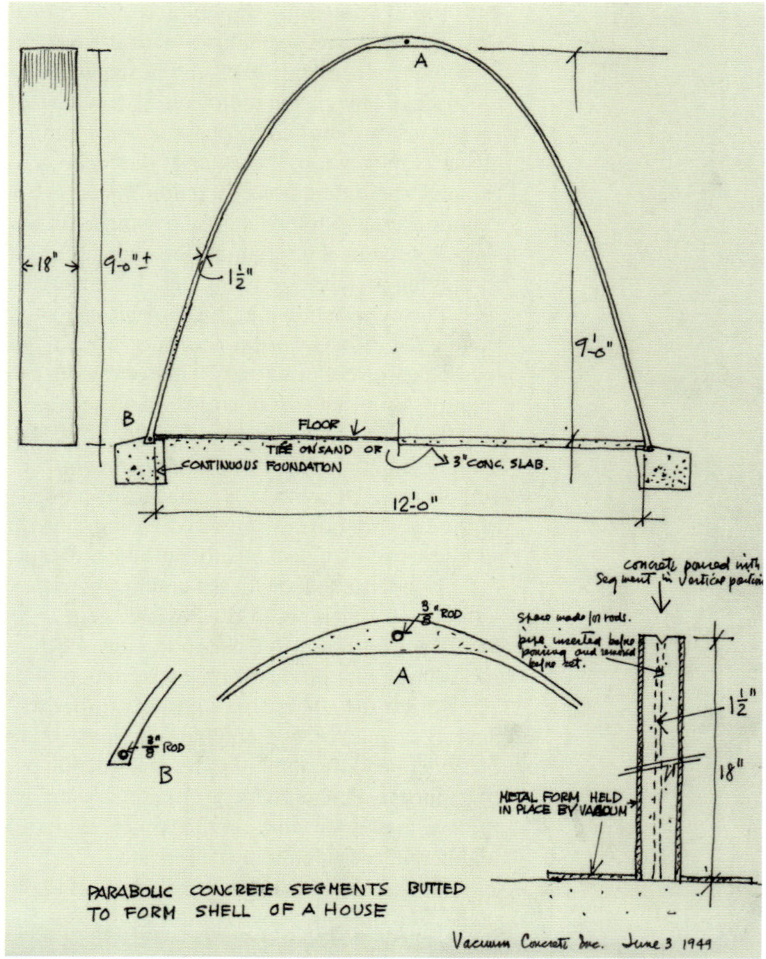

The members of the Israel Housing Committee also visited the Neshar workshop in Hafia, specializing in cement production - as well as the cement block factory, again in Hafia[53] - and sites using the tilt-up construction system, in which walls were cast on site in full height forms with doors and windows and then raised with windlasses.[54]

The summary of the observations and proposals of the Israel Housing Survey Committee is prepared in June 1949, when the various members have already left Israel. The "Draft of Report of Housing Survey for Israel" contains the suggestion for the construction systems for five types of housing units: the "semidetached house 2 units" with walls in "concrete cast in situ"; the "terrace house 4 units" with walls in "hollow concrete blocks"; the houses of "2 stories" with walls in "hollow concrete blocks" and slabs in "reinforced concrete"; the "block of flats 3 stories" with "R.C. skeleton" and walls in "silicate bricks" and "partitions" in "hollow block"; finally, the "terrace house 4 units" with walls in "wooden frame board and Masonite sheeting."[55] In the chapter of the report on the "expansible house" indications are made of the potential of concrete, in the form of blocks or poured in place (the roof, too, can be formed of "pre-cast roof slabs"; "concrete pipes" are also suggested, to reduce the use of metal).

Kahn's report stands out from those of the other members for its proposal of certain original construction systems, also documented through drawings, and based on the premise of the development of "*new* techniques based on the wider range of concrete products [...]."[56] He examined some of these systems together with Billner.

Kahn envisions the building of houses with three construction systems, two based on the integration in two successive phases of wood and masonry, as discussed with Sharon and Barnett, and one based on the use of concrete alone. The first two systems are composed of an "inner shell" and an "outer shell" to be built in two phases.[57] In the first case, the "inner shell" is made first, with vertical posts and horizontal boards (including the roof), and only later are the "masonry walls" added; in the second case the "outer shell" in masonry (including the roof) is built first, followed by the internal cladding, indicated as the "fur interior" (no longer as the "interior shell," since it has become masonry cladding, to apply with different materials).[58]

While the first two systems are directly based on the logics of Israeli construction of houses in wood or in masonry, the third – based on the construction of houses as "concrete shells" – more directly reflects Kahn's initial research on new methods of deployment of concrete. It is no coincidence that this system is developed, also in its formal conception, when Kahn returns to America (there are no traces of the "concrete shell" in the documents related to his stay in Israel).

The concrete shell has paraboloid section and is built with a series of juxtaposed "segments"; only in a second phase is an "insulating layer" added that can also be a "gunite exterior."[59] The idea of using a concrete shell probably relates to the transportable barracks in corrugated sheet metal with a semi-circular section, known as the Quonset Hut, which had been produced in great numbers by the George A. Fuller Company as temporary shelters for the Navy and the Marines during World War II.[60]

But Kahn's "concrete shell" has a monumental character not found in the Quonset hut. It is significant that Kahn, speaking of his houses for Israel, mentions Fred N. Severud, the engineer of Norwegian origin, founder of the firm Severud, Elstad, Krueger Associates which is working with Eero and Eliel Saarinen on the project for the grand arch of the Jefferson National Expansion Memorial in St. Louis, Missouri, winner of the two-phase competition in September 1947 and February 1948 – the construction of the arch in reinforced concrete clad in steel does not happen until 1963-65. In 1947 Kahn

[53] Albert Meyer, "Report of Albert Meyer," n.d., [1949], LIK, 030.II.A.35.33.
[54] David Rose, "Report of David Rose," 26 May 1949, LIK, 030.II.A.35.33. See also [Israel Housing Survey Committee], "Draft of Report of housing survey for Israel," n.d., [June 1949], LIK, 030.II.A.35.33.
[55] [Israel Housing Survey Committee], "Draft of Report of housing survey for Israel" cit.
[56] Louis I. Kahn, "Preliminary report on housing in Israel," 7 June 1949, LIK, 030.II.A.35.33.
[57] See the graphic diagrams taken from, Kahn, "Preliminary report on housing in Israel" cit.
[58] In the report Kahn explains the logic of the "first stage of construction" that serves to guarantee "the minimum acceptable for shelter" and the minimum of services; he recommends, regarding the "frame construction," that "until more advanced production systems are developed in masonry and concrete, frame construction be used to prime the building program by providing enclosure quickly." Masonry construction, to be done in a second phase by the immigrants themselves, and placed over the wooden structure – "construction masonry walls on the exterior against the studs" –, ensures "an insulating air space, and adds permanency to the construction." In the case of "masonry construction" done in the first phase thanks to machinery for the mixed design of concrete blocks or silica brick, the plan is to later add what Kahn calls "fur masonry walls on the interior with board or sheet material" (Kahn, "Preliminary report on housing in Israel" cit.). See also the draft for the "Preliminary report on housing in Israel," typewritten and handwritten pages, without title or date, LIK, 030.II.A.35.30.
[59] See the graphic diagrams taken from, Kahn, "Preliminary report on housing in Israel" cit.
[60] See Lou R. Crandall, *The George A. Fuller Company, War and Peace, 1940-1947*, New York, George A. Fuller Company, 1947.

23-26. Study drawings for Israel housing (L. I. Kahn, "Preliminary report on housing in Israel," 7 June 1949, LIK, 030.II.A.35.33).

lets Saarinen know that he admired the "physical monumentality" of that arch, with its profile similar to the one he later used for his concrete shell.[61]

The details of the concrete shell system, including the idea of its construction in segments, are developed by Kahn in collaboration with Billner.[62] The technicians of Vacuum Concrete Inc. prepare the design of "pre-cast houses" starting with the shell principle;[63] then they suggest its construction in segments.[64]

For the construction of arches, use would be made of the concrete platform previously placed on the lot to outline the footprint of the house, as in the "tilt-up" system. The layout of 13 x 30 feet is calculated to be able to contain the successive fabrication of all the 17 arches that form the concrete shell, simplifying the worksite and reducing the movement of pieces to a minimum. The individual arches, each with a width of 18 inches, are poured in a "metal form held in place by vacuum" and arranged in a vertical position - "concrete poured with segment in vertical position."[65] Placed side by side, the segments are connected together thanks to a special "galvanized rod" passing through the hole placed at the top of each one. Kahn also studies the various systems of lot layout for the concrete shell houses and their extension, and goes into the technical detail of the proposal of the "parabolic shaped shell forms," listing their characteristics:

"Each shell is composed of segments fitted and tied together to form an enclosure of required length and about twelve feet wide. Enough segments are cast on the house floor slab to complete the shell of the house. It is assumed that the mixing of concrete is done by hand and the erection of the structure is done by the use of a windlass. Six ordinary laborers can, after being given proper instruction, set the vacuum forms, mix and place the concrete and apply the vacuum for one complete structure in one day. The vacuum forms can be released five minutes after the concrete is in place. With proper scheduling, the same form can be reused

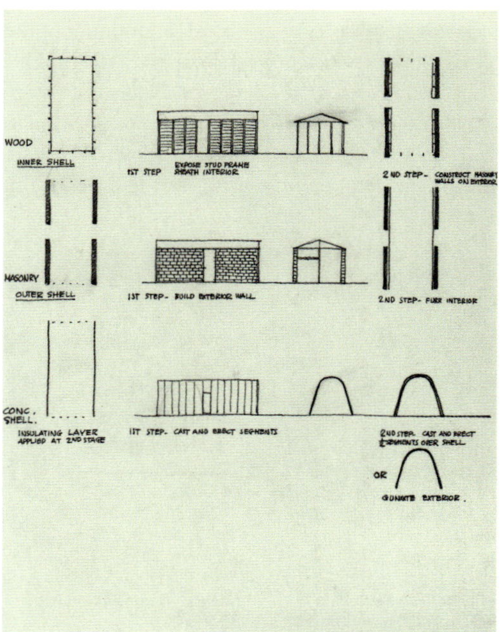

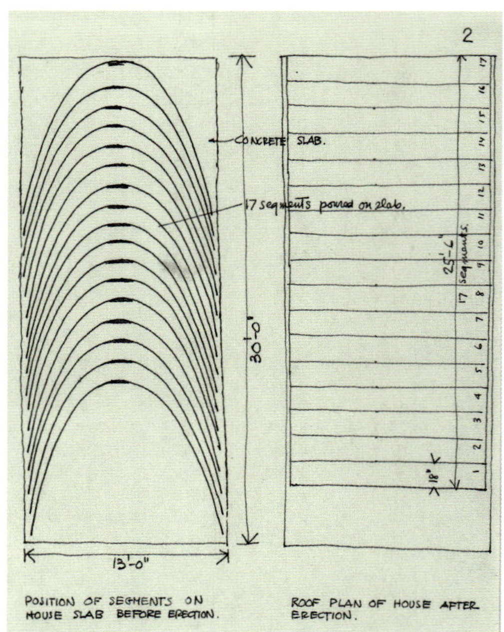

every 20 minutes. There are 17 to 20 segments to one complete shell, therefore all these segments can be made by six men in one working day. [...] since the machinery can be used for the production of 3 houses per day, three sets of forms should be required. It is calculated that the forms and machinery are good for 10,000 houses."[66]

Kahn expresses the hope that the Israel Housing Survey Committee will undertake or sponsor "a project to erect an experimental enclosing concrete shell structure" in terms that offer a glimpse of the crucial role that, in his research on concrete structures, would be played by collaboration with engineers. This was already implicit in *Monumentality*, since he states, in the report, that "such outstanding concrete engineers as Carl Billner and Severud" should "be retained to work with a U.S. contractor and a U.S. architect together with a chosen architect and contractor from Israel on the design development and erection of such structure."[67]

Two other proposals of Kahn, not described in the report but documented by graphic diagrams, call for the use of prefabricated concrete elements and differ from the previous construction systems because the single pieces, in the case of the two variations of the concrete shell, are shaped to have a section with two inclined sides closer to the walls of a house, and because – in the case of the rectangular panels measuring 8 feet x 30 inches with which to make the walls and roof – the prefabrication permits the creation of

[61] Louis I. Kahn, letter to Eero Saarinen, 21 October 1947, cit. in Eeva-Liisa Pelkonen, ed., *Eero Saarinen: shaping the future*, New Haven, Yale University Press, 2006, p. 226.
[62] Kahn, "Preliminary report on housing in Israel" cit. The drawing by Kahn of the details is dated 3 June 1949 and made based on the drawing of Vacuum Concrete Inc., also dated 3 June 1949, (LIK, 030.II.A.35.30). See Karl Pauli Billner, letter to Louis I. Kahn, 28 June 1949, LIK, 030.II.A.61.10.
[63] See the axonometric "PRE-CAST Houses, Vacuum Concrete Inc., 4210 Sansom St., Philadelphia," LIK, 030.II.A.61.10.
[64] Kahn, "Preliminary report on housing in Israel" cit.
[65] See the graphic diagram "Vacuum Concrete Inc." dated 3 June 1949, attached to Kahn, "Preliminary report on housing in Israel" cit. On the same diagram we read: "Concrete poured with segment in vertical position."
[66] Kahn, "Preliminary report on housing in Israel" cit.
[67] *Ibid.*

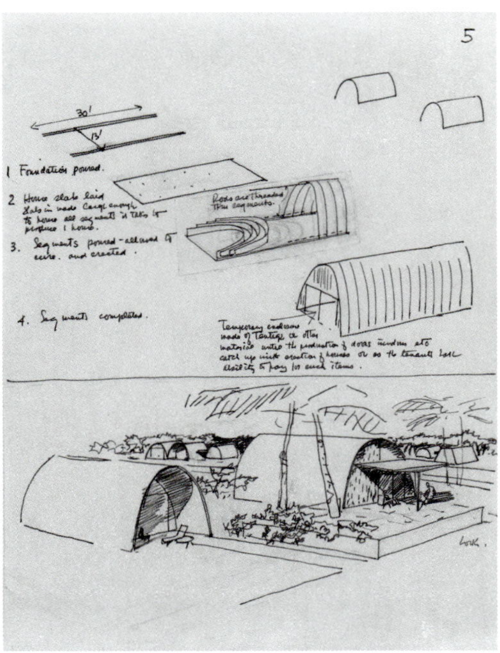

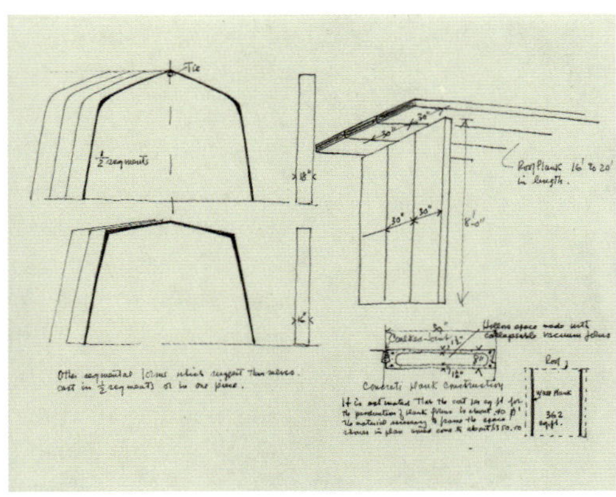

an insulating cavity, as in the pre-cast "hollow-cored units" of Atterbury. This cavity is obtained, as Kahn writes in a note, "with collapsible vacuum forms."[68]

In all the observations of the local technicians and members of the Israel Housing Survey Committee we can see that in Israel concrete works are systematically concealed by claddings, due to the low quality of the fabrication of the material, whether poured in place or used in the form of blocks produced by machinery. During his stay in Israel, Kahn expressed his own views regarding the sight, during construction, of rugged concrete structures before they were hidden by plaster. In the final report he goes back to the question of exposed concrete and its formwork, which not coincidentally becomes an essential issue for his architecture. That skeleton left exposed, without claddings, announced in *Monumentality*, is now put into the context of international reinforced concrete construction. So after the hints in *Monumentality* and in the "Outline Report," Kahn's final report on the prefabricated houses for Israel adds further, significant information on the progressive definition of his idea of "exposed concrete," whose examples include, significantly enough, works of engineering like bridges:

"The potentialities of advanced design in concrete are evident in the new systems of forming, stripping and curing by use of vacuum and boiling. In the vacuum system, the same equipment used for applying vacuum to the forms is also being employed for the lifting of castings, which simplifies the handling of large castings and provides greater possibilities in concrete design. In France, Belgium, Switzerland and South America, where concrete has been used extensively for buildings and bridges, concrete is asserting itself as a material of beauty born of necessity and economy. In Israel, excellent work is performed in concrete finishing, making worktop slabs, floor tile, handrails, and terrazzo work. Because forms are expensive, structural concrete work is done roughly, and is covered with cement plaster. Skeleton forms of exposed concrete - a logical perfection in the use of concrete - are not seen in Israel."[69]

27. Samuel Radbill Building, Philadelphia Psychiatric Hospital, Philadelphia, Pennsylvania, 1949-54. Detail of the drawing "First Floor Plan. Column Schedule, etc.," 16 February 1950, detail (LIK, 030.I.C.180.003).

The reference to France might lead us to assume that Kahn was aware of the work in progress in Marseille for the construction of the Unité d'Habitation of Le Corbusier; but no documents confirm this hypothesis.

Following the studies on the possible applications of concrete for the construction of low-cost housing in Israel, Kahn would state that due to the scarcity of wood and steel, the Israelis "must find the glory of man in the use of concrete, with all the technical know-how available."[70]

The research on prefabrication of concrete conducted for Israel often leads Kahn to take a position in favor of experimental initiatives for low-cost construction in the United States. "Suggested," it is reported regarding one of his contributions in 1950, "desirability of having demonstration of new methods of concrete construction in various localities to advance the use of systems which promise greater economy."[71]

[68] See the graphic diagram attached to Kahn, "Preliminary report on housing in Israel" cit.
[69] Kahn, "Preliminary report on housing in Israel" cit. It is hard to ascertain whether at this time Kahn is familiar with images of the worksite of the Unité d'Habitation in Marseille by Le Corbusier. Only Scully states "Kahn visited Le Corbusier's Unité d'Habitation at Marseilles when it was under construction" (Vincent J. Scully, Jr., *Louis I. Kahn*, New York, G. Braziller, 1962, p. 18).
[70] Harold Feldman, *Inter-Views: Philosophic Architect Would Like to Redesign Center City*, in, "Jewish Exponent," 13 February 1959, Jewish Exponent-LIK, Box 66 (now cit. in Solomon, *op. cit.*, p. 28).
[71] Cit. in, R.J. Wadsworth, "Advisory Committee, Summary Notes of Conference - Longfellow Building Washington," 3-4 February 1950, LIK, 030.II.A.61.11.
[72] The design of the canopy includes the participation of Anne Griswold Tyng (see curriculum vitae of Anne Griswold Tyng, n.d., LIK, 030.II.A.45.64).
[73] See the drawing "Second Floor Plan," 16 February 1950, LIK, 030.I.C.180.002.

Canopy and sunshades of the Samuel Radbill Building

The pursuit of exceptional figures for contemporary structures, as it emerges in the "tubular forms" of *Monumentality* and with the concrete shells for housing in Israel, finds confirmation in certain elements of a sculptural character Kahn develops for the Samuel Radbill Building, designed in 1949-50 as an annex to the Philadelphia Psychiatric Hospital: the canopy and the protruding sunshades, which run along the main facade, animate a concrete skeleton with infilling slate panels. The construction is done by the Master Masons Construction Company of Philadelphia; the concrete is supplied by the Keystone Concrete Company; Alex Guerrina & Sons of Philadelphia is the company hired for the masonry work; and John Briggs is the engineer who works with Kahn (the building is completed in 1954).

The triangular canopy seems to be an interpretation of the sculptural canopy of the Unité d'Habitation in Marseille, but with a geometry that reveals structures to come, also partially expressing that complexity seen in the figures published in *Monumentality*[72]. The triangular slab in "exposed concrete," as indicated in the drawings,[73] rests on three pairs of pillars tapered towards the bottom, for which Kahn invents a load-bearing

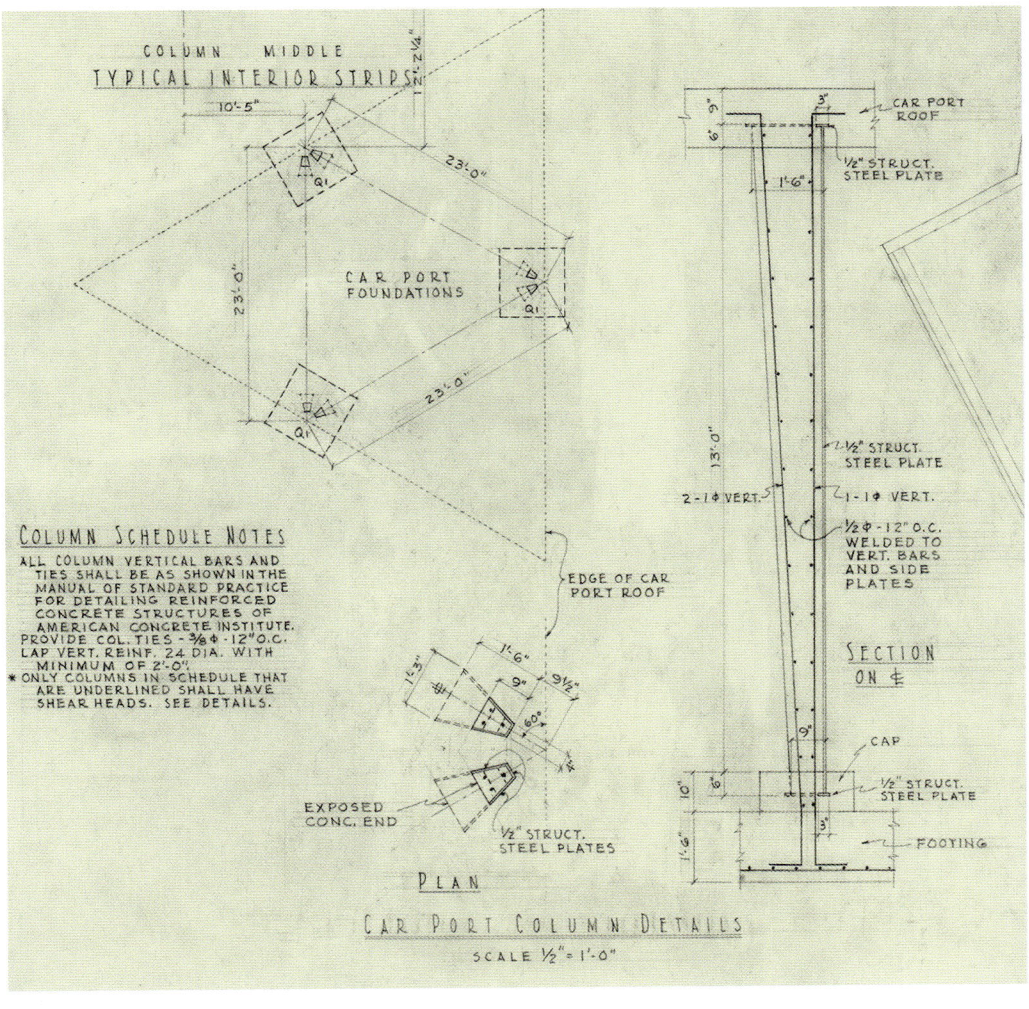

28-30. Samuel Radbill Building, Philadelphia Psychiatric Hospital, Philadelphia, Pennsylvania, 1949-54.

28. Main facade with the canopy [1954 circa] (D. B. Brownlee, D. G. Delong, *Louis I. Kahn: In the Realm of Architecture*, New York, Rizzoli International Publications, 1991, p. 42).

29-30. Details of the brise-soleil and bricks wall (photographs by the author, 2010).

structural system that is also the form for the pouring of the concrete: a thick piece of steel sheet bent in a V, for each pillar, with an open part that leaves the reinforced concrete core visible.[74] Precisely this metal pillar, its hollow configuration and tapering towards the bottom, can be seen as a variation on the "tubular forms" published in *Monumentality*. With the solution of the metal bent into a V, Kahn also demonstrates the will to control the finish of the more sculptural pillars of the Samuel Radbill Building, without having to make use of wooden formwork – otherwise, the profile of the pillars would have required a complex assembly of wooden boards. Not all of the reinforced concrete structure of the building, in fact, is left exposed, as demonstrated by the rear infill where the sheets of slate are arranged so that the pillars will form uninterrupted strips, without visibility of the ends of the slabs in reinforced concrete.

The sunshades are another invention that sets out to explore the expressive potential of exposed concrete, in the ways that are becoming fa-

mous thanks to Le Corbusier and his Brazilian followers. Kahn makes overhanging planes perforated by rows of deep square-section tubes – the number of hollow tiles varies from level to level of the building.[75] The rays of sunlight cross the tiles generating a pattern of lights on the facades.[76]

On the gable walls of the building, Kahn erects walls in concrete cinder blocks, with Nazareth dark cement-base mortar, demonstrating chromatic sensitivity to variations of gray that would later also be seen in other projects.[77]

"Exposed finished concrete" and "democratic taste"

Though limited in number, there are increasingly telling clues of Kahn's progressive focus on techniques that permit building with exposed concrete. "Section IV. Concrete Work" in the specifications prepared in April 1950 for the annexes to the Philadelphia Psychiatric Hospital, of which the Samuel Radbill Building is also a part, contains useful indications to understand how Kahn conceives of the deployment of exposed concrete with the typical formwork and finishes of American "architectural concrete," prior to learning of the famous statements of Le Corbusier on *béton brut* that should not be worked after stripping the formwork, in spite of possible defects in its finish.[78]

The material of the formwork for the "exposed finished concrete" of the new buildings at the Philadelphia Psychiatric Hospital is typically American plywood, which Kahn indicates should be made with "highly water-resistant glue," "coated before using" and selected "in as large units as practicable"; furthermore, the panels should be "laid out in a regular and symmetrical pattern [...]."[79] The recommendations for the implementation of plywood for the "exposed finished concrete" range from checking of joints to the coplanar arrangement of the panels, to obtain a clean concrete without flaws – a "smooth surface" (as the surface of the concrete for the staircases is defined).[80]

"Edges of units shall be tightly butted together with straight, clean joints; any appreciable space at joints shall be in filled flush and smooth as approved. The material for 'special surface' form shall be in flat sheets, handled, treated and nailed according to manufacturer's recommendations. Special care shall be exercised in the construction and erection of wood

[74] See the drawing "First Floor Plan. Column Schedule, etc," 16 February 1950, LIK, 030.I.C.180.003. For the reinforcement of the canopy see the drawing "Second Floor Plan. Fire Tower Detail," 16 February 1950, LIK, 030.I.C.180.003. In the project by Stonorov and Kahn for the expansion of the hospital, dating back to 1946, the structure of the building was envisioned with a steel "frame" and "reinforced concrete floor slabs" (Oscar Stonorov, Louis I. Kahn, "Outline specifications for the Philadelphia Psychiatric Hospital. Hospital Building, Convalescent Pavilion, Nurses' and Doctors' Building," May 1946, LIK, 030.II.A.20.8). These "Outline specifications" also indicate that the "concrete blocks are to be laid in a smooth joint for painting instead of plastering."

[75] "The tiles for sun shades will act as forms for holes in sun shade. This Contractor shall work with the Tile Contractor in placing the tiles securely in place in the forms" (Louis I. Kahn, "Additions to Philadelphia Psychiatric Hospital. [...]. Bulletin No. 6, "Make the changes, deletions and additions to drawings and specifications described in Item 63 to 178," 26 April 1950, LIK, 030.II.A.20.7).

[76] The drawing "Concrete layout at first floor plan showing placement of sunshade tile at overhang" is dated 18 December 1950, LIK, 030.I.C.180.003.

[77] David P. Wisdom, letter to Master Masons, 28 September 1950, LIK, 030.II.A.20.26. See also Martin B. Schwartz, letter to David P. Wisdom, 15 May 1951, LIK, 030.II.A.20.21. For the mortar of the masonry it is indicated that Nazareth cement should be used (Alex Guerrina & Sons, of Philadelphia, letter to Master Masons, 7 September 1950, coll. 20.26). The same materials are also specified, still in 1949, for the Pincus Therapy Building (David P. Wisdom, letter to Building Committee, Philadelphia Psychiatric Hospital, 26 May 1949, LIK, 030.II.A.20.32).

[78] On the *béton brut* of Le Corbusier see Roberto Gargiani, Anna Rosellini, *Le Corbusier, Béton Brut and Ineffable Space, 1940-1965. Surface Materials and Psychophysiology of Vision*, Lausanne, EPFL Press, 2011.

[79] Kahn, "Additions to Philadelphia Psychiatric Hospital" cit.

[80] *Ibid.*

forms for finished concrete surfaces, such as exposed to view, exterior and interior columns, pilasters, beams, girders, etc., to ensure straight, plumb, level exposed surfaces with all corners and angles sharp and true to line so as to require no other finish."[81]

The operations recommended after stripping, in the section "Patching – Finished Concrete Surface" of the specifications, bear out Kahn's desire to intervene with painstaking work on the surfaces of the exposed concrete, to remove any type of irregularity. It is only after Le Corbusier explains the artistic nature of *béton brut* and the untouchability of the surfaces once the forms have been struck that Kahn modifies, during the course of the 1950s, the specifications for his own works in concrete to forbid any intervention on the concrete after the removal of the formwork.

For the buildings of the Philadelphia Psychiatric Hospital, after the stripping of the formwork he urges that flaws "such as voids, stone pockets, tie holes, fins, ridges and pour joints or other irregularities as remain shall be patched or removed while concrete is green"; all "patches" have to be "kept moist for at least seven days."[82] "Ridges and irregularities shall be ground with special steel grinders or honed with carborundum blocks."[83] The tie holes are also to be carefully sealed with cement, disguising a feature of the formwork system that would later, instead, in subsequent works, become an essential earmark of Kahn's concrete surfaces.[84] The finishing operations for the concrete after stripping also have to do with the ceilings of the concrete slabs, where suspended ceilings are not stipulated - "the filling of air bubbles and the rubbing with the carborundum rock."[85] For the exposed concrete of the columns, a sample is ordered, to assess "the finish on the exposed concrete column surfaces."[86]

All the finishes and their colors combine to a fundamental extent to create what Kahn calls the "democratic taste" of the Philadelphia Psychiatric Hospital. In this perspective, exposed concrete with its gray color and its texture is destined to play an increasingly decisive role in the selection of other materials and their meanings, in keeping precisely with that concept of democracy invoked for the hospital, which would reach its representative apotheosis in the concrete applied in Dhaka.

"Democratic taste," Kahn writes to the president of the hospital, Samuel Radbill, "is needed for most of the interior spaces of your hospital, because one must satisfy a variety of personalities in and out of the hospital. The usual aesthetic decision involving democratic taste is qualified by restraint and reasonable uniformity. Finally the choice of materials, colors, textures, and forms should not shock one but be unobtrusively there. Too much arbitrary variety and too definite a pattern can grow tire-

[81] *Ibid.*
[82] *Ibid.*
[83] *Ibid.*
[84] "All holes resulting from form ties, rods or sleeve nuts," it is explained in the specifications, "shall be solidly packed completely through all exterior walls by pressure grouting from back of wall with Alemite gun or similar compacting device after the holes have been flushed with water. Cement grout shall be used for the plugging material. The grouted holes of finished surfaces shall be acreeded off flush and ground to match adjoining surfaces. The finished exposed concrete surfaces, when finished, shall be left free of defects with corners, jambs, arrises and angles, straight, plumb, true and level. Upon completion of the work, the exposed concrete surfaces shall be thoroughly cleaned and inspected. All remaining blemishes shall be removed at this time and all finished surfaces thoroughly washed down with a solution of caustic soda in water" (Kahn, "Additions to Philadelphia Psychiatric Hospital" cit.).
[85] *Ibid.*
[86] Martin B. Schwartz, letter to Louis I. Kahn, 14 August 1951, LIK, 030.II.A.20.20.

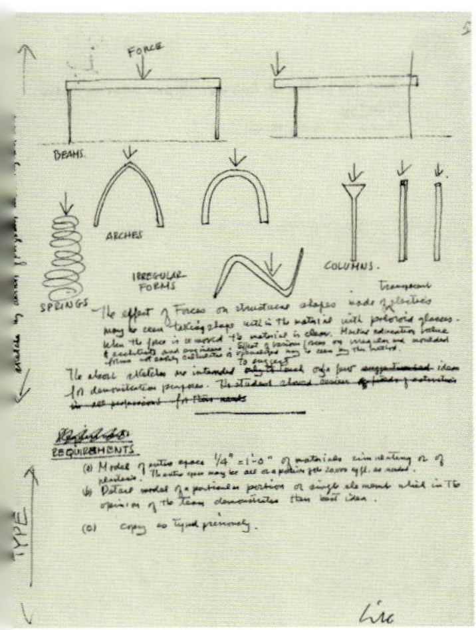

31. "Collaborative problem - 1949. An idea center for plastics" (LIK, 030.II.A.61.27).

some quickly (as the vogue changes!). A building should not be furnished and decorated in the same spirit as a woman buys a new dress which is meant to be worn only occasionally and be discarded as the new styles come in. My efforts and interests have always extended through the finishing phases of any building I have designed."[87]

Only in the light of statements regarding "democratic taste" to also be achieved through the choice of materials can the meaning Kahn assigns to the nature of concrete become fully comprehensible. This is to be seen not only as an aspect of the pursuit of an idea of construction, but also as the expression of the processes of fabrication, the constitution of the concrete with economical components, all that way to the structures which that material, used in all its forms, from the block to the brick, the skeleton to the wall, permits to be made for the construction of an architecture for the democratic society. Precisely the color of concrete guarantees, for Kahn, that "reasonable uniformity" that has to form the backdrop of a "variety of personalities," about which he writes with regard to the Philadelphia Psychiatric Hospital. Jasper Johns would begin his obsessive research on the values of gray in the mid-1950s.

Yale Art Gallery: beyond the Flexible Ceiling

The invention of the first exceptional exposed reinforced concrete structure by Kahn happens in the context of a university institute for artistic research, where Kahn and Howe are both professors.

Starting in 1947, Kahn is invited to take part in the Department of Architecture of the Yale School of Fine Arts, at Yale University in New Haven, as a critic, "senior critic" and then, in 1950, as "chief critic."[88] Certain programs of study proposed by Kahn in agreement with the new director of the Division of the Arts, Charles H. Sawyer, have to do with the creation, for the school year 1949-50, of plastic objects to be displayed in particular lighting conditions, and to be achieved by "dramatizing the visual and structural potentialities of plastics."[89] As in the projects of the 1940s or the canopy of the Radbill Building, in the case of plastic Kahn continues to pursue a structure with exceptional geometrics, as he had suggested in *Monumentality*.[90] A series of his sketches demonstrate to students the criteria to discover special structural forms through the deformation of the material, starting with familiar figures like the trilith, the arch, the spring, the column with various sections, from the H to the tube, and the "irregular forms" which are those of the structures he published in *Monumentality*[91]. Another meaningful example of the creation of structures, and in particular of reticular structures with a triangular grid, in the exercises at the Yale School of Fine Arts done under

Kahn approves the column sample on 17 August.

[87] Louis I. Kahn, letter to Samuel Radbill, 19 September 1952, LIK, 030.II.A.21.19. See also [Louis I. Kahn], "Design definition," 10 February 1950, LIK, 030.II.A.63.40.

[88] Patricia Cummings Loud, *The Art Museums of Louis I. Kahn*, Durham, Duke University Press., 1989, p. 53.

[89] Louis I. Kahn, sheets of handwritten notes, "Collaborative problem - 1949. An idea center for plastics," LIK, 030.II.A.61.27.

[90] Kahn, *Monumentality* cit., p. 586.

[91] Louis I. Kahn, sheets of handwritten notes, "Collaborative problem - 1949. An idea center for plastics," LIK, 030.II.A.61.27. See also Charles H. Sawyer, "Yale University - Division of the Arts. Proposal for collaborative problem to be discussed at meeting Saturday, July 9, 1949," LIK, 030.II.A.61.27. Louis I. Kahn, sheet of sketches and notes, LIK, 030.II.A.61.27.

32. Roman ruins, 1951 (J. Hochstim, *The Paintings and Sketches of Louis I. Kahn*, New York, Rizzoli International Publications, 1991, p. 242, No. 332).

Kahn's guidance in 1949, is the scholastic project for the National Center of Unesco by a group of students calling themselves Team 14.[92]

The fact that Kahn points students towards the invention of structures starting with the potential of material is essential to understand the reasons that prompt him to experiment with an utterly exceptional form for reinforced concrete, created for the famous institution based in New Haven.

The event that emblematically demonstrates the new orientation of the Yale School of Fine Arts, and in which the structure invented by Kahn takes its rightful place, is the creation in 1950 of the Department of Design, entrusted to the leadership of Josef Albers.[93]

When he finds himself at the American Academy in Rome at the start of December 1950, where he remains for about two months before going to Egypt and Greece and then returning to the United States at the start of March,[94] Kahn asks himself the same question approached in *Monumentality* about the role of constructed masonry forms for contemporary materials. As soon as he arrives in Rome, he identifies "translating the masonry construction into steel and concrete" as his field of research, which not coincidentally has an impact on his concrete masterpieces.[95]

The project for the expansion of the headquarters of the School of the

[92] Team 14 is composed of William Guy Garwood and William Sharp Kirkpatrick, architects, Howard John Besnia and Edith Ayrault Rose, painters, and Evans Woollen, sculptor. The subject is proposed by Kahn with Louis-Henri Jean Charlot, an artist of Russian origin of the Colorado Springs Fine Arts Center (*Student architects, painters, sculptors design together*, in "Progressive Architecture," vol XXX, April 1949, pp. 14, 16, 18, fig. p. 14).

[93] Charles H. Sawyer, letter to Louis I. Kahn, 7 July 1950, LIK, 030.II.A.61.27.

[94] David P. Wisdom, letter to H. D. Palmer, 22 February 1951, LIK, 030.II.A.107.45.

[95] Louis I. Kahn, letter to office staff in Philadelphia, 6 December 1950, in, Eugene J. Johnson, Michael J. Lewis, *Drawn from the source. The Travel Sketches of Louis I. Kahn*, Cambridge - Massachusetts, London, MIT Press, 1996, p. 72.

33. Acropolis of Athens, 1951
(J. Hochstim, *The Paintings and Sketches of Louis I. Kahn*, New York, Rizzoli International Publications, 1991, p. 273, No. 376).

34. Pyramids of Giza, 1951
(J. Hochstim, *The Paintings and Sketches of Louis I. Kahn*, New York, Rizzoli International Publications, 1991, p. 298, No. 418).

35. Column capitals, Karnak, 1951
(J. Hochstim, *The Paintings and Sketches of Louis I. Kahn*, New York, Rizzoli International Publications, 1991, p. 293, No. 407).

Fine Arts, known later as the Yale Art Gallery, is assigned by the Division of the Arts to Kahn on 8 January 1951,[96] when he is still at the American Academy. His collaborators on the project are Douglas W. Orr, an architect from New Haven, and Anne Griswold Tyng, who works in his office. Kahn's remark "brick and stone are wonderful," written to Howe from Rome during the exchange of letters in the wake of this new commission, suggests the potential of the historical character of the materials that would be used in the Yale Art Gallery.[97]

The program for the addition to the building is issued in January and calls for Museum Functions, to host the artworks of the collection, and Teaching Functions, for art education ranging from studios to drawing classrooms, a library and offices.[98] For both types of functions, the need is emphasized to identify a system capable of offering "maximum flexibility."[99] The program leaves the questions of the material of the structure – in steel or reinforced concrete – and the physical plant elements open, to be resolved based on costs. A few weeks after the issuing of the program, the ideal type of structure for the Yale Art Gallery is defined as a "'loft-type' structure."[100]

Right from the first project versions dating back to March-April, Kahn considers a system of reinforced concrete structure to be left partially exposed only inside the building. Stone slabs are to be used to cover the concrete pillars around the perimeter, except along Chapel Street, where there is a wall without windows, perhaps evoking that "incomplete architectural composition" made of "derelict buildings," and the "abandoned palace" that is discussed in the preliminary phases of the program for the Yale Art Gallery.[101]

Already in the preliminary version of the Yale Art Gallery, Kahn introduces a particular vision in American contemporary architecture of what was becoming, thanks to the Unité d'Habitation in Marseille, the international aesthetic of *béton brut* (the work by Le Corbusier was inaugurated on 14 October 1952). But the *béton brut* of Kahn takes on the form of a structural nucleus in exposed reinforced concrete, inside a stone envelope, almost like the Roman technique of *opus caementicium*.

Thanks to the floor system, Kahn gives the Yale Art Gallery a specific formal character. The floor system contains the physical plant elements to guarantee the differentiated and flexible functioning of the "'loft-type' structure" in a way intended by Kahn as an alternative to the American contemporary tradition of the Flexible Ceiling in which artificial light, air conditioning and acoustic panels form a ceiling suspended from the structure.[102]

But starting with the first projects the integration of the structure and the physical plant systems raised difficulties. The second design, dating back to April 1951, is significant in this sense (the building is called the "Yale Laboratory of Design"). The fact that it has a reinforced concrete structure explains the appearance of the definition "concrete loft building."[103] While the reinforced concrete of the pillars, arranged in squared spans, is visible everywhere in the interiors, the finish of the "flat floor slab,"[104] also in reinforced concrete, varies from space to space: in the "Architectural School Space" the concrete is indicated as "exposed" and to be painted after stripping; in the "Gallery Space," on the other hand, "acoustical plaster ceilings" are specified to conceal the lighting and air conditioning equipment, based on the model of the Flexible Ceiling. The facade towards the garden has large openings, framed by beams and pillars clad in slabs of black slate, while the front along Chapel Street is a blind wall in Aquia Freestone, to match the finish of the contiguous building. Only as an alternative, Virginia

36-37. L. I. Kahn, D. W. Orr, Yale University Art Gallery and Design Center, New Haven, Connecticut, 1951-53.

36. Perspective drawing (LIK, 030.IV.A.370.2).

37. Perspective of galleries, 1951 (Yale University Art Gallery Collection, New Haven).

[96] Charles H. Sawyer, letter to Louis I. Kahn, 8 January 1951, LIK, 030.II.A.107.42. The contract is drawn up on 22 June 1951. For the Yale Art Gallery, besides the already mentioned essays on Kahn see also Cummings Loud, *op. cit.*, pp. 52-99.
[97] George Howe, letter to Louis I. Kahn, 22 January 1951, LIK, 030.II.A.107.42.
[98] "Art Gallery Wing," 6 January 1951, LIK, 030.II.A.107.42.
[99] *Ibid.*
[100] "Yale University - School of the Fine Arts Department of Architecture," 6 February 1951, LIK, 030.II.A.107.42.
[101] "Art Gallery Wing," 6 January 1951, LIK, 030.II.A.107.42.
[102] *The flexible ceiling*, in, "Architectural Forum," vol. XCIII, July 1950, No. 1, pp. 71-72.
[103] H. D. Palmer, letter to Noble, 10 May 1951, LIK, 030.II.A.107.45.
[104] Office of Douglas W. Orr, report attached to project submitted on 10 May 1951, LIK, 030.II.A.107.45. The internal walls are to be made with Haydite type concrete blocks. The floors of the exhibition spaces and the main staircase are to be clad in slate (in the case of the staircase an alternative in marble is proposed); instead, for the secondary staircases terrazzo is indicated.

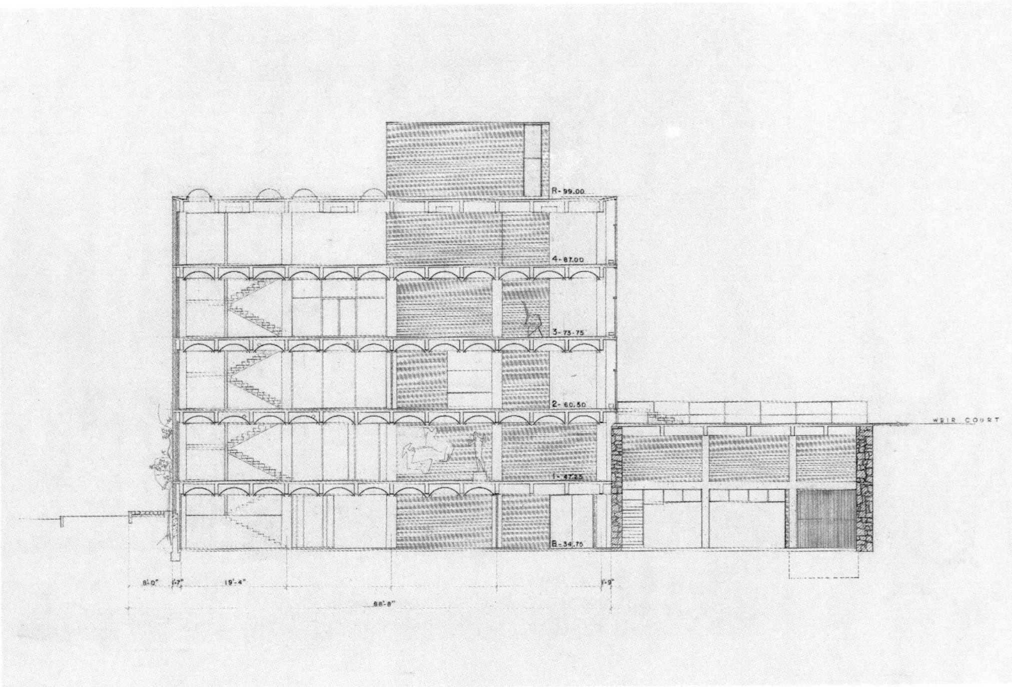
R-99.00
4-87.00
3-73.75
2-60.50
1-47.25
B-34.75
WEIR COURT
8'-0"
1'-7"
19'-4"
1'-9"
88'-8"

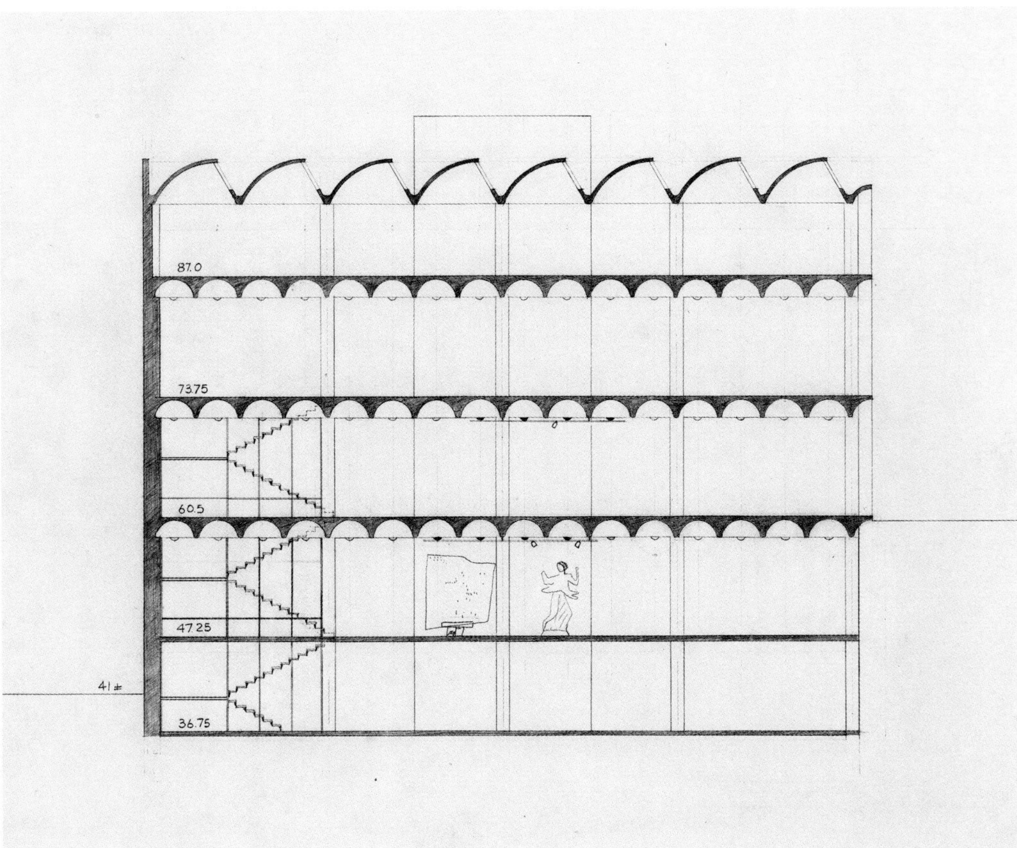
87.0
73.75
60.5
47.25
36.75
41±

38-39. L. I. Kahn, D. W. Orr, Yale University Art Gallery and Design Center, New Haven, Connecticut, 1951-53.

38. Cross-section of design laboratory, 25 September 1951 (Yale University Art Gallery Collection, New Haven).

39. Cross-section of design laboratory (Yale University Art Gallery Collection, New Haven).

type brick is suggested for this wall, but still without the white horizontal bands, corresponding to the floors, found in the final version.

In the project in May the crucial modification has to do with the span of the structure, which becomes rectangular (20 x 40 feet), with the long side parallel to Chapel Street, freeing up the spaces from the pillars. The floor structure developed in August with the help of the engineer Henry Pfisterer of New Haven is now formed by a series of parallel beams. The main physical plant conduits run between the pairs of beams, at the center of the span. The need to light the artworks induces Kahn to insert, between each pair of beams, small steel vaults, for diffused artificial light. The association of the figure of the vault with light would return in later exposed concrete works. The conduits run in the interspace between the reinforced concrete slab and the steel vaults. Thus the structure of the slab, at the first four levels, remains visible only as the intrados of the beams. The roof slab on the upper level has no vaults, since there are skylights – only along the middle beams of each span are flat ceilings introduced, to conceal the conduits.

The Tetrahedron Floor System

A breakthrough in the design of the floor system for the Yale Art Gallery is made by Kahn probably during the first days of March 1952, with the invention – in collaboration with Tyng – of a type of structure with a tetrahedron geometry, based on the solutions of Richard Buckminster Fuller for the Geodesic Domes and of Konrad Wachsmann for the various types of Space Frames. The result is the "new concrete floor system" of the Yale Art Gallery, called the Tetrahedron Floor System,[105] Tetrahedral Floor System,[106] or "space slab".[107] This type of space slab has a pattern of triangular base pyramid-shaped cavities to contain the physical plant systems and artificial lighting, and has such plastic force that it can transform the whole reinforced concrete structure into the "gigantic sculptural forms" envisioned in *Monumentality*.

40. Geodesic dome realized by the students of R. Buckminster Fuller (LIK, 030.V.F.5.7).

41. A. G. Tyng, project for an Elementary School, 1950-51. Model of classroom unit (A. G. Tyng Archive, AAUP, 074.IV.B.14).

The choice of the tetrahedron pattern for the floor system permits alternation of closed pointed coffers and cavities through which to run the various conduits and in which to house the lighting fixtures. A fundamental step to achieve this alternation is the passage from metal, normally used for the rods of such structures, like the Space Frame of the design by Tyng for the prototype of an Elementary School in 1950-51, to reinforced concrete.

[105] H. D. Palmer, "Memo Yale Art Gallery. Meeting March 11, 1952," LIK, 030.II.A.107.42.

[106] *Building engineering*, in, "Architectural Forum," vol. XCVII, August 1952, No. 5, (pp. 148-152), pp. 148-149. On the meanings of the concrete of the Yale Art Gallery see Goldhagen, *op. cit.*, pp. 59-63; Legault, *Louis Kahn and the Life of Materials* cit., pp. 220-221, and Leslie, *op. cit.*, pp. 48-83.

[107] Louis I. Kahn, letter to Anne G. Tyng, 3 September 1954, in, Anne Griswold Tyng, ed., *Louis Kahn to Anne Tyng: The Rome Letters, 1953-54*, New York, Rizzoli, 1997, p. 174.

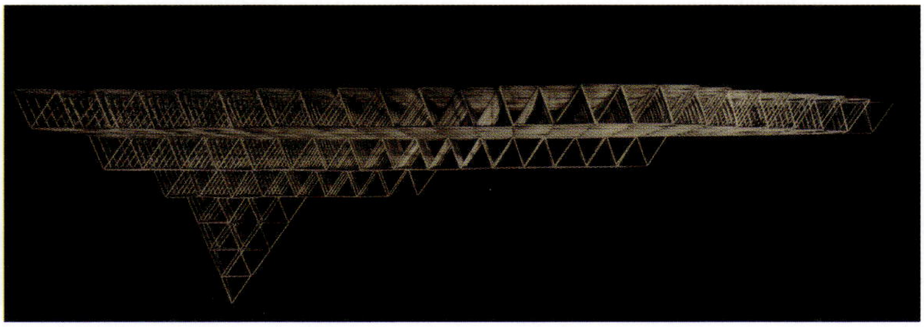

We should remember that in 1951 the geodesic domes of Buckminster Fuller are also seen as ideal structures for removable scaffolding. "Most domes are very difficult to build in concrete, the usual material for them, because they call for complicated formwork," is a current comment.[108]

If we consider Kahn's resolve expressed during his stay in Rome to "translate masonry construction into steel and concrete," we cannot rule out the hypothesis that in the tetrahedral structures he also sees the possibility of creating, in contemporary reinforced concrete, like those built by Pier Luigi Nervi, a sort of flat vault of coffers comparable to those made in the domes and vaults in *opus caementicium*.

The Tetrahedron Floor System and the Flexible Ceiling have some similar characteristics. "Except to people looking directly upward, lighting equipment is concealed yet accessible," reads the commentary in 1952 on Kahn's system, with an observation that is also true for the Flexible Ceiling with perforated grids.[109] Both the Tetrahedron Floor System and the Flexible Ceiling ensure uniform, free utilization, ready for transformation over time, of the floor surface. As in department stores or offices, in the Yale Art Gallery the works of art can be grouped in sections thanks to easily moved panels, with a special system to attach them to the floor and ceiling.

In the transformation of the floor of the Yale Art Gallery, from the early designs to the definitive version, a crucial role is probably played by examples covered in American magazines. "Architectural Forum" offers articles and publicity on different types of complex rectangular grid ceilings used to screen artificial light. One of the most important examples is the "cellular ceiling" created for an office building at 383-5 Madison Avenue in New York, whose hexagonal coffers are like forebears of the geometry of Kahn's floors system.[110] Examples of ceilings with lights built into them in an artistic way, which were familiar to Kahn and may have influenced the formal choices for the floors system of the Yale Art Gallery, include the "lunar ceilings" made by the American designer and sculptor Isamu

[108] *The dome is its own scaffolding*, in, "Architectural Forum," vol. XCV, February 1951, No. 2, p. 148. A statement by Buckminster Fuller indicates encounters between him and Kahn in the preliminary phase of the design of the floors system of the Yale Art Gallery, during the course of 1951 (Richard Buckminster Fuller, letter to John Entenza, 5 April 1965, LIK, 030.II.A.55.42). According to an article published in "Perspecta," Kahn opted for the "space frame" structure following lectures given by Buckminster Fuller at Yale (*cf. Order and Form. Yale Art Gallery and Design Center designed by Louis I. Kahn*, in, "The Yale Architectural Journal," No. 3, pp. 47-58, p. 51).

[109] *Building engineering* cit.

[110] See *Lobby of light*, in, "Architectural Forum," vol. XCVI, January 1952, No. 1, pp. 118-120. See also the "translucent ceiling" of the building at 600 Fifth Avenue, New York (*New Skyscraper*, *ibid.*, pp. 121-125).

[111] Documentation of the project for the slab of the American Stove Company and the mural sculpture *Lunar Landscape* are shown at the Museum of Modern Art of New York in 1951 (Ana Maria Torres, *Isamu Noguchi. A Study of Space*, New York, The Monacelli Press, 2000). It is in the context of the reasoning on hidden light sources explained at the conference that Kahn makes reference to the ceilings of Noguchi (Louis I. Kahn, *The Relation of Light to Form*, North Carolina State Conference, 28 February 1953, LIK, 030.II.A.56.86).

42. Cellular ceiling of the office building, 383-5 Madison Avenue, New York (*Lobby of light*, in, "Architectural Forum," vol. XCVI, January 1952, No. 1, p. 120).

43. Detail of the concrete vault of the Basilica of Maxentius, Rome (photograph by the author, 2009).

44. I. Noguchi, ceiling for the American Stove Company Building, Saint Louis, Missouri, 1948 (A. M. Torres, *Isamu Noguchi. A Study of Space*, New York, The Monacelli Press, 2000, p. 273).

Noguchi in 1947-48 for the Time-Life Building at Rockefeller Center in New York, and for the American Stove Company in Saint Louis[111].

The change introduced by Kahn in the configuration of the floor system does not involve only the geometry of the structure and the insertion of the various physical plant systems. With the Tetrahedron Floor System, Kahn chooses to erect, inside the body of the building, a powerful reinforced concrete structure to be left exposed, without any kind of ceiling. The choice of this material also leads Kahn to adapt the structure of the floor system, abandoning the earlier metal vaults and taking the tetrahedral geometry as the most suitable solution to bring out the potential of the new material. So Kahn's instruction to the students working on the exercise on plastic, urging the "dramatizing [of] the visual and structural potentialities," is applied in this case to reinforced concrete.

The model for the powerful concrete structure of the Yale Art Gallery can only be the platform of the Unité d'Habitation.[112] The fact that the modification of the design happens in March can be seen in relation to the publication, that same month, of an article on the Unité d'Habitation in "Architectural Forum."[113] It is no coincidence that the platform, from Kahn's viewpoint, seems like the first fundamental example of the insertion of physical plant systems inside cavities made in an exposed concrete structure, a solution he is seeking to get beyond the criteria of the Flexible Ceiling and the Cellular Ceiling.[114] The reasons that prompt Kahn to make this radical breakthrough in his design can be found in the intuition of attempting to create the first American work capable of being a counterpart to what is destined to become the founding example of *béton brut*. With the exposed concrete pillars, beams, walls and floors system of the new project for the Yale Art Gallery, Kahn undertakes what can be considered the pursuit of his own personal vision of the *béton brut* of Le Corbusier, which will continue all the way to his last works. With the Tetrahedron Floor System patterned with cavities, Kahn manages to translate the great urban potential of the platform of the Unité d'Habitation, its way of becoming an artificial ground, into a structure on the scale of spaces for exhibitions and cultural and artistic activities. The Tetrahedron Floor System in exposed concrete asserts itself as a powerful alternative to the various American decorative approaches to concrete floor slabs still accepted in the 1940s for monumental buildings, and covered in the term "architectural concrete slabs."[115]

Only the discovery of the potential of the platform of the Unité d'Habitation in *béton brut* and the typical issues of the Flexible Ceiling, with air conditioning conduits, artificial lights and screening, can explain Kahn's insight that makes him turn to the geometrics of Buckminster Fuller and Wachsmann, also appreciated by Tyng, to find a possible solution for an exposed concrete floors system with compartments for physical plant elements.[116]

[112] *Le Corbusier's Marseilles Apartments*, in, "Architectural Forum," vol. XCVI, March 1952, No. 3, pp. 142-145. Eero Saarinen visits the Unité d'Habitation in September 1951, when the project for the Yale Art Gallery is still being developed (*cf.* Gargiani, Rosellini, *op. cit.*, p. 60, note 228).

[113] *Le Corbusier's Marseilles Apartments* cit.

[114] "Corbusier," Kahn writes in his notebook, "feels what a space 'wants to be,' passes through order impatiently and hurries to form. In Marseilles order was string. [...]. At Ronchamp order is only dimly felt in form born of dream" (Louis I. Kahn, notebook, LIK, 030.VII.4).

[115] *Architectural concrete slabs*, in, "Architectural Forum," vol. LXXII, February 1940, No. 2, pp. 101-106.

[116] See Tyng, *op. cit.*, p. 47.

45-46. L. I. Kahn, D. W. Orr, Yale University Art Gallery and Design Center, New Haven, Connecticut, 1951-53. Study drawings of the effects of light on the surfaces of tetrahedra (LIK, 030.IV.A.370.2).

47. Mr. and Mrs. Morton Weiss House, Philadelphia, Pennsylvania, 1947-50. Study for the mural, 1955 (J. Hochstim, *The Paintings and Sketches of Louis I. Kahn*, New York, Rizzoli International Publications, 1991, p. 300, No. 422).

48. Study for a mural based on Egyptian motifs, 1951 (J. Hochstim, *The Paintings and Sketches of Louis I. Kahn*, New York, Rizzoli International Publications, 1991, No. 425, p. 302).

49. J. Albers, study for a structural constellation, 1950 circa (F. A. Horowitz, B. Danilowitz, *Joseph Albers: to Open Eyes, The Bauhaus, Black Mountain College, and Yale*, London, Phaidon Press Limited, 2009, p. 120).

50. Paper construction, exercises by the students of J. Albers at the Yale University (F. A. Horowitz, B. Danilowitz, *Joseph Albers: to Open Eyes, The Bauhaus, Black Mountain College, and Yale*, London, Phaidon Press Limited, 2009, p. 146).

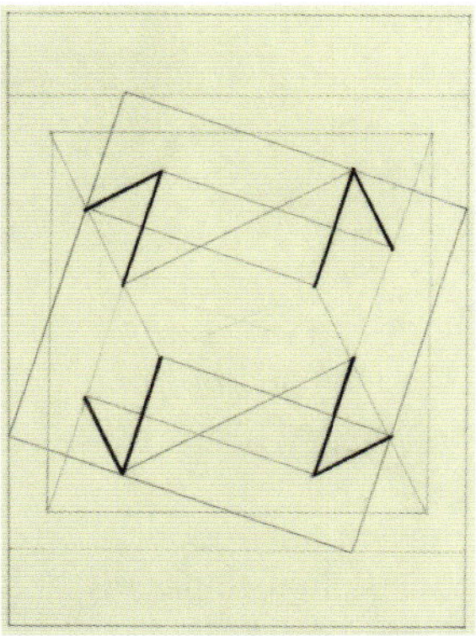

The use of brick cladding for the front along Chapel Street and exposed concrete in the interior forecasts a line of research in Kahn's future architecture aimed at uniting what were the two basic materials of the Roman technique of *opus caementicium*.

But Kahn's tetrahedral structure also has another, more important value, quite apart from the research of Buckminster Fuller, Nervi or Tyng. Imagined in the space of the gallery, floating over the artworks, it seems like a variation of the Structural Constellation developed by Albers with paper models, already seen by Kahn in 1949.[117] The multiplicity of planes of which the Tetrahedron Floor System is composed, and the effects of light reflecting on its many inclined surfaces, make it into an application of the exercises on vision assigned by Albers to the students in the art courses of Yale University. Precisely in the light of the artistic values of models like those of Albers, Noguchi, Buckminster Fuller, Tyng and Le Corbusier, Kahn's floor system for the Yale Art Gallery becomes the emblem of an art school striving for didactic renewal in keeping with the principles of the avant-garde vision, and transforms the geometrics of artistic exercises done on paper into a monumental structural device in *béton brut*. From this standpoint, the Tetrahedron Floor System can even be seen as the transposition in load-bearing structure of the contemporary graphic studies by Kahn himself for artworks, like the murals for the Weiss House.[118]

The transformation of the volume of the staircase, in the project that includes the Tetrahedron Floor System, into a hollow cylinder with three ramps around a triangular void, covered by a triangular beam, reflects the pursuit of a structure that can express an approach to art education focused on geometric abstraction, and the figure of the triangle – an emblem, at this point, of the entire Yale Art Gallery.

From the model to the definitive project: tetrahedral geometry and construction in conflict

Studies on the colors of the concrete of the Yale Art Gallery,[119] of the wall infills ("iron spot brick - light gray"[120]), the window casements ("steel, galvanized, painted [Hope]"[121]), the type of stone for the cladding of the pillars (Aquia Freestone), and the type of brick for the wall along Chapel Street ("Virginia hand made [Locher]"[122]) have been under way since the start of 1952.

The floor structure is discussed in a meeting on 11 March by Howe, Solomon, Pfisterer, Kahn and Orr. The name of the Tetrahedron Floor System, which arises on this occasion, emphasizes the fact that the structure has now been freed of any contemporary ceiling type.[123] Nevertheless, the opinions lean towards criticism of Kahn's floor system, due to the increased quantity of steel, the construction costs, the complexity of the formwork and the implementation, and the need to make a prototype for structural testing in order to obtain a building permit.

In spite of the negative views and the official decision made at the end of the meeting "to go back to the original simple beam and floor

[117] *Cf.* Frederick A. Horowitz, Brenda Danilowitz, *Josef Albers: To Open Eyes*, London, Phaidon Press Limited, 2006, p. 45.
[118] Robert McCarter, *Louis I. Kahn*, Paris, Phaidon, 2007, p. 53. For the dating of the first studies for the mural to 1951, see E. Bittermann, letter to Louis I. Kahn, 15 June 1951, LIK, 030.II.A.60.34.
[119] "Laboratory of design. Notes on the meeting at Mock Up building - January 23, 1952," LIK, 030.II.A.107.42.
[120] Douglas W. Orr, Louis I. Kahn, "Design Laboratory - Yale University. Memorandum description and list of materials," 20 March 1952, LIK, 030.II.A.107.45.
[121] *Ibid.*
[122] *Ibid.*
[123] Palmer, "Memo Yale Art Gallery. Meeting March 11, 1952" cit.

51-52. L. I. Kahn, D. W. Orr, Yale University Art Gallery and Design Center, New Haven, Connecticut, 1951-53. Cardboard model (LIK, 030.V.A.11.3; LIK, 030.V.A.11.4).

system,"[124] Kahn intends to proceed with development of his Tetrahedron Floor System and to make the structure in exposed reinforced concrete. In the meantime, also in March, the contractor is selected, the George B.H. Macomber Company of Boston, selected over others for its "extensive experience with reinforced concrete construction."[125]

The quality of the concrete to be left exposed has already been studied in detail by Kahn and his staff in relation to the design of the formwork. In the documents we find references to a kind of architectural concrete to be obtained not with the typical American plywood formwork, but with wooden boards.

"Because of the nature of the structure," Orr writes in a letter on 21 March, "the character of the concrete work is of extreme importance, as much of it will be left exposed. The quality of the formwork desired for columns and beams is narrow vertical boarding rather than ply form and must be very carefully studied and executed. The manner of pouring each floor with particular reference to the exposed concrete will be a matter of extreme interest and will require the utmost cooperation from the contractor to produce 'Architectural' concrete."[126]

The choice of boards instead of plywood must be carefully considered. These boards are precisely the most telling clue regarding Kahn's desire to give the pillars and beams a formal character in line with the *béton brut* of the Unité d'Habitation of Marseille. Of equal importance is the fact that thanks to the dense pattern of imprints, Kahn can make lines on the concrete surfaces in tune with the graphic pattern of the tetrahedral floor system.

Towards the end of March and in April, Kahn makes a cardboard model of the Tetrahedron Floor System representing a "parallel chord, multi-planar, truss system (space frame) of equilateral triangles with alternate inclined triangles, in each of the three directions, also made solid."[127] After having found that the quantity of steel required to make the structure is too large with respect to the quantity needed for a normal concrete slab, and that the hierarchy of the structural members is not clear, Pfisterer proposes modifications to Kahn's model in order to "express the structural skeleton more significantly and provide a ceiling pattern retaining the tetra-

hedron form with alternate openings."[128] The transformation of the structural functioning of the faces of the tetrahedra, which for reasons of construction cannot all be made as load-bearing parts, is discussed. The crucial question has to do with the making of the concrete. In fact, the tetrahedral pattern is so complex that certain parts of the formwork have to be stripped from above, since it is not possible to simultaneously pour the horizontal slab to obtain a continuous structure.[129]

Precisely due to the need to divide the pouring of the Tetrahedron Floor System into two phases, in order to remove certain formwork of the pyramids and add the horizontal ones of the top slab, the structure of the floor is changed, introducing a hierarchy that winds up reducing the tetrahedron floor to a conventional system of parallel beams with non-load-bearing crosspieces. Kahn, in any case, has no intention of giving up the tetrahedral image, and even insists on keeping the thickness of all the

[124] *Ibid.*

[125] Charles H. Sawyer, letter to Laurence G. Tighe, 18 March 1952, LIK, 030.II.A.107.42. In March 1952 three construction companies are contacted: McGinn & Company of Naugatuck, Dwight Building Company of New Haven, and the George B. H. Macomber Company of Boston (H. D. Palmer, "Memo Yale Art Gallery. Meeting March 14, 1952," LIK, 030.II.A.107.42).

[126] Douglas W. Orr, letter to G. Vincent Maconi, 21 March 1952, LIK, 030.II.A.107.42.

[127] "The floor system is a modification of a design concept developed in model form fastening a continuous plane element to the apices of open-base, hollow, equilateral tetrahedrons which were joined at the vertices of the base triangles in the lower plane" (Henry A. Pfisterer, "Design Laboratory - Yale University. Description of the structural System," n.d., [1952], LIK, 030.II.A.107.42). Kahn's model is critically analyzed by Pfisterer (Henry A. Pfisterer, "Memorandum," 8 April 1952, Louis I. Kahn Collection, LIK, 030.II.A.107.42).

[128] Pfisterer, "Design Laboratory - Yale University. Description of the structural System" cit.

[129] "Unfortunately, to permit removal of the interior formwork the top slab must be poured separately from the rest of the structure, necessitating a construction joint with unavoidable loss of strength at the critical connection between the top of each pyramid and the slab. Inability of concrete to transfer horizontal shear at joints required a modification of the original scheme. One face of each pyramid was made part of a 5" thick sloping joist running continuously across the entire floor" ([Verner Read], *Tetrahedral Floor System*, in, "Architectural Forum," vol. XCVII, November 1952, No. 5, p. 149).

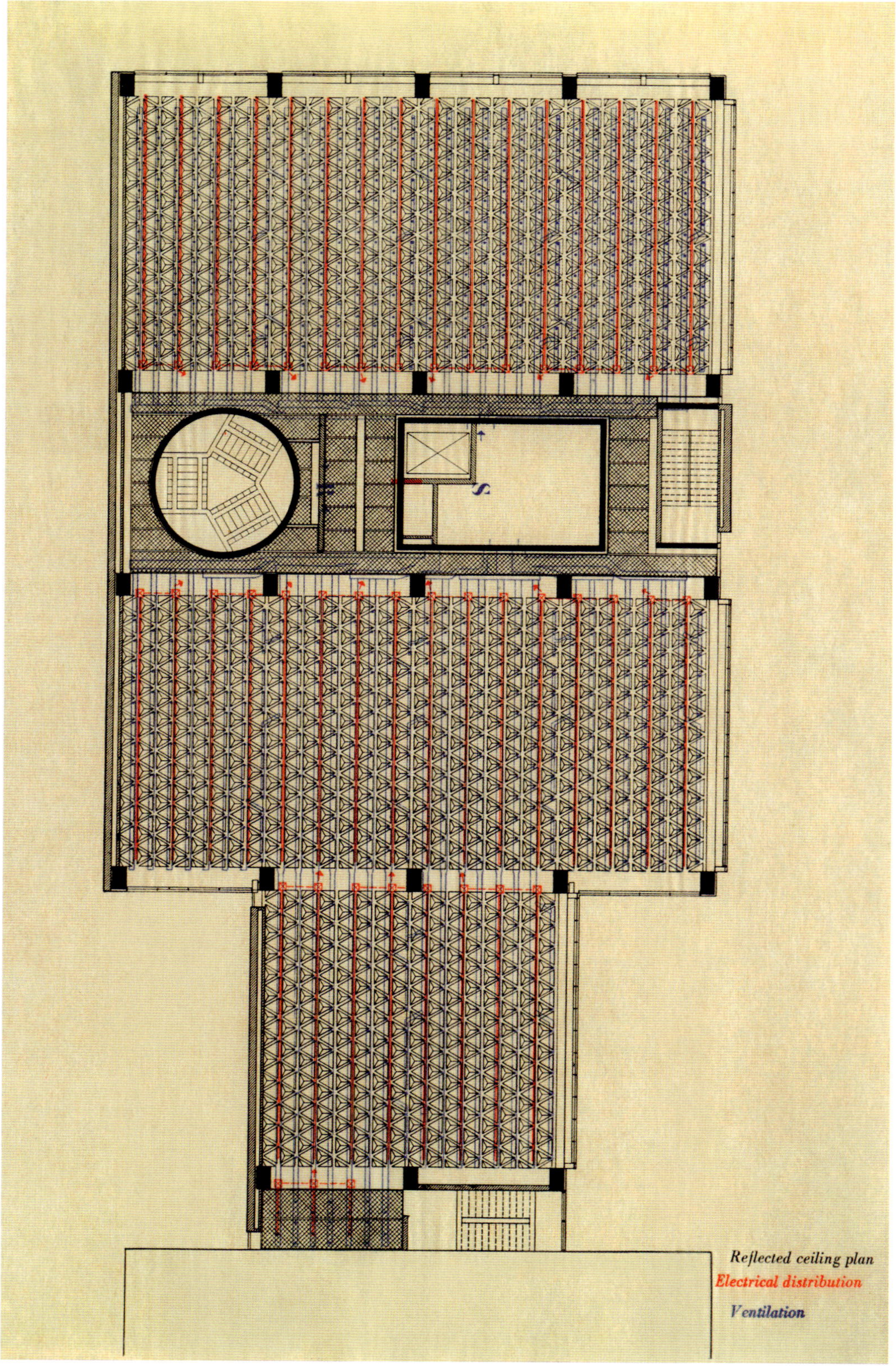

Reflected ceiling plan
Electrical distribution
Ventilation

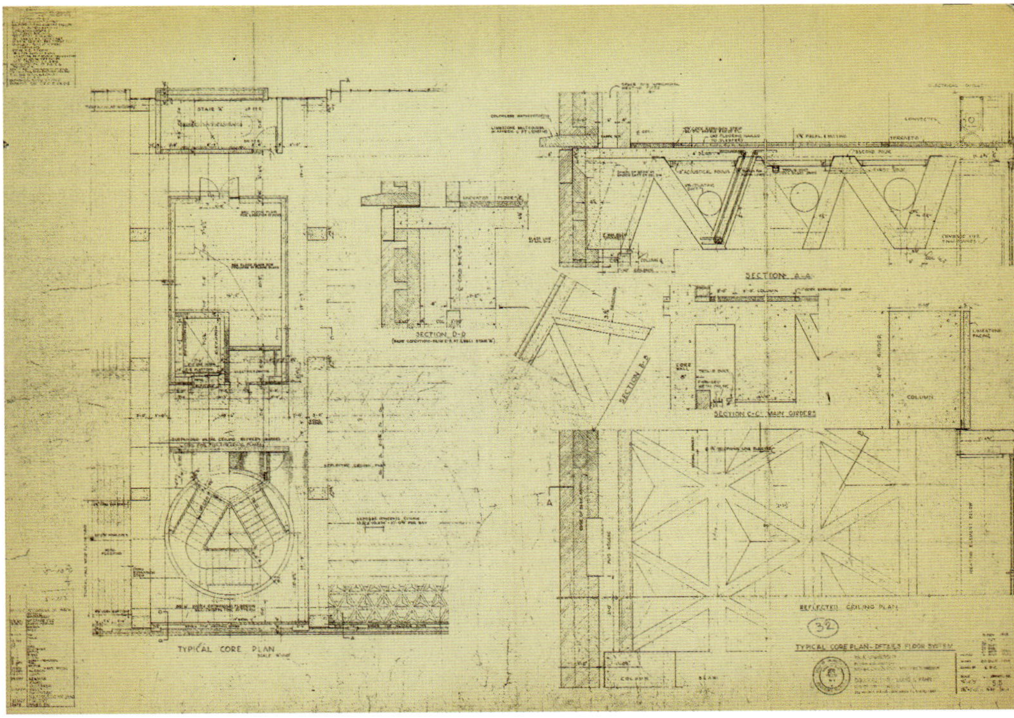

53-55. L. I. Kahn, D. W. Orr, Yale University Art Gallery and Design Center, New Haven, Connecticut, 1951-53.

53. "Reflected ceiling plan. Electrical Distribution. Ventilation" ("Perspecta", [1955], No. 3, p. 49).

54. "Typical core plan-details floor system," 20 May 1953 (LIK, 030.I.C.370.001).

55. Isometric view of the Tetrahedron Floor System (LIK).

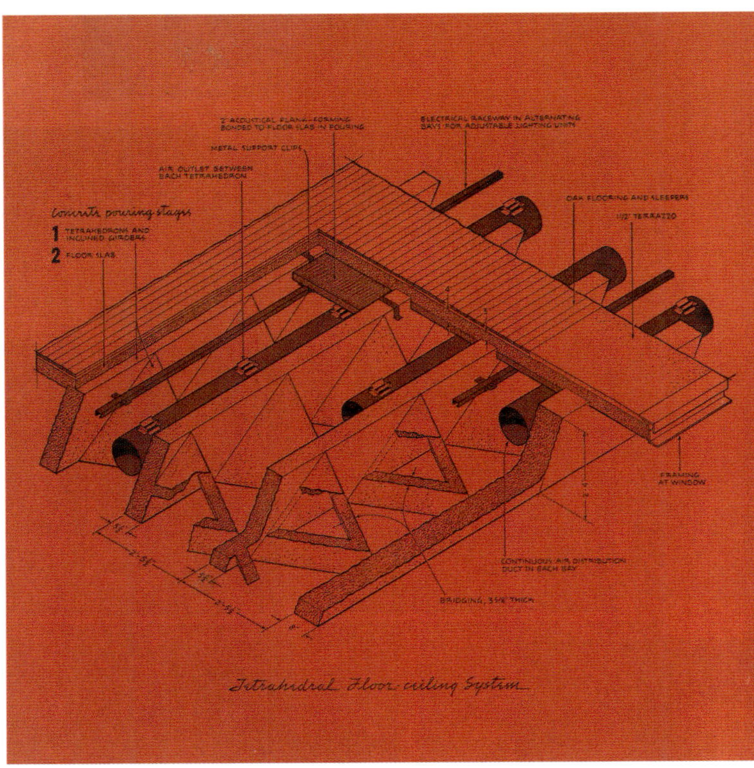

members constant. At this point, for him the tetrahedral pattern determines a floor system capable of generating a space that is no longer unidirectional like the one he had designed with the solution of parallel beams and small vaults. Rather than pursuit of truth of the individual parts of the structure, what interests him are the spatial qualities of that pattern, its meaning as a symbolic icon of the School of Fine Arts, its potential spherical form based on its derivation from the Geodesic Dome, and the integration of the physical plant system with the rhythm of the openings for lighting and ventilation. It is clear that Kahn is willing to develop a floor pattern that is a cross between structure and suspended ceiling, a "three-dimensional concrete ceiling structure" as the floor system of the Yale Art Gallery is significantly defined in 1953.[130] But this vision clashes once again with that of Pfisterer, who instead wants to make visible the radical transformation of the Tetrahedron Floor System following the introduction of a hierarchy among the constituent parts of the pattern.[131] The terminology used in the documents bears witness to the desire to distinguish the homogeneous pattern of Kahn in load-bearing members, known as "girds" or "ribs," and crosswise elements, known as "joists" or "diaphragms." Precisely the appearance of girds and joists in the pattern of the tetrahedra leads, according to Pfisterer, to the need for variation of the thickness of the parts to make their different structural role visible - 4 inches for the girds and 2.5 inches for the joists; the main border beams should present a "solid soffit." In this phase of the project, through Pfisterer's intervention, originates the contradiction that survives in the built work, between the idea of a continuous pattern and a structure with ranked elements.

Pfisterer's criticism of Kahn's model with members of uniform thickness is expressed in no uncertain terms: "The finished construction will not express the relative structural significances of the various elements. We will see a uniform pattern of ribs in 3 directions. The ribs in only one or, at most, two directions will be the primary lead members with the others being of secondary importance structurally."

The geometric complexity of the ceiling requires careful study of the formwork at the points of contact between the border beams and the tetrahedral pattern. "Molds," Pfisterer writes, "require layout study to properly meet exterior wall beam parallel to main joists similar on each side. Main design objection is discontinuity at girders necessitating partial triangles or zigzag pattern of girder soffit. However, this same problem exists with the original concept."[132]

[130] Caption of the photograph of a portion of the floor system, LIK, 030.IV.310.125.
[131] Pfisterer, "Memorandum" cit.
[132] *Ibid.*
[133] On 18 April the design of the floor system with the rods of different thicknesses is prepared (Anne Griswold Tyng, study drawing for the development of the tetrahedron floor, LIK, 030.IV.A.370.2.11). See also Leslie, *op. cit.*, p. 62.
[134] Pfisterer, "Memorandum" cit.
[135] Robert B. Newman, letter to H. D. Palmer, 17 April 1952, LIK, 030.II.A.107.45.
[136] "The sound absorbing material," Newman writes, "could either be placed in the forms before pouring the upper slab or could be attached to the slab through the triangular openings after the formwork has been removed. You might, for example, consider the use of a 1" thick material such as Porex, Insulrock, Durisol or the like. These materials can be painted without any loss of efficiency but they will not give a high degree of light reflection even when painted white. Alternatively, you might cement to the slab after the forms have been removed a standard perforated acoustic tile which has a better light reflectivity and possibly a slightly better appearance. If the upper space is to be dark, this is probably not such a problem" (Newman, letter to Palmer, 17 April 1952, cit.).

Also considering the static behavior of the floor system, which varies if the upper surface is poured at the same time as the small beams or afterwards, Pfisterer would even prefer to create a floor system that is completely different from the Tetrahedron Floor System, where instead of the inclined elements there would be vertical beams, to make it possible to use prefabricated parts in the pouring of the concrete.[133]

To gain the approval of Yale University officials for the high cost of the Tetrahedron Floor System, Kahn is asked to specify to what extent the characteristics of his configuration are capable of granting "a greater sense of space," also taking the limited height of the floors into account; he is also asked to explain the system's "acoustical properties" and the type of lighting system it would permit.[134]

The company Bolt Beranek and Newman, based in Cambridge, Massachusetts, hired for the study of the acoustics, reports that with the presence of brick walls and glass partitions, the Tetrahedron Floor System of Kahn, that ensured "diffusion and break-up of sound," would not be able to control reverberation without the use of sound-absorbing materials like Porex, Insulrock or Durisol panels, or perforated tiles.[135] The position recommended for the panels or tiles is in the openings of the Tetrahedron Floor System. The different nature of the two types of acoustic materials implies different modes of installation, either during the process of construction of the floor system or after the pouring has been completed. Bolt Beranek and Newman also suggests the possibility of inserting the panels during floor construction, attaching them over the formwork, or in the case of the use of tiles, applying them through the tetrahedral openings. Nevertheless, the solution of stripping the formwork after the pouring of the slab seems impossible, precisely due to the configuration of the floor. Bolt Beranek and Newman also provides indications on the different light reflection qualities of the various materials in the openings – a fundamental question for a floor system that also has to contain artificial lighting fixtures for artworks.[136]

The complexity of the tetrahedral pattern prevents Kahn from reaching the goal of an integrated system of structure and physical plant extending into every sector of the Yale Art Gallery. Kahn has to sacrifice that geometry in the floor system of the narrow service spaces, crossed by the volumes of the staircases, where the pattern would have been too fragmented. The choice of the type of ceiling for these spaces, in any case, sheds light on the desire to create a connection: the ceiling is made with a grid of lozenges that leaves a view of the physical plant systems and the reinforced concrete slab, acting as a substitute for the contiguous tetrahedral floor. The breaking up of the Tetrahedron Floor System caused by the middle strip of services reveals an unresolved contradiction between indications of continuity, comparable to those of the Flexible Ceiling, and the ranked organization of spaces.

The construction of the floor system prototype

The Tetrahedron Floor System of the Yale Art Gallery is so unusual that structural calculations are difficult using the parameters of the building regulations in New Haven. Also for this reason, Kahn contacts the engineer William H. Gravell of the Keast & Hood Company. Nicholas L. Gianopulos and Thomas J. Leidigh of Keast & Hood, who take the place of Gravell after his death, bring new solutions to the project, such as that of shifting the top of the pyramids to leave a large cavity for the conduits between the tetrahedra and the top slab.[137]

To obtain the approval of the Building Department, it is decided to make a half-span prototype of the Tetrahedron Floor System at the worksite, with four inclined beams, to conduct load tests.[138] The prototype also offers an opportunity to study the formal effects of the use of the formwork, to achieve a refined surface of the exposed concrete.

The construction of the prototype of the floor system is done during the week from 28 July to 1 August 1952, by the Macomber Company. The load testing is done by the engineering firm Thompson and Lichtner of Brooklyn.[139]

The formwork for the pillars, like those of the beams, are made with panels composed of wooden boards held in position by braces perpendicular to their length. The boards are joined using the system of "tongue and groove floor boards." After assembly on a workbench, the panels are laid out to be sprinkled on the side that comes into contact with the concrete with a product that facilitates stripping; they are then assembled with the others to create the formwork.

The difference between the two types of formwork – wooden boards for the pillars and beams, metal for the tetrahedra – depends on the differing geometry of the structural elements. The choice of metal reveals Kahn's concern with obtaining the most perfect possible making of his Tetrahedron Floor System.

[137] Nicholas L. Gianopulos, interview with Anna Rosellini, Philadelphia, July 2010.
[138] See Henry G. Falsey, letter to Henry A. Pfisterer, 18 June 1952, LIK, 030.II.A.107.45. For the construction of the prototype the formwork system and the type of concrete compound are also studied, including the addition of an "air-entraining agent" to improve workability (*cf.* Pfisterer, "Design Laboratory - Yale University" cit.). On 20 May the phases of the design, construction and testing of the prototype, to be built in two weeks, are planned – excavation is slated for 10 June (H. D. Palmer, "Yale Art Building," 21 May 1952, LIK, 030.II.A.107.42).
[139] Henry A. Pfisterer, letter to Miles Claire, 1 July 1952, LIK, 030.II.A.107.45.

56-59. L. I. Kahn, D. W. Orr, Yale University Art Gallery and Design Center, New Haven, Connecticut, 1951-53.

56-57. Construction of the prototype (LIK, 030.IV.A.370.1.32 and LIK, 030.IV.A.370.5).

58. Formwork for the construction of the prototype (LIK, 030.IV.A.370.1.27).

59. View of the extrados of the tetrahedra with air outlet pipes and some acoustical planks (LIK, 030.IV.A.370.1.3).

60-62. L. I. Kahn, D. W. Orr, Yale University Art Gallery and Design Center, New Haven, Connecticut, 1951-53.

60. Photograph of the prototype (LIK, 030.IV.A.370.13).

61. Kahn during the visit of the prototype ("Architectural Forum," vol. XCVII, November 1952, No. 5, p. 149).

62. View of the prototype during load testing and contruction of the columns formwork (LIK, 030.IV.A.370.1.15).

The parallel inclined beams and their connection in the form of "diaphragms" have to be poured simultaneously to obtain the image Kahn wants: "a desired finished ceiling effect of exposed concrete."[140] Therefore the formwork system for the tetrahedral pattern is made in metal, with different workmanship and implementation. For the hollow pyramid-shaped openings with a triangular base a formwork composed of two pyramids is created: the inner pyramid is in contact with the poured concrete that will be left visible, and is made of welded sheet metal; the outer pyramid to contain the poured concrete is in sheet metal bolted on special border tabs and is formed by just two sides (one is left open). Both pyramid shells are bolted onto the horizontal support of plywood panels, to prevent movement and imprecise angles during the pouring. For the inclined beams sheet metal formwork reinforced by a frame are used. Wooden dowels keep the two parallel formwork of each beam in place. To allow the workmen to move around during

the pouring, boards are installed leaving only slim openings of the beams through which to pour the concrete. The concrete is vibrated, also to make it penetrate all the crevices of the pyramids. Once the pouring and stripping operations of the tetrahedral system are finished, horizontal acoustic panels of the Tectum type, produced by the Tectum Corporation of Columbus, Ohio[141], are installed, forming the permanent formwork for the pouring of the top slab. The panels remain visible through the gaps of the tetrahedral structure. Precisely the complex form of the tetrahedra and the technical need, dictated by the formwork, of pouring the concrete in two phases, confirm the transformation into a system of two not perfectly correlated parts, which in Kahn's original idea would have formed a single structure instead.

[140] Henry A. Pfisterer, "Design Laboratory - Yale University," 21 May 1952, LIK, 030.II.A.107.45. "Pouring of the floor construction," Pfisterer writes, "will be done in two stages. Ribs and diaphragms will be poured in the first operation to the bottom of the top slab using metal forms. Side forms for ribs and top forms for diaphragms will be stripped after two days to allow installation of the circular sheet metal ducts. Top slab will then be poured as soon as possible using acoustical soffit forms which will remain in place. Bond between ribs and top slab, to provide for horizontal shear and develop T-beam action will be obtained by doweling of shear reinforcement in ribs and by transverse keys formed in the top of the ribs. Bottom forming for floor system will be constructed to maintain shoring under ribs at mid or third points of span for not less than two weeks after slab is poured" (*ibid.*).

[141] See Robert E. Rosene, letter to Louis I. Kahn, 14 February 1956, LIK, 030.II.A.65.36; "How prominent architects use TECTUM Modern architecture at Yale art gallery and design center by Douglas Orr, F.A.I.A – Louis I. Kahn, F.A.I.A," 14 February 1956, LIK, 030.II.A.65.36.

63. L. I. Kahn, D. W. Orr, Yale University Art Gallery and Design Center, New Haven, Connecticut, 1951-53. Photograph of the metal formwork for the tetrahedra (LIK, 030.IV.A.370.14).

The load tests are done at the end of August.[142] The experiences with the prototype make it possible to set the thickness of the parts of the floor system[143] and to solve problems such as that of ensuring the stiffness of the beams and the inclined triangular elements, avoiding the raising of the formwork during the pouring phases. To prevent this from happening, a new mix-design is developed, with a way of working the concrete to permit a "minimum of internal vibration."[144] The Macomber Company is authorized to construct other formwork for the tetrahedra, so as to have the right number to pour, in the final phase, a floor sector corresponding to a span of four pillars.[145]

The quality of the concrete obtained in the prototype with the metal formwork for the tetrahedra meets with positive reactions: "it was judged that these locked as well as expected, although there are still some bugs to be worked out."[146] But the view of the prototype must have raised some doubts regarding the finishing of the works on the part of certain members of the institution, since Orr and Kahn hear talk of the hypothesis of eventually painting certain parts of the floor system. "Painting Tetrahedrons?" is the question at the center of a letter of the office Orr-Kahn to Sawyer, in which explanations are demanded regarding what parts should be painted – directly on the concrete of the tetrahedra, or the conduits and acoustic panels visible through the openings – in order to make decisions about the working methods: in the first case, it would be necessary to alter the process of the tetrahedra, making it more economical, because if it is worked "to be exposed" the concrete "is expensive," and it would make no sense to paint it; in the second case it would be necessary to paint the panels and conduits before their installation, because once the concrete had been poured the operation of painting would be "nigh impossible."[147]

At the meeting on 23 September, with Kahn, Palmer, Howe and Sawyer, among others, it is decided to leave all the materials of the floor system exposed, both the concrete and the panels, discarding the idea of painting and confirming an aesthetic that aims to achieve *brut* truth and an artistic effect based on a vision of the textures of materials used without any finishing treatment.[148]

Even prior to the making of the work, American architecture magazines, and especially "Architectural Forum" (Kahn is contacted by the editorial staff) and "Progressive Architecture" (the editors contact Pfisterer), are interested in publishing the phases of the design and the construction process of a floor system that promises to be exceptional for the complexity of the structure, as already shown by the prototype.[149]

In the report following a visit to the prototype of the "new concrete floor system based on tetrahedral shapes," Verner Read, a contributor to

[142] Burton H. Holmes, letter to Douglas W. Orr, 7 August 1952, LIK, 030.II.A.107.45. A document shows that the load trials were planned for 28 August (Douglas W. Orr, letter to Burton N. Holmes, 12 August 1952, LIK, 030.II.A.107.45).
[143] The report indicates that "the concrete placing experience indicated that thicknesses used are a reasonable minimum for cast-in-place work of this type" (Pfisterer, "Design Laboratory - Yale University" cit.).
[144] Pfisterer, "Design Laboratory - Yale University" cit.
[145] H. D. Palmer, "Yale Design Laboratory," 11 August 1952, LIK, 030.II.A.107.45.
[146] *Ibid.*
[147] H. D. Plamer, letter to Charles H. Sawyer, 19 September 1952, LIK, 030.II.A.107.45. Kahn reiterates the decision to avoid painting the concrete structures, also in the case of the aluminium grille ceiling applied to the flat slab in the zone of the service spaces: "The ceilings and walls above the mesh will not be painted [...], and the painting of the ducts above the mesh will be quoted as an alternate" (H. D. Palmer, letter to George B.H. Macomber

"Architectural Forum," underlines the characteristics of a structure capable of incorporating and making available for maintenance the conduits of the lighting and air conditioning systems, but also remarks on the contradiction between a pattern based on the geometrics of Buckminster Fuller done with slender metal rods, and its translation into reinforced concrete (the weight of the Tetrahedron Floor System is 60% more than that of a conventional slab of the same size).

"Its floor design," Read writes, "was motivated by Fuller's theories which convinced Kahn of the efficiency of tetrahedron as a basic structural frame. While Fuller concentrates upon pure structure, striving to carry the greatest load with the least material, Kahn saw the possibility of adapting Fuller's three-dimensional metal space-frame to a fireproof concrete structure."[150]

Read makes direct comparison between the reinforced concrete floor of Kahn and the metal version of Buckminster Fuller, known as the "Octet truss" made by students at North Carolina State College, in Raleigh, under the guidance of James W. Fitzgibbon, a collaborator of Buckminster Fuller.[151]

After reading Read's piece, Kahn advises that its publication be postponed; "such a negative article" is his drastic assessment.[152] Kahn provides the definition, added by hand to Read's text, of the "breathing ceiling" regarding the floor system of the Yale Art Gallery, where the air, as he explains, runs in the conduits and "filters down through the openings."[153]

In the photographic views of the floor shot at the prototype, pictures of details made "by looking directly upward" prevail, with the aim of showing the functioning of the cavities and the pyramids.[154] A note in September about one photograph of the prototype is significant: "The photographer was directly underneath and aimed his camera straight up at the underside. The geometrical pattern of the reinforced concrete construction conceals the two pipelines that run through the entire floor. The smaller pipe at the top is for electrical wiring and electrical fixtures will be plugged in, and lights will be placed within the recesses. The large pipeline, running horizontally through the middle of the photograph, is for air conditioning units."[155]

The first American construction in *béton brut*

Based on the construction of the prototype of the Tetrahedron Floor System a report is drawn up containing the specifications for the concrete and the formwork of pillars, beams, walls and floors, to prepare for the construction of the Yale Art Gallery. "Forms for tetrahedron system shall be of not less than 14 gauge metal with angle or bar stiffeners as required to prevent bulging," the report indicates for the floor system.[156] Then all the parts to be included in the floor system are decided on, from the pipes to the

Company, 4 June 1953, LIK, 030.II.A.107.45). "Walls & ceiling above mesh: Not to be painted" is repeated in another document, also dated 4 June (Office of Douglas W. Orr, "Addendum 1 to painting specifications," 4 June 1953, LIK, 030.II.A.107.45).

148 H. D. Palmer, "Design Laboratory - Yale University," 23 September 1952, LIK, 030.II.A.107.42. On 15 October an addition cost of 6000 dollars is recorded for "changed formwork" ("Design Laboratory Yale University," 15 October 1952, LIK, 030.II.A.107.45).

149 *Cf.* Burton H. Holmes, letter to Douglas W. Orr, 7 August 1952, LIK, 030.II.A.107.45, and Douglas W. Orr, letter to Burton N. Nolzes, 12 August 1952, LIK, 030.II.A.107.45.

150 [Verner Read], *Tetrahedral Floor System*, in, "Architectural Forum," vol. XCVII, November 1952, No. 5, p. 149.

151 *Ibid.* See *Space frame discussed*, in, "Architectural Forum," vol. XCVIII, February 1953, No. 2 (pp. 154-160), p. 158.

152 Louis I. Kahn, draft of letter, n.d., [1953], attached to typewritten manuscript of the article, LIK, 030.II.A.107.44.

153 *Ibid.*

154 Caption of the image of a prototype detail, in, [Read], *op. cit.*, p. 148.

155 Note attached to image of the prototype, September 1952, LIK, 030.II.A.107.41.

156 "Division. Concrete. Y.U. Design Lab.," 4 August 1952, LIK, 030.II.A.63.33.

pivots of the doors "which have to be accurately located before the slab can be poured."[157] The discussion on one detail of the pillar formwork reveals what would become a constant concern in Kahn's subsequent works. This is the assembly with the corner splint, known as a chamfer, of the panels for the pouring of the concrete of a pillar. The specification that the "corners" should be "chamfered with as small a chamfer as possible" points to Kahn's goal for his architectural concrete.[158]

Work on the structure begins in October 1952. As in the prototype, the concrete is poured in two phases, first in the tetrahedra and then in the top slab. After the stripping of the tetrahedra, the conduits, the lines of the physical plant and the Tectum panels are installed, followed by the pouring of the top slab[159]. The plywood deck on which the metal formwork for the tetrahedra are attached leaves uncontrolled imprints on the rods of the floor, at the position of the panel joints. In later works, Kahn will no longer leave anything to chance.

At the moment of construction of the first floor system precautions are taken in the making of the formwork and the pouring, bringing about perceptible improvements in the quality of the finish.[160] In the early pouring phases delays are caused by the lack of experience of the workmen in the coordination of the problematic implementation of the various integrated parts. Eight weeks are needed for the construction of one entire floor system.[161]

The passage between different surface textures is usually emphasized by a recessed joint that creates a shadow line, and is therefore called a "shadow joint" by Kahn.[162] This joint takes on multiple constructive meanings in his later exposed concrete works, with different definitions according to the role it plays in the fabrication process.

"One thing I learned from that job," Tyng recalls, on the subject of the Weiss House, "was what Lou called the 'shadow joint' detail. He be-

64-65. L. I. Kahn, D. W. Orr, Yale University Art Gallery and Design Center, New Haven, Connecticut, 1951-53.

64. View of the Tetrahedron Floor System before the construction of the curtain walls (LIK, 030.V.A.11.23).

65. Photograph taken during the implementation of the columns cladding (LIK, 030.IV.B.370.2.3).

came very famous for that. The shadow joint is the idea of separation between materials so that if one has a door jamb in wood against rough stone-work, rather than to trying to scribe the wood to the stone shape, one actually lets it be separate, with its own straight edge away from the irregular edge of stone."[163]

The material for the cylindrical staircase goes through different variations, even when the work is already in progress. In November 1952 it is decided to give up the original ordinary brick, replacing it with cement bricks – "from chocolate brick to cement brick."[164] The definitive solution, decided in 1953, calls for the construction of the wall of the staircase in concrete poured into formwork made with wooden boards, like those of the pillars and beams, with the aim of creating another sculptural feature in *béton brut*. In May the staircase has been completely poured.[165] Horizontal grooves identify the pouring lines, the passage from one formwork system to another in the vertical sequence corresponding to the floors, while the

[157] H. D. Palmer, letter to Charles H. Sawyer, 19 September 1952, LIK, 030.II.A.107.45.
[158] H. D. Palmer, "Yale University Design Laboratory," 23 October 1952, LIK, 030.II.A.107.45
[159] Pfisterer, "Design Laboratory - Yale University" cit.
[160] "The methods now adopted on the form placement and concrete pouring are apparently developing very satisfactory finished concrete (at least at those points where forms have been removed for observation), but it is also proving to be a slower and more costly operation" (Douglas W. Orr, "Memorandum," 24 November 1952, LIK, 030.II.A.107.45).
[161] Leslie, *op. cit.*, p. 74.
[162] H. D. Palmer, letter to George B.H. Macomber Company, 25 February 1953, LIK, 030.II.A.107.43.
[163] *Anne Griswold Tyng, Interview Alessandra Latour, Philadelphia, Pennsylvania, May 2, 1982*, in, Alessandra Latour, *Louis I. Kahn. L'uomo, il maestro*, Roma, Kappa, 1985 (pp. 39-55), p. 43.
[164] H. D. Palmer, "Yale University - Design Laboratory," 28 October 1952, LIK, 030.II.A.107.42; H. D. Palmer, letter to George B.H. Macomber Company, 12 November 1952, LIK, 030.II.A.107.45. Still in February 1953 there are no definitive drawings for the staircase (Charles H. Sawyer, letter to Douglas W. Orr, 19 February 1953, LIK, 030.II.A.107.42).
[165] See the back of the worksite photograph of the staircase, LIK, 030.IV.A.370.1.8.

66-69. L. I. Kahn, D. W. Orr, Yale University Art Gallery and Design Center, New Haven, Connecticut, 1951-53.

66. Photograph of the construction of the stair (LIK, 030.IV.A.370.1.10).

67-68. Views (photographs by the author, 2010).

69. Detail of the brick wall (photograph by the author, 2010).

surface is marked by the holes left by the tie cones, arranged in a regular way (the metal pipes of the spreaders remain in the concrete). The form of the cylindrical wall in exposed concrete with the imprints of the boards and the pattern of the holes can be considered an initial achievement of Kahn's architectural concrete.

The Yale Art Gallery is opened on 31 October 1953. Thanks to its Tetrahedron Floor System, the Space Frame has now become a structure capable of generating spatial effects in the perspective view of the multiple and thick inclined planes that vary according to one's vantage point, always offering new, powerful lines of perspective – an effect similar to the one depicted by Mondrian in *Composition with Grid 4 (lozenge)*.

The Tetrahedron Floor System is simply Kahn's starting point for the invention of an extraordinary series of floors, always with different structures, in exposed reinforced concrete, all capable of housing physical plant conduits and lines.

Vincent J. Scully, Jr., describes the qualities of Kahn's concrete floor, directly comparing it to the concrete of Le Corbusier: "The ceiling canopy is heavy; it has real visual weight. Its concrete is rough, unfinished, and rich in texture, like the concrete of Le Corbusier's Unité d'Habitation at Marseille. The insistent pattern of its crystalline triangles becomes some-

thing somber and of an archaic dignity in rough concrete. Like the Unité d'Habitation itself, this building is not a triumph of modern industrial techniques – few buildings are – it is instead, like the Unité, the result of a moving and expressive tension between geometric will, structural imagination, and rough structural techniques. It is, in intent, as strict and as ordered as the Medici-Riccardi Palace, turned now into one rough new volume of space."[166]

Due to its "stimulatingly controversial" structure, the Yale Art Gallery was also to be considered a model of the "possibilities and the limitations of in situ concrete to future Architects from Yale."[167]

But the most significant testimony to grasp the cultural value of Kahn's choices regarding formwork is a statement made by the president of George B.H. Macomber Company Builders, C. Clark Macomber, who at the start of 1954 underlines the particular character of the boards preferred by Kahn, which he significantly defines as "methods of the past." What Macomber says allows us to determine the degree of technical "outdatedness" with respect to American construction of the boards of the formwork of the Yale Art Gallery, and therefore to fully comprehend the extent to which Kahn, with that choice, intends to pay direct homage to the *béton brut* of Le Corbusier in the version generated by French techniques at the end of the 1940s. It is worth emphasizing that the person closest to the artistic and technical sensibilities of Le Corbusier, for the choice of the materials of the formwork, is Macomber, not Kahn.

"We think the tetrahedron construction is practical and as economical as standard construction, plus furred ceiling and the like, for this type of problem," Macomber states. "The tetrahedron forming and pouring were complicated to work out. The heavy rib forms lifted, pivoting on the cross-rib form; the crossing of rods in three directions caused serious voids. The forms had to be tied down, a novel problem; and hand tamping by special tools was required in place of mechanical vibration. It gave all of us in this organization a challenge and a feeling of accomplishment when we obtained control of the false work and the pouring operation. As builders we do not see eye to eye with Kahn in the special usage of structural false work to obtain artificial results. The narrow faced (two inches wide) boards used

166 Vincent J. Scully, Jr., *Somber and Archaic; Expressive Tension*, in, *The New Art Gallery and Design Center*, theme issue of "Yale Daily News," 6 November 1953, p. 10.

167 Pfisterer, "Design Laboratory - Yale University" cit.

for column, beam, and silo forming to obtain close parallel lines on the concrete surfaces we do not consider functional. They perhaps recall methods of the past but do not provide the functional boldness or sincerity of structure developed by the normal large sheet of plywood or metal form. This narrow-faced formwork, the special sizes of concrete blocks, the floor-to-ceiling structural channel frames, and the like are the kinds of things that cost excessively; and in our opinion go into the field of special veneers rather than honest fundamental expressions of structural elements used for finish surfaces."[168]

[168] C. Clark Macomber, *Making theory work*, in, *Opinions of the Yale Art Gallery and Design Center*, in, "Progressive Architecture," May 1954, (pp. 15-16, 22, 24), pp. 22, 24. In 1954 the office of Marcel Breuer, working on the design of the building at St John's Abbey at the Benedictine Monastery in Minnesota, asks Kahn for images of the "space-frame ceiling and roof construction" of the Yale Art Gallery (Hamilton Smith, letter to Louis I. Kahn, 18 February 1954, LIK, 030.II.A.66.9). On the floor system of the Yale Art Gallery see Anna Rosellini, *Le Tetrahedron Floor System de Kahn à la Yale Art Gallery*, in, Roberto Gargiani, L'architrave, le plancher, la plate-forme, Lausanne, PPUR, 2012, pp. 798-808.

70-74. L. I. Kahn, D. W. Orr, Yale University Art Gallery and Design Center, New Haven, Connecticut, 1951-53.

70. View, 1953 circa (LIK, 030.IV.A.310.1.28).

71-74. Details of the Tetrahedron Floor System, blocks wall, columns and wooden flooring (photographs by the author, 2010).

75-78. L. I. Kahn, D. W. Orr, Yale University Art Gallery and Design Center, New Haven, Connecticut, 1951-53. Details of the structure and views (photographs by the author, 2010).

79-82. L. I. Kahn, D. W. Orr, Yale University Art Gallery and Design Center, New Haven, Connecticut, 1951-53. Views and details of the stair (Photographs by the author, 2010).

83. Card of the Alumni Day at the Yale University, 1958 (LIK, 030.II.A.54.25).

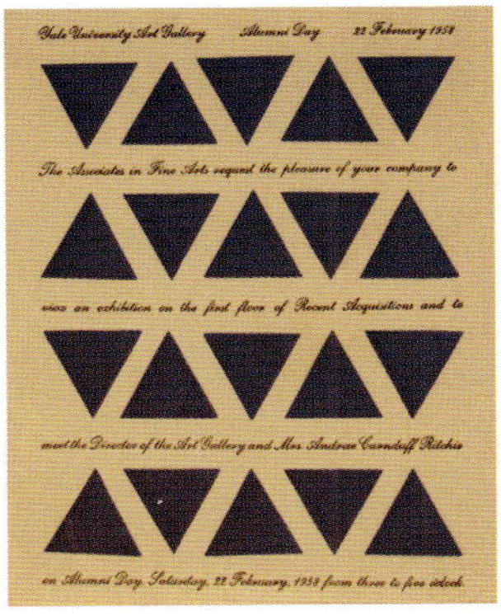

chapter two
hollow stones and hollow columns

Tetrahedral Space Frame for the City Hall Building

The construction of the Yale Art Gallery marks the beginning of the refining of the concepts implied in the structure of that work and connected with different ways of using concrete. The discovery of the tetrahedron floor with openings for ventilation and lighting gear leads Kahn to investigate all the potentialities in order to define a complete structural form in exposed reinforced concrete. What is at stake is the civic and democratic meaning Kahn has proven he wants to attribute, at this point, to concrete.

The design process of the Yale Art Gallery necessarily excludes a configuration of pillars integrated into that of the tetrahedral floor system. Precisely such an integration becomes the goal pursued by Kahn in his research starting in 1952, with the City Hall Building, in the context of the "Plan for Midtown Philadelphia," all the way to the project in 1954 for the Adath Jeshurun Synagogue, also in Philadelphia. The arrival point of the research would be the ideal project of the City Tower, developed with the support of the Universal Atlas Cement Company, as the evolution of the City Hall Building. With these projects tetrahedral structures take on an expressive symbolism of the various communities they shelter and represent: that of citizens and that of religious Jews. The ideal material for these structures with their symbolic function remains exposed reinforced concrete, which is emerging by now as a radical alternative to the steel of skyscrapers with their glass curtain walls that are becoming the expression of American financial capitalism, also thanks to speculative operations like the extension of Rockefeller Center in New York and the skyscrapers of Emery Roth & Sons, Mies van der Rohe or Skidmore, Owings & Merrill.

84, 86, 88-90. Project for City Hall Building, First Scheme, Philadelphia, Pennsylvania, 1952-53.

84. Esplanade plan based on a triangular grid (LIK, 030.I.A.395.1.4).

86. Plan (LIK, 030.I.A.385.3).

88. Elevation (LIK, 030.I.A.385.1).

89. Section (LIK, 030.I.A.385.5).

90. Plan (LIK, 030.I.A.385.4).

85. A. G. Tyng, Walworth Tyng House, Cambridge, Maryland, 1950-53 (A. G. Tyng Archive, AAUP, 074.IV.B.52).

87. D'Arcy W. Thompson, *On Growth and Form*, Cambridge, Cambridge University Press, 1944, p. 222.

In the discussions in the 1940s and 1950s on the fate of the 19th-century City Hall of Philadelphia, built straddling Market Street, the positions waver between demolition and the safeguarding of only the bell tower, which has become a symbol on the skyline of the city.[1]

In Kahn's initial project the whole Esplanade between Pennsylvania Boulevard and Market Street, starting from the 19th-century City Hall, is carpeted with buildings of different heights and configurations, but based on a geometric tetrahedron structure derived from the Tetrahedron Floor System of the Yale Art Gallery, to create a complex destined to pay tribute to the epic of the Space Frame in reinforced concrete for a democratic community.

In the first studies of Kahn and Tyng for the tetrahedral structure of the City Hall Building, to be built in the area near the Schuylkill River, the Space Frame generates floor plans with variously inclined sides; but the facades remain vertical. The connections between the tetrahedra of the floor system and the inclined structural frames are assigned to the geometry of inclined rods, as demonstrated in the section of the second variant of the City Hall Building. It is in these studies that Kahn and Tyng introduce the hexagonal plan of the imposts based on the geometry of the tetrahedra.[2]

In the version from 1953, Kahn and Tyng modify the configuration of the volume of the City Hall Building to adapt it to the Space Frame. The plan of the new versions comes from the grouping of three hexagons. For the floor, the Tetrahedron Floor System of the Yale Art Gallery returns, while the pillars form a great grid of inclined rods with a triangular pattern – a "space frame tower," as Kahn puts it.[3] The geometry of the structure

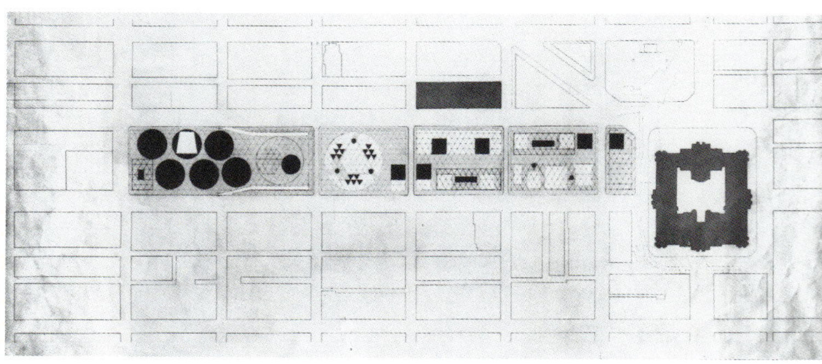

[1] See Goldhagen, *op. cit.*, p. 237, note 51.
[2] See Heinz Ronner, Sharad Jhaveri, Alessandro Vasella, ed., *Louis I. Kahn: Complete Work 1935-1974*, Zürich, ETHZ, 1977, p. 74.
[3] [Louis I. Kahn], *Proposed City Hall Building*, in, "Perspecta," n.d., [1953], No. 2, p. 26.

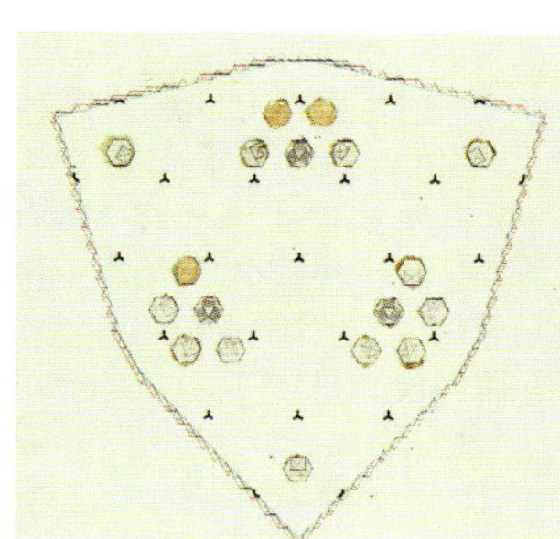

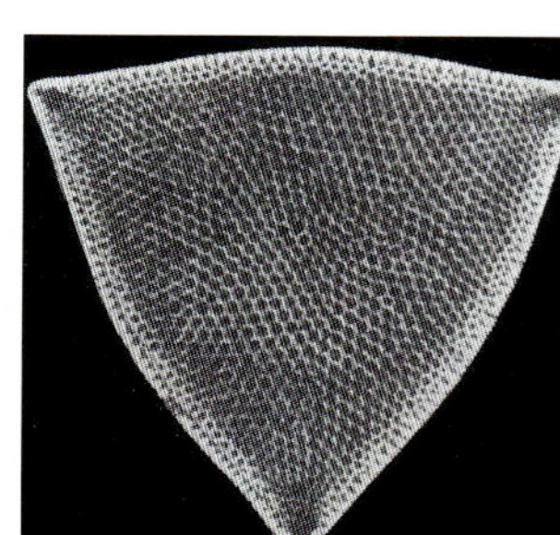

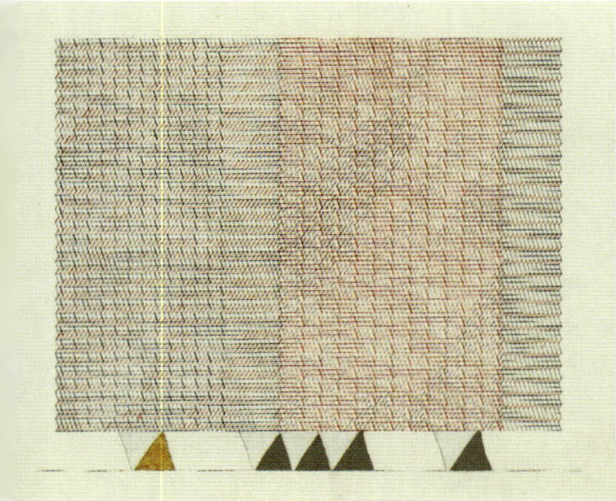

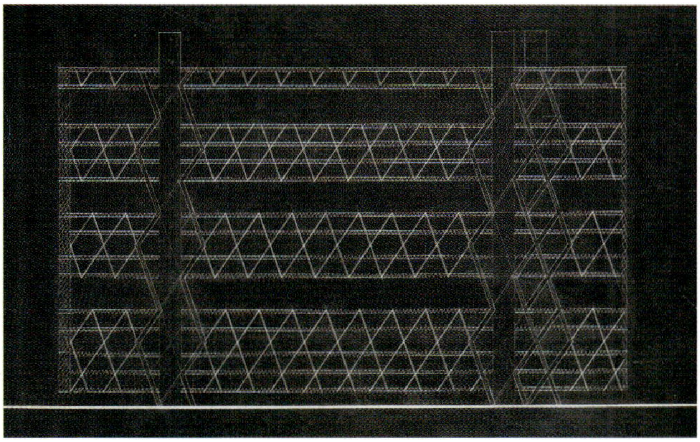

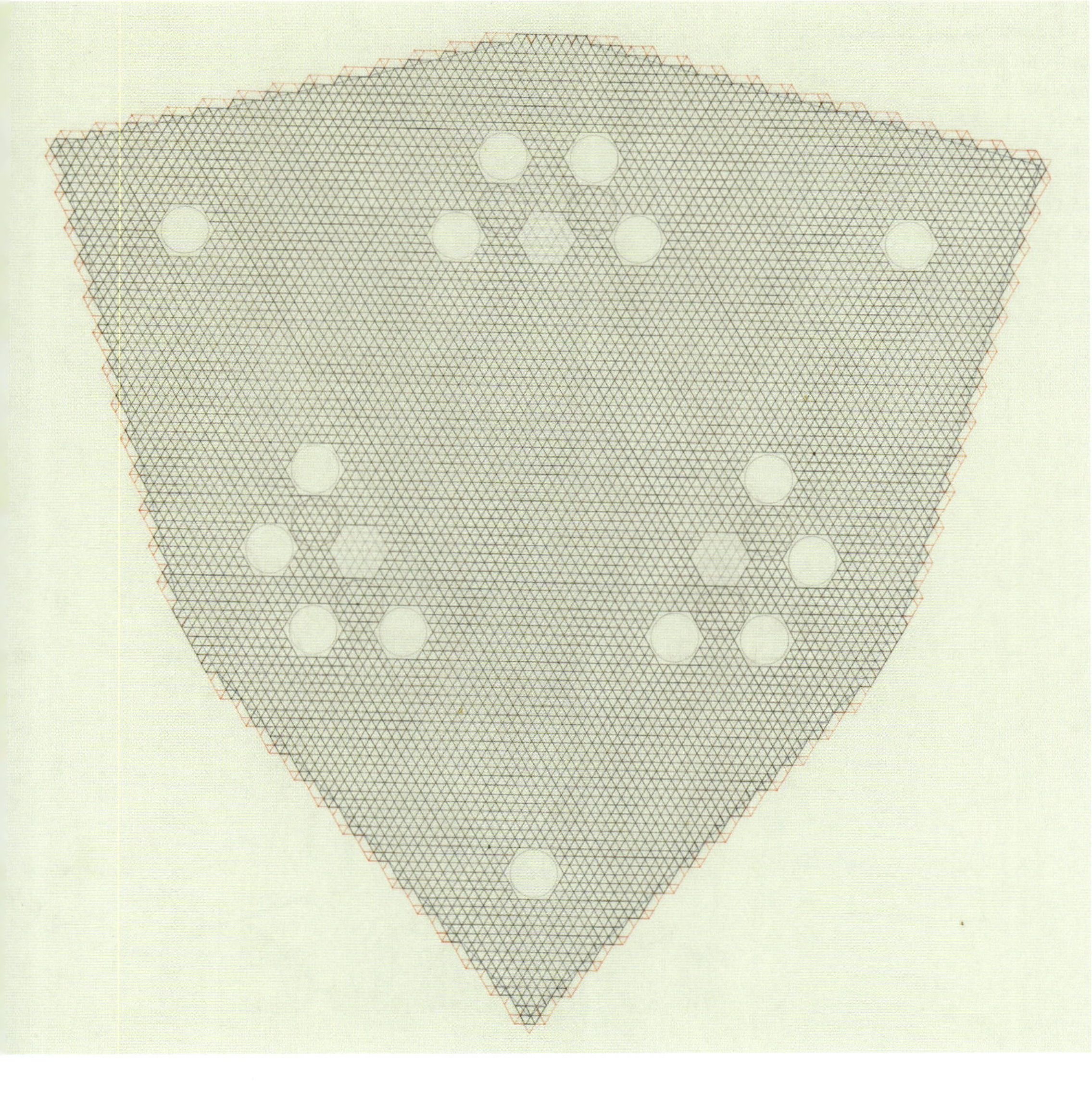

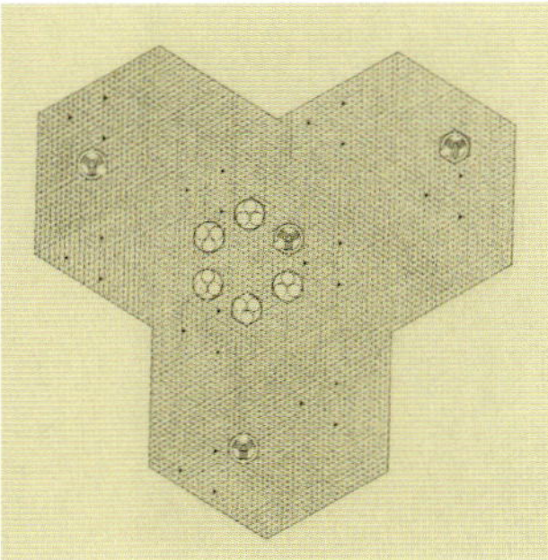

generates sequences of staggered slabs depending on the slope of the pillars, rejecting the prismatic image of the skyscraper.

The model built by Tyng is an initial ideogrammatic representation of the Space Frame – it "shows only the basic geometry of the structure"[4] – while Kahn's sketches made on the basis of the model depict the tower standing in the heart of Philadelphia, crossed by a vital impulse, like a spiral column, the new symbolic bell tower of the community.

All the parts of the structure, from the pillars to the floor system, are envisioned in exposed reinforced concrete. The construction of the prototype of the floor system of the Yale Art Gallery, done in the summer of 1952 to test the technical and load-bearing characteristics of the Tetrahedron Floor System, serves to refine the technical details of the City Hall Building. So the project developed by Kahn and Tyng is a first critical revision of the construction under way in New Haven.

91-95. Project for City Hall Building, Second Scheme, Philadelphia, 1954.

91. Plan (LIK, 030.IV.A.385.18).

92. Section (LIK, 030.I.A.385.7).

93. Elevation (LIK, 030.I.A.385.6).

94. Perspective study ("Perspecta", [1953], No. 2, p. 26).

95. Model (LIK, 030.IV.A.385.2.15).

96. P. L. Nervi, Exhibition Hall C, Turin, 1950. Process of manufacturing prefabricated units for the four-sided dome (*The Works of Pier Luigi Nervi*, London, The Architectural Press, 1957, p. 73).

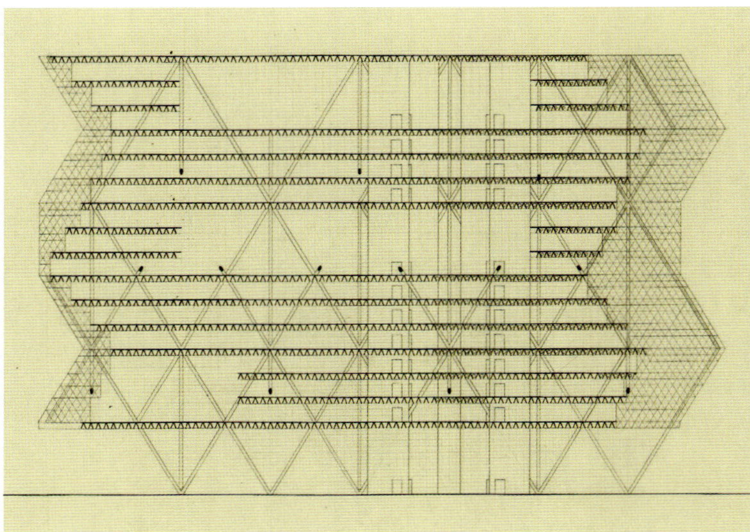

The documents from 1953 indicate that the reinforced concrete of the "space frame tower" is prefabricated and prestressed.[5] The individual tetrahedra of the floor system are to be prefabricated in "light-weight concrete"; once installed, the formwork would be assembled, identical to those of the floor system of the Yale Art Gallery and made with "insulation panels which absorb the sound"; the reinforced concrete top slab would then be made on these formwork.[6] Again, as in the floor system of the Yale Art Gallery, ducts are to be installed between the tetrahedra, before the "pouring of the ceiling structure."[7] The idea of transforming the Tetrahedron Floor System, which in New Haven was poured in place, into a grouping of prefabricated hollow parts is the response of Kahn and Tyng to the costs and difficulties of the making of the floor system of the Yale Art Gallery, and probably reveals even some influences by Nervi's researchs about rhomboid precast units in "ferrocemento." The "triangulated frame" is also to be made like the tetrahedra: "precast, prestressed concrete struts."[8] It is not until the years around 1960 that certain American universities conduct studies to test the possibility of prefabrication of the type of floor system invented by Kahn.

Had it been built, the City Hall Building would have been the first large civic monument in exposed concrete in the United States, and Kahn would have immediately entered the ranks of the protagonists of the international spread of *béton brut* for the construction of public buildings, on a par with Perret's Le Havre and Le Corbusier's Chandigarh. That animated tower would have demonstrated the potential of the extension of "a triangular space frame system in a vertical direction"[9] and been built in the landscape of the Civic Center by the Schuylkill River, amidst a forest of other tetrahedral public facilities that would have translated the metal "tubular forms" of the portico published in *Monumentality* into the new geometry of reinforced concrete.

[4] *Ibid.*, p. 23.
[5] *Ibid.*, p. 26.
[6] *Ibid.*
[7] *Ibid.* Regarding the buildings of the Philadelphia center, Kahn puts such emphasis on the slab that he defines them as "slab office towers" (Louis I. Kahn, *Toward a Plan for Midtown Philadelphia*, *ibid.*, pp. 10-27, p. 21).
[8] *Spatial triangulation. City Hall, Philadelphia, Pennsylvania*, in, "Progressive Architecture," vol. XXXV, June 1954, No. 6, p. 102.
[9] [Kahn], *Proposed City Hall Building* cit., p. 26.

Theoretical principles for concrete: hollow stones

In the lectures and writings of 1953 on the technical and design questions that emerge following the construction of the Yale Art Gallery and the Space Frame studies with the tetrahedron motif, Kahn outlines the principles of his own vision of architecture, principles that depend on concrete construction, one decade after his first important theoretical assessment made in *Monumentality*.[10] These principles are indicated against the backdrop of the passage from the continuity of tubular steel structures without joints, made possible by welding and discussed in *Monumentality*, to the lack of continuity in reinforced concrete construction, poured in place or prefabricated, and made clear by the pouring joints or the connections between the pieces.

The characteristic of the Tetrahedron Floor System in reinforced concrete, in the two versions envisioned by Kahn – poured in place, as in the Yale Art Gallery, or prefabricated, as in the City Hall Building – is that it possesses a series of functional cavities of a size that makes the floor system comparable to other American products of the concrete industry. Precisely those cavities take on particular significance for Kahn in the context of his reflections on steel and concrete. Thus the earlier "tubular forms" are now transformed into what Kahn, starting in January 1953, calls "hollow stones."

The use of the adjective hollow, as we have seen, is widespread in the technical literature familiar to Kahn, to describe prefabricated concrete parts. We should note that in the essay by D'Arcy Wentworth Thompson there is also a chapter on "hollow structures."[11] But apart from these precedents, the motivation behind the invention of the term "hollow stones" lies in Kahn's intuition of being able to prefabricate each single pyramid of the Tetrahedron Floor System planned for the City Hall Building. The stones of Greek temples and Gothic cathedrals discussed in *Monumentality* become hollow and are made of reinforced concrete, for the construction of the contemporary structure, reflecting the intent of "translating masonry construction into steel and concrete" indicated by Kahn as a research hypothesis during his stay in Rome. Once it has been admitted that the "hollow stones" are the individual prefabricated tetrahedra, it seems logical for Kahn to associate them with the "space frame," as we can see from the corrections, in July 1953, to the text of his lecture at the School of Design of North Carolina State College held in February: "space frames or structures of hollow stones."[12] The fact that his reasoning about "hollow stones" takes its cue from Gothic construction is further proof of how the "space frame tower" constitutes an evolution of the "tubular forms" that in *Monumentality* were compared to the flying buttresses and arches of Gothic cathedrals.

"In Gothic times, architects built in solid stones," Kahn explains. "Now we can build with hollow stones. The spaces defined by the members of a structure are as important as the members themselves. These spaces range from the voids of an insulation panel, voids for air, lighting and heat to cir-

[10] The fundamental texts of Kahn are the lecture at the School of Design of North Carolina State College, Louis I. Kahn, *The relation of light to form*, 23 January 1953, LIK, 030.II.A.56.87; and the article, [Kahn], *Proposed City Hall Building* cit., p. 23.
[11] D'Arcy W. Thompson, *op. cit.*, vol. II, pp. 934-957.
[12] Louis I. Kahn, telegram to Mary Ann Brown, 22 July 1953, LIK, 030.II.A.56.86. The modifications requested by Kahn have to do with the text, the summary of his lecture at North Carolina State College, [Louis I. Kahn], "Summation of the Conference," n.d., LIK, 030.II.A.56.86. On space frames and hollow stones see Goldhagen, *op. cit.*, p. 236, note 48.
[13] [Kahn], *Proposed City Hall Building* cit. The text of the conference includes this passage: "We know that in Gothic days they built

culate, to spaces big enough to walk through or live in. The desire to express voids positively in the design of structure is evidenced by the growing interest and work in the development of space frames."[13]

The fact that the "hollow stones" are constituent parts of the Space Frame leads to a new vision of concrete. While the leading protagonists of the Space Frame, from Wachsmann to Mies van der Rohe, come to grips with metal structure, Kahn attempts to translate those same figures into the prefabrication of reinforced concrete to be left exposed, because only with that material can he manage to express a monumental force that can be compared to that of the antique stone structures. In spite of the affinities with the "hollow siliceous tubes" of which Robert Le Ricolais writes in that same year, 1953, regarding spatial or geodesic structures[14], Kahn never returns to the metal structure of the "tubular forms" outlined in *Monumentality*.[15] The path he is taking from metal to concrete is revealed when he states that he wants to conserve, also in the new "hollow stones," the conquests of construction in steel and glass: "In my opinion a great problem is that of preserving the discovery of such a miracle as the tenacity of steel and the transparency of glass."[16] In this sense, his specification of the way of prefabricating the single pieces of the Tetrahedron Floor System for the City Hall Building should not be seen as a mere technical note, since it expresses the goal of concrete: "the walls of which [tetrahedra] are just thick enough to cover the reinforcing."[17]

If Kahn has reached – through the study of the Tetrahedron Floor System – the formulation of the concept of "hollow stones" it is also because in his work, and to a significant extent starting in the 1940s, he has made use of that special form of application of concrete widespread in the United States, namely concrete blocks, which, when not load-bearing, are made with holes. Only by taking the concept of "hollow stones" and the hypothesis of prefabrication of the tetrahedra into account can we better understand why Kahn is increasingly oriented towards the use of exposed concrete block walls. Those walls are not at all a corollary of his vision of exposed concrete.

It is not surprising that Kahn, after having defined the "hollow stones," is struck by the phrases written in *The Book of Tea* by Kakuzo Okakura on the concept of the vacuum formulated by Lao Tse.[18] Kahn transcribes the whole passage on the vacuum, and sends it in January 1954 to Tyng who is in Rome (he explains that the "vacuum" corresponds to "space"). Precisely the statements of Lao Tse can be seen as a fundamental stimulus for Kahn and for the subsequent evolution of the "hollow stones" into entities no longer limited to the single prefabricated concrete piece, but becoming instead true inhabitable places.

"Lao Tse claimed that only in vacuum lay the truly essential," we read in the passage transcribed by Kahn. "The reality of a room, for instance, was to be found in the vacant space enclosed by the roof and walls, not in

in solid stones. Now we can build with hollow stones. Those hollow stones, those spaces, are important to us. Less material supports more. Air, lighting, heat have a place to circulate. Structural members should reflect our new knowledge of strength, and ducts, pipes, and wires are designed and arranged to express their function. The awareness of these hollow spaces in the design of structure is evidenced by the growing interest and work in the development of space frames" (Louis I. Kahn, "The relation of light to form," 23 January 1953, LIK, 030.II.A.56.87).

[14] Goldhagen, *op. cit.*, p. 236, note 48.

[15] Goldhagen believes that for Kahn the hollow stones "meant the steel or aluminium cylinders" (*ibid.*, p. 73).

[16] [Kahn], "Summation of the Conference," n.d., LIK, 030.II.A.56.86.

[17] [Kahn], *Proposed City Hall Building* cit., p. 26.

[18] Kakuzo Okakura, *The Book of Tea*, New York, Duffield and Company, 1931.

the roof and walls themselves. The usefulness of a water pitcher dwelt in the emptiness where water might be put, not in the pitcher or the material of which it was made. Vacuum is all potent because all containing. In Vacuum alone motion becomes possible."[19]

Joints of formwork versus monolith, or "how it is built"

Still in the context of the reflections on the potential of the Tetrahedron Floor System built for the Yale Art Gallery, Kahn develops awareness of the existence of a concept of ornament that differs from the decoration based on a process of stylization. The assembly of elements and their overlapping and joining form the basis of his idea of contemporary ornament. Precisely the awareness that the expressive potential of ornament lies in the assembly prompts him to examine the processes of fabrication of concrete, also thanks to already completed experiences, and in the wake of the success of *béton brut*, as a new era of the art of building. The whole question of ornament revolves around the "joint," which is there even when the construction takes on the appearance of a monolith.

The first evidence of Kahn's idea that concrete, its formwork and the phases of pouring constitute the origin of the new ornament can be traced back to the lecture at the School of Design of North Carolina State College and to the article *Proposed City Hall Building* written for "Perspecta." Kahn formulates his defense of the joint, which he will reiterate often in other lectures and texts. He does so by focusing the reasoning on the role of drawing, which remains that outlined in 1931, when he wrote about the drawings of Auguste Rodin that they "embody the hidden potentialities of his medium," as if Rodin were "thinking in terms of his chisel and hammer."[20] That joint becomes, in the words spoken by Kahn in 1953, the distinctive sign of the construction of concrete, which has to guide the design:

"The feeling that our present day architecture needs embellishment stems in part from our tendency to fair joints out of existence, to conceal, in other words, how parts are put together. If we were to train ourselves to draw as we build, from the bottom up, stopping our pencils at the joints of pouring or erecting, ornament would evolve out of our love for the perfection of construction and we would develop new methods of construction. It would follow that the pasting on of lighting and acoustical material, the burying of tortured unwanted ducts, conduits and pipe lines would become intolerable. How it was done, how it works should filter through the entire process of building, to architect, engineer, builder and craftsman in the building trades."[21]

The two crucial terms for the mode of construction that leads to Kahn's ornament, "pouring or erecting," refer to concrete: the pouring of the material and the assembly of the formwork. With the term joint, as he explained

[19] Louis I. Kahn, letter to Anne G. Tyng, 8 January 1954, in, Tyng, *op. cit.*, (pp. 88-90), p. 89. Kahn possesses the copy of a document on the building regulations of Philadelphia, significantly entitled "Cavity walls and hollow walls. Philadelphia building code amendments, March 12, 1956" (LIK, 030.II.A.63.25). For the concepts of "solid stones" and "hollow stones" and of "building elements of solids and voids," seen as "inherent in steel and concrete," see [Louis I. Kahn], "A space in architecture shows how it is made," n.d., LIK, 030.II.A.62.44.

[20] *The Value and Aim in Sketching*, in, "T-Square Club Journal," vol. I, May 1931, No. 6, now in, Alessandra Latour, *Louis I. Kahn: Writings, Lectures, Interviews*, New York, Rizzoli, 1991, (pp. 10-12), p. 11.

[21] Louis I. Kahn, *The relation of light to form*, conference held at the School of Design of North Carolina State

in 1953, Kahn is not referring to the assembly of "tongue and groove floor boards" used for the boards of the formwork, but to the horizontal border of the panel formed by the boards themselves, which sets the limit of the pouring, as in the cylindrical wall of the staircase of the Yale Art Gallery. That type of joint becomes part of the "shadow joints," though it indicates not the passage from one material to another, but a phase of the concrete construction process. We can say, after the statements of 1953, that Kahn is starting to think, on the worksite of the Yale Art Gallery, about concrete poured in place as a material that does not make it possible to obtain a continuous construction, because it is separated into phases of pouring and different types of joints divide it into sectors, as if it had been laid like masonry.

Kahn clarifies in greater depth the meaning of the joint and ornament when he discusses the role of the formwork used in the Yale Art Gallery. We can attribute to Kahn the content of the captions of the article by George A. Sanderson, published in "Progressive Architecture" in May 1954 and underlining the meaning of the visible signs on the "exposed concrete":

"Everywhere, construction method is documented and revealed. [...] marks forming are left on concrete columns and girds; holes that received lugs in formwork for the stair silo remain – as decoration, as well as to show how it was built. [...]. Joints of erecting, far from being faired out of sight, are boldly exploited with 'shadow joints.' In the tetrahedral ceiling 'lung' all elements are left exposed exactly as they set – unpainted and unadorned."[22]

Kahn's account of the construction of the Yale Art Gallery, in a lecture, starts precisely with the formwork, because he intends to indicate some of the technical issues to then "develop new methods of construction." The explanation clearly shows that all the lines of the construction produce a logical design only if they are directly derived from the construction process. Precisely because the joint discussed by Kahn is that of the borders of the individual panels of the formwork, it seems evident that the wooden boards are not only a corollary, but even become unsuited to the idea of concrete fabrication. Those boards represent a more than eloquent sign of the admiration for the way in which Le Corbusier makes the concrete in the Unité d'Habitation in Marseille; but, not coincidentally, they are destined to be exiled from the main future worksites of Kahn.

"The formwork was made from floor to floor," Kahn explains regarding the Yale Art Gallery, "and this line was accented in the design; because what we tried to do in the expression of the building was to show in every way how it was built. This formwork was made of small floor boards, and the little holes that you see there indicate the through-ties in the formwork. These were left in as holes in the concrete so that in every way, how it was made is apparent. We accentuated the struggle of building; of building from

College, 23 January 1953, LIK, 030.II.A.56.87. The text in "Perspecta" is perceptibly different: "If we were to train ourselves to draw as we build, from the bottom up, when we do, stopping our pencil to make a mark at the joints of pouring or erecting, ornament would grow out of our love for the expression of method. It would follow that the pasting over the construction of lighting and acoustical material, the burying of tortured unwanted ducts, conduits, and pipe lines, would become intolerable. The desire to express how it is done would filter through the entire society building, to architect, engineer, builder and craftsman" ([Kahn], *Proposed City Hall Building* cit., p. 23; Latour, *Louis I. Kahn, l'uomo, il maestro* cit., p. 57). It is significant that speaking of ornament Kahn writes "would evolve," in the text of the conference, and "would grow" in the text for "Perspecta."

[22] George A. Sanderson, *Extension: university art gallery and design center*, in, "Progressive Architecture," vol. XXXV, May 1954, No. 4, (pp. 88-101), p. 100.

floor to floor, because that joint is a critical thing in the construction. When you take the forms off, something always happens in a very ugly soupy way. Whereas, if you actually know that, and you place in there something which is very positive; you put this in it (pointing to the form joint) so you can really see it, then it sets up its own pattern. And I believe that these joints are the beginning of ornament. It's true certainly of the old buildings. It's when you change your form of construction. It's true of anything I know, that the ornamentation is really the joint problem…"[23]

Having developed, after the building of the Yale Art Gallery, the principles for a theory of meaning of the joint for concrete construction is what sets Kahn's position apart from that of the architects and theorists of *béton brut*, including Le Corbusier. What Kahn has to say about the joint is of maximum importance because those phrases contain the foundations of his exceptional position in the vast panorama of research on exposed concrete, due to the radical rejection of any pursuit of imprinted decoration as an end in itself, obtained through manipulation of the components of the formwork in order to invent an artistic pattern. By now Kahn has created the theoretical premises for a different interpretation of architectural concrete, in which the fundamental imprints are only those that express the growth of the building and not the constitution of the arrangement of the individual parts of the formwork. While Le Corbusier explores all the potential of the bonding of the formwork, not coincidentally reaching the definition of the "nouvelle stéréotomie," Kahn on the other hand moves towards erasure of any effect of contrived pattern, trying to discover the nature of concrete itself, not the imprint.

The ideas of joint and ornament are utterly unique in Kahn's concrete fabrication because they are backed by a vision of the construction of a building as a process of organic growth, like that of a living being. The metaphors of skeletons, embodied in the "tubular forms" without joints, precisely due to the technical characteristics of the fabrication of concrete take on a more pregnant meaning in Kahn's vision. In the mid-1950s Kahn begins to make frequent use of the concept of "growth" in his writings and lectures. Already, when he describes the ornament generated by a construction method, he uses the verb "grow out."[24] The particular feature of Kahn's outlook with respect to that of his contemporaries, from Le Ricolais to Buckminster Fuller, who are also interested in the relationship between "growth and form" forcefully described by D'Arcy Wentworth Thompson,[25] is that he associates the concept of growth with "construction" and not just with structure, so much so that for Kahn ornament is the particular sign of a recognizable phase in the process of the building of a work. So while Le Corbusier sees the pouring of concrete into formwork as an operation similar to the casting of bronze, Kahn proceeds towards a completely different idea of concrete, which already at the end of 1953, precisely through the concept of growth, takes the form of the apodeictic statement, uttered in November of that year, that "buildings are not monoliths."[26] The "building

[23] Louis I. Kahn, *This Business of Architecture*, in, "The Student Publication of the School of Architecture of Tulane University," 1955, n.p. (now in, Latour, *Louis I. Kahn: Writings, Lectures, Interviews* cit., pp. 62-66, p. 63).

[24] [Kahn], *Proposed City Hall Building* cit.

[25] See Goldhagen, *op. cit.*, pp. 64-74.

[26] Louis I. Kahn, in "Architecture and the University Conference, December 12, 1953, School of Architecture, Princeton University, Princeton, New Jersey," LIK, 030.II.A.55.75 (now in, Latour, *Louis I. Kahn:*

is constructed or built-up, not (like the Lever Building) a monolith," he explains during that same month.[27] In the notes in the form of aphorisms published in 1954, Kahn outlines the relationships among "form," "growth" and "construction": "Form emerges out of a system of construction / Growth is a construction." The practical examples of the concept of growth, significantly enough, refer to works of engineering: "Nervi grows an arch / Fuller grows a dome."[28]

An image, that of the section of a building on multiple levels taken from a patent of Edison and published in Zucker's anthology,[29] may have helped Kahn to reach the conclusion that the construction of concrete through formwork, even when the latter is as high as one floor of the building, as in the "tilt-up" or the IBEC method, cannot produce a monolith, unless an exceptional method is put into practice, namely the one invented at the start of the 20th century by Edison, which called for the assembly of the formwork of the entire building, including the roof, and then a single pouring of the concrete, into multiple specially positioned channels, as if the building were a bronze casting to make an enormous sculpture. Halfway through the 1950s, Kahn is aware that not only gold, but also the possibility of making a monolith, belong to the sculptor. Through the discovery of the joint and of growth as they manifest themselves on concrete worksites, the value assigned by Kahn to welding to make joints invisible becomes a thing of the past. The whole evolution of the modes of deployment of concrete invented by Kahn takes place in terms of the desire to make the growth phases of the building visible, to the point of insisting that the individual parts of the formwork be the largest possible plywood panels, as already indicated in the specifications for the Samuel Radbill Building – "in as large units as practicable" – prior to undergoing the passing fascination with the small boards of Le Corbusier.

"Grow out" and "joints" in the towers of the Mill Creek Redevelopment Area

The questions of the expression of the construction and the resulting meaning of the joint, as discussed in 1953, are central to the making of the skeleton in reinforced concrete of the towers of the Mill Creek Redevelopment Area in Philadelphia, earmarked like their neighboring buildings for low-cost housing organized by the Philadelphia Housing Authority (for this project Kahn works in partnership with Louis E. MacAllister and Kenneth Day). In the design of the skeleton formed by "columns" and "slabs," developed to "free the apartment spaces of columns,"[30] Kahn inserts large balconies that form "exposed concrete T-shaped column walls,"[31] and the central wall nucleus with the vertical access elements.

On the Mill Creek worksite, as in that of the Yale Art Gallery, certain significant joints, in the columns of the skeleton at the positions of the various slabs, indicate the process of construction and growth of the building, the deployment of the formwork "from floor to floor." The meaning of these few visible signs is discussed at the end of 1953, when Peter Blake, for

Writings, Lectures, Interviews, cit., pp. 54-56, p. 55).
[27] Louis I. Kahn, letter to Anne G. Tyng, 18 December 1953, in, Tyng, *op. cit.*, (pp. 76-79), p. 78.
[28] Louis I. Kahn, *Order is*, in, "Perspecta," n.d., [1955], No. 3, p. 59.
[29] Zucker, *op. cit.*, p. 179.
[30] "The concrete structure is designed to free the apartment spaces of columns and shaped to brace the building, support the balconies and enclose the central core" (*Millcreek Redevelopment Project*, in, "Perspecta," n.d., [1953], No. 2, p. 20).
[31] Typewritten version of *Millcreek Redevelopment Project* for "Perspecta," LIK, 030.II.A.62.45.

97-98. Mill Creek district, Philadelphia, Pennsylvania, 1951-56.

97. Perspective (H. Ronner, S. Jhaveri, eds., *Louis I. Kahn Complete Work 1935-1974*, Basel, Birkhäuser, 1987, p. 43).

98. View during the construction (LIK, 030.V.B.30.16).

"Architectural Forum," and Roger Schafer, secretary of Kingsview Homes, Inc., a redevelopment company in New York, turn to Kahn to find out ways of giving greater formal character to low-cost buildings like the Mill Creek towers. "[...] the problem of making the exterior of our cooperative projects more attractive" is the subject of the discussions between Blake and Schafer. "It was Peter's thought," Shafer writes to Kahn, "that you might either have or know of photographic material which might give us an idea of what could be done in the way of interesting design in low-cost construction."[32]

The color slide taken on the Mill Creek worksite with its three towers, one already faced-off with infills, the others simply skeletons of reinforced concrete standing out against the sky, is perhaps the most expressive document, in Kahn's view, of the meaning of "growing out," of "joint" and of his idea that "buildings are not monoliths." He draws the towers himself as if they were composed only of the "*construction frame* from all sides without windows, curtain walls, etc."[33] "In effect," Kahn writes, "it is a picture of the building before the enclosing material is delivered to the goals of have found that the frame must be in itself aesthetically satisfying before I can place the enclosing material alongside it or over it."[34]

In the notes written in response to Schafer, Kahn makes reference to a type of photograph like this color slide. It is no coincidence that Kahn explains the way in which he designed the skeleton of the towers, if we consider his statements in 1953: "to draw as we build, from the bottom up, stopping our pencils at the joints of pouring or erecting." The response to Schafer's requests confirms the theoretical questions of concrete construction with which Kahn is engaged in this moment:

"I have taught myself to make drawings of a building choosing only the structure frame – without windows, enclosing walls, etc. I have even gone so far as to develop these drawings in imitation of the building process – starting my drawings from the bottom up and making special emphasis on the construction joints. On the way up this representation of the grow out, the expression of the building process (how it is done), I

have found the basis of scale and proportion. [...] I chose exposing the concrete frame and enclosing the space between with glass and brick. I see no reason however why the concrete frame may be covered with the same brick – but how? – and still retain the underlying proportions. I am to concluded one way, line one close near another but certainly this is no aesthetic formula. It is possible to use the lined up joint for flush panels of brick confined to the width of the columns – spandrel beams windows areas and enclosing wall areas. All these panels flush with the others but distinguished from one another by only a camber or joint of separation. May be a bit delicate but at least comes like the veneered wall it is. I believe and I know your architects will agree that the entire space and mechanical plan must be affected in a design approach which considers the cladding frame as the basis of scale and proportion. I do not believe that this approach should add to the cost."[35]

"I have also found that by expressing the structure one does not arrive at the bulky monolithic appearance."[36]

In spite of the international success of the Yale Art Gallery, the towers of Mill Creek constitute the first skeletons by Kahn in entirely exposed concrete, those that achieve the hope expressed in *Monumentality* for a "skeleton of the structure" that asserts "its right to be seen." It should be mentioned that in any case Kahn is able to do without the cladding, which instead he had to apply to the columns of the Yale Art Gallery, only because the Mill Creek towers are part of a low-cost development operation. In spite of his defense of the concrete skeleton with the pouring joints, written to Schafer, the Mill Creek worksite reveals the technical difficulties involved in achieving high quality for exposed concrete. The workmanship leaves him unsat-

[32] Roger Schafer, letter to Louis I. Kahn, 28 December 1953, LIK, 030.II.A.62.34.
[33] Louis I. Kahn, handwritten notes for the letter to Roger Schafer, LIK, 030.II.A.62.34.
[34] *Ibid.*
[35] *Ibid.*
[36] *Ibid.*

isfied, as he admits on 1 January 1954: "I am trying to do a good concrete job but it is practically impossible with such deliberately grasping people."[37] When he visits the worksite again in March, he instead notices the quality of the materials of the low buildings, and once again, as in the case of the Samuel Radbill Building, he associates the range of shades of gray with the concept of uniformity, which in the context of the Mill Creek operations even takes on communitarian and civic overtones: "The white brick and the gray cement are likely to look very good indeed. The bricks (poorly laid) are going up in the quads. The Formigli joists are up (very good and clear) and the whole idea of the cement beams in white brick and mortar I believe is going to blend to uniformity appropriate to the low-cost materials."[38]

Untouchable and scarface: visual illusion of values

The ineptitude of the workmen assigned the task of pouring the concrete for the Mill Creek towers makes Kahn begin to wonder about the perception of the defects caused by imprecise deployment of concrete. As an artist sensitive to chromatic and graphic effects, the signs left on the concrete by the formwork system become, for him, not just precious evidence of the process of fabrication of the building. If the growth of the construction is displayed by the joints between the large formwork panels, the patterns generated by the holes of the ties also have another role in the image of the concrete, as their pattern creates a rhythmical chiaroscuro motif capable of attracting the gaze and of preventing it from freely scanning the surface in search of the unpredictable effects generated by the flaws. The many considerations on the perception of points or lines on a surface, the optical rules for their balance, contained in *Language of Vision* by Gyorgy Kepes, can help us to understand what Kahn sees on the concrete surface.[39]

99. G. Kepes, scheme of the interactions of different fields (G. Kepes, *Language of Vision*, Chicago, P. Theobald, 1944, p. 29).

Kahn would return to the question of the perception of defects in the future. We cannot rule out the idea that he may already have made observations of this type regarding the cylindrical wall of the Yale Art Gallery. But in the case of the construction of the Mill Creek towers, we see the first evidence of him taking a position in meaningful terms on this aspect of exposed concrete: "we are using screw ties throughout – they (the screw ties) leave a hole about 1" deep after the forms are removed. This gives a strong construction rhythm to the concrete and tends to erase some of the pour disfigurings. Now they come to the penthouse and the contractor decides to form up without those ties (they use snap ties which have to be snapped off resulting in a nasty surface scar)."[40]

If we take into account that starting with the Yale Art Gallery Kahn orders, in the specifications, that the concrete not be repaired after stripping, it becomes easier to understand – precisely in the light of his statements about the Mill Creek towers – the meaning of the obsessive presence of the pattern of holes made by the ties. This practice becomes an outstanding feature of his works to come, as a secret optical correction for in-

[37] Louis I. Kahn, letter to Anne G. Tyng, 1 January 1954, in, Tyng, *op. cit.*, (pp. 86-87), p. 86.
[38] Louis I. Kahn, letter to Anne G. Tyng, 19 March 1954, *ibid.*, (pp. 118-119), p. 119.
[39] For the relationship between the two artists see, Gyorgy Kepes, letter to Louis I. Kahn, 27 January 1956, LIK, 030.II.A.62.34.
[40] Louis I. Kahn, letter to Anne G. Tyng, 10 July 1954, in, Tyng, *op. cit.*, pp. 159-160.
[41] Louis I. Kahn, letter to Anne G. Tyng, 2 December 1953, *ibid.*, (pp. 69-70), p. 69.
[42] *Ibid.*

evitable defects. On the question of the meaning attributed to flaws and their perception, and the modes of their optical correction, from the early 1950s Kahn demonstrates a precisely opposite approach with respect to the pictorial operations done by Le Corbusier on *béton brut* for the "camouflage" of defects, with strong color contrasts. Kahn, unlike the apostle of *béton brut*, always entrusts the control of the perception of defects only to the means of construction of the formwork. The calculation of the form of the cones to obtain a given depth of the holes is not a mere question of perfection of the construction, for Kahn, but the clue of an artistic discipline of control of the "visual illusion of values," a true expedient of a "language of vision."

"Column joints" or "column capitals," "column clusters" and "hollow columns"

The process of development for the structure of the City Hall Building does not end with the project of 1953. The main question remains the form of the column most suitable for the new type of Tetrahedron Floor System. The versatility of shaping concrete grants Kahn the possibility of exploring different hypotheses that are more complex than simple inclined rods. Starting in December 1953 and throughout 1954 different solutions and definitions are investigated in succession for the exposed reinforced concrete column of the Tetrahedron Floor System, with the goal of achieving that Space Frame which Kahn now, in general, also identifies with the term "3D" accompanied by "truss," "structure" or "construction."

Against the backdrop of Kahn's reflections on the joint as a symbol of growth of a building, it should come as no surprise that the evolution of the design of the City Hall Building also includes the definition of similar details, as if the joint were taking on an entirely different configuration with respect to that of the Yale Art Gallery and the Mill Creek towers.

Immediately after publication of the City Hall Building in "Perspecta" in 1953, Kahn starts to work on the connections between the columns and slabs, which he significantly defines as "column joints."[41] While he is convinced that he knows "all we need to know for the moment about floors," on the other hand "the connection and the nature of the column I believe we know but too little," as he writes regarding the City Hall Building in December 1953 to Tyng.[42] Kahn's intuition, destined to influence the evolution of his tetrahedral Space Frame, consists in the introduction in the junction points between columns and slabs of true hollow capitals or, as he calls them, "open star-like capitals (as the Greek) which may be the shear head transition (capital is the shear head expression) to the column and that the column may express its bending (L/R) in a graphic way."[43] Kahn encourages Tyng to contact Nervi to get information on how to resolve the questions of the tetrahedral structure, after having learned of the interest expressed by Nervi himself in the Space Frame of the City Hall Building, which Tyng showed him on the pages of "Perspecta."[44]

[43] Louis I. Kahn, letter to Anne G. Tyng, 2 December 1953, *ibid.*, pp. 69-70.

[44] "What you write about Nervi is really fun and I am all for the idea of working together on the space frame [...]" (Louis I. Kahn, letter to Anne G. Tyng, 2 December 1953, *ibid.*, pp. 69-70, p. 69). Since 1952, Tyng had requested the possibility of an apprenticeship with Nervi (see Tyng, *op. cit.*, p. 58).

The fact that Kahn considers the system outlined during the course of 1953 with the studies for the City Hall Building, with the "capital" as well, to be a fundamental evolution of the Yale Art Gallery, is borne out by his own words, still at the end of the year, in reaction to the possibility of designing a building similar to the one in New Haven at Durham, North Carolina: "I lost the job at Durham too because I told them (the committee) that I should like to produce a perfected construction similar to the Gallery (with 'newly' conceived column idea). They misunderstood – thinking I was going [to] give them a second hand job."[45]

The search for the ideal column for the Space Frame is also conducted by Mies van der Rohe, who precisely in the wake of the success of the Tetrahedron Floor System of the Yale Art Gallery begins, in September 1953, the project for the Convention Hall in Chicago, equipped with a great metal roof structure created first with the students of IIT and then in collaboration with the engineer Frank Kornacker. From the rods of the structure others descend to form, only along the perimeter, truss columns similar to those made by Le Ricolais.[46] Kahn is aware of this project of Mies van der Rohe: "I keep dreaming about space frames which are more and more important. Evidence of their importance is appearing in several places. I believe I mentioned that Mies van der Rohe put out an analyzed [tetrahedron] roof frame 600' x 600' with no columns on the inside. [...]. My only objection is that he does not use the cantilever."[47]

At the end of 1953, Kahn envisions the construction of a model of the City Hall Building "at a large enough scale to show the floor construction too."[48] In the early days of 1954, when the municipal administrators of Philadelphia are considering the possibility of constructing the new facility as an annex to the City Hall on Reyburn Plaza, Kahn sketches the City Tower with three hexagons on one block, and explains the spatial characteristics of the floor system just completed in New Haven, now defined as a "cellular floor structure" to the authorities, including its "wide flexibility of the space for varied uses."[49] Kahn and Tyng work independently on the detail of the "column joints." "I am working on some joints too," Tyng informs Kahn on 1 January.[50]

The definition of the structure with the Tetrahedron Floor System happens not only in the context of the project for the City Hall Building, but also in the design of other commercial buildings. To develop the structure, Kahn involves the two leading figures of the construction of the Yale Art Gallery, the Macomber Company and Pfisterer. Robert Venturi, on Kahn's office staff, takes part in the preparation of drawings and models, while Tyng continues to be consulted on questions related to the "3D structure." Kahn and Tyng, in any case, are still looking for an engineer to study the complex structure. Tyng leans toward calling in Nervi; Kahn underlines the qualities of Le Ricolais and Felix J. Samuely.[51] But all attempts in this direction lead nowhere. "The development of so unique a structure," Kahn later recalls, "would require funds which I don't have and was unable to

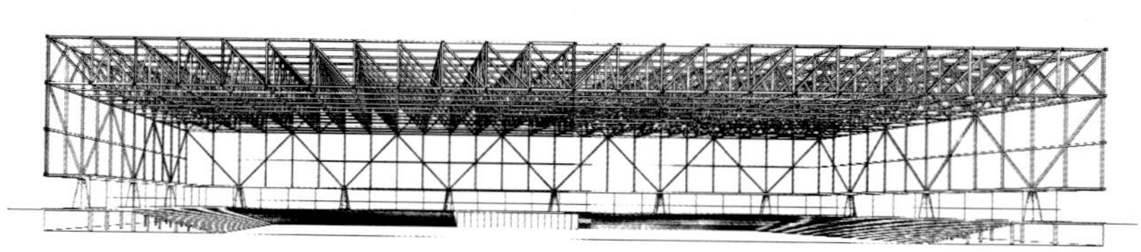

100. R. Le Ricolais, lattice structure (G. R. Le Recolais Archive, AAUP, 086.III.A.29a).

101. L. Mies van der Rohe, Convention Hall, Chicago. Perspective (P. Carter, *Mies van der Rohe at Work*, London, Phaidon, 2005, p. 104).

obtain. Luigi Nervi was extremely interested in the idea behind this structure and, for a while, I thought he might help to develop the engineering."[52]

For a shopping complex to be built in Washington, Kahn proposes "the use of a 3D Truss." "The Truss would be built of 'toothpick' members with reflecting panels for acoustical absorption and light. Harbored of course would be the air conditioning ducts and electrical sources. The whole area would have no more than 4 columns."[53] In certain studies Kahn envisions a column, like those of Le Ricolais and Mies van der Rohe, equipped with a lower vertical segment and branches that connect to the geometry of the Tetrahedron Floor System. Regarding this new solution, he writes: "I will be able to make it pure as we progress through the confidence with Macomber's cost [...]. I believe the column cluster idea as in I is a good idea."[5]

Also in the work on the City Hall Building to be erected next to the City Hall, Kahn studies a radical modification of the form of the building, starting precisely with the idea of its construction in two phases, as planned for the tower with three hexagons. Now he envisions two towers with a square plan, but with an unconventional arrangement of the columns – the larger tower has three resting points and one cylinder; the smaller tower has only two resting points.[55] Kahn calls the load-bearing system of the towers "a new column spacing."[56]

The development of the "column cluster idea" emerges in the definition of another form of column even more dependent on the Space Frame of the tetrahedral floor system. This column is the arrival point of Kahn's thinking about the connections between the column and the Tetrahedron Floor

[45] Louis I. Kahn, letter to Anne G. Tyng, 24 December 1953, in, Tyng, *op. cit.*, (pp. 80-81), p. 81.
[46] *Cf. Space frame discussed*, in, "Architectural Forum," vol. XCLVIII, February 1953, No. 2, (pp. 154-160), p. 155.
[47] Louis I. Kahn, letter to Anne G. Tyng, 26 December 1953, in, Tyng, *op. cit.*, p. 83.
[48] *Ibid.*
[49] "In my recently completed work, the Yale University Design Center, which is essentially a loft space, I have developed wide flexibility of the space for varied uses. This was accomplished by a distributed integration of the air conditioning, heating and lighting requirements with a cellular floor structure of acoustical quality" (Louis I. Kahn, letter to Francis J. Lammer, Executive Director of the Redevelopment Authority of the City of Philadelphia, 30 January 1954, LIK, 030.II.A.56.87). For the purpose of the letter – to offer services for a "study of the New City Building, citing the Gallery as an example of the kind of economic integration" – see Louis I. Kahn, letter to Anne G. Tyng, 8 January 1954, in, Tyng, *op. cit.*, (pp. 88-90), p. 88.
[50] Louis I. Kahn, letter to Anne G. Tyng, 1 January 1954, *ibid.*, (pp. 86-87), p. 86. Kahn refers to the question of "joint type A or type B" in Louis I Kahn, letter to Anne G. Tyng, 24 January 1954, *ibid.*, (pp. 93-94), p. 93.
[51] Louis I. Kahn, letter to Anne G. Tyng, 6 June 1954, *ibid.*, p. 146. See Felix J. Samuely, *Space Frames and Stress Skin Construction*, in, "The Journal of the Royal Institute of British Architects," Third Series, vol. LIX, March 1952, No. 5, pp. 166-173.
[52] Louis I. Kahn, letter to Carmine Pacileo, 25 February 1959, LIK, 030.II.A.65.18. See Carmine Pacileo, letter to Louis I. Kahn, 26 January 1959, LIK, 030.II.A.65.18. Nervi does not remain impervious to the Tetrahedron Floor System and its developments in the City Hall Building, as demonstrated by the exposed reinforced concrete structure with triangulated rods of his George Washington Bridge Bus Station, in Manhattan, designed and built in 1959-63.
[53] Louis I. Kahn, letter to Anne G. Tyng, 21 February 1954, *ibid.*, (pp. 105-106), p. 105. The architect John Hans Graham and the local architects Sweet and Schwartz are involved in the project, for which Kahn also envisions working with the Macomber Company and Pfisterer, with the aim of "working on the 3D truss" (Louis I. Kahn, letter to Anne G. Tyng, 27 February 1954, in, Tyng, *op. cit.*, pp. 107-111, p. 107).
[54] Louis I. Kahn, letter to Anne G. Tyng, 7 March 1954, *ibid.*, pp. 111-115, with sketches.
[55] See Louis I. Kahn, letter to Anne G. Tyng, 7 March 1954, *ibid.*, pp. 111-115, with sketch (p. 108).
[56] Louis I. Kahn, letter to Anne G. Tyng, 16 March 1954, *ibid.*, (pp. 116-117), p. 117.

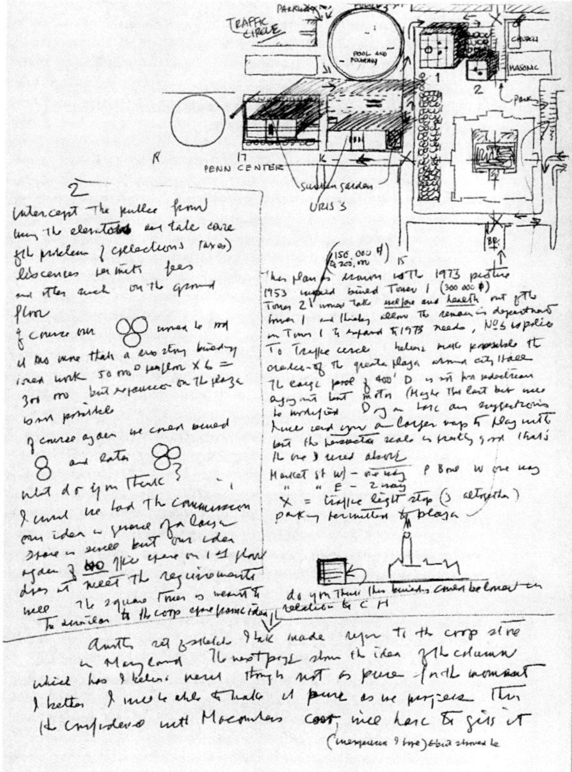
TRAFFIC CIRCLE
POOL AND FOUNTAIN
CHURCH
MASONIC
PENN CENTER
sunken garden
URIS 3
2
Intercept the gallery from
using the elevators and take care
of the problem of collections (taxes)
licences permits fees
and other such on the ground
floor
(150,000 #)
Thes plan is drawn with the 1973 position
1953 original bound Tower 1 (300 000 #)
To Traffic circle
Market St W) – one way
X = traffic light stop (3 altogether)
parking permitted to plaza
do you think these buildings could be closed in relation to C H
Another set of sketch I have made refer to the corp store in Maryland

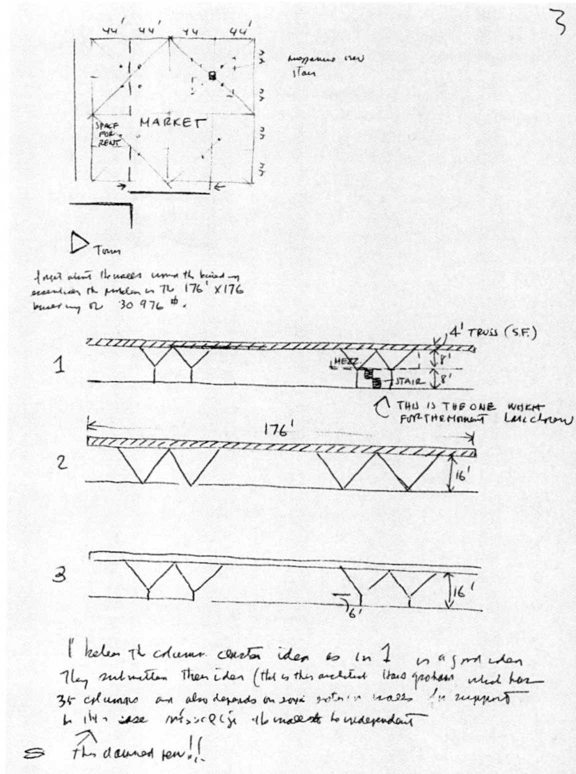
3
44' 44' 44 44'
MARKET
SPACE FOR RENT
Tower
4' TRUSS (S.F.)
MEZZ
STAIR
THIS IS THE ONE WHICH FURTHERMORE
176'
16'
6'
1
2
3
I believe the column cluster idea as in 1 is a good idea
This damned pen!!

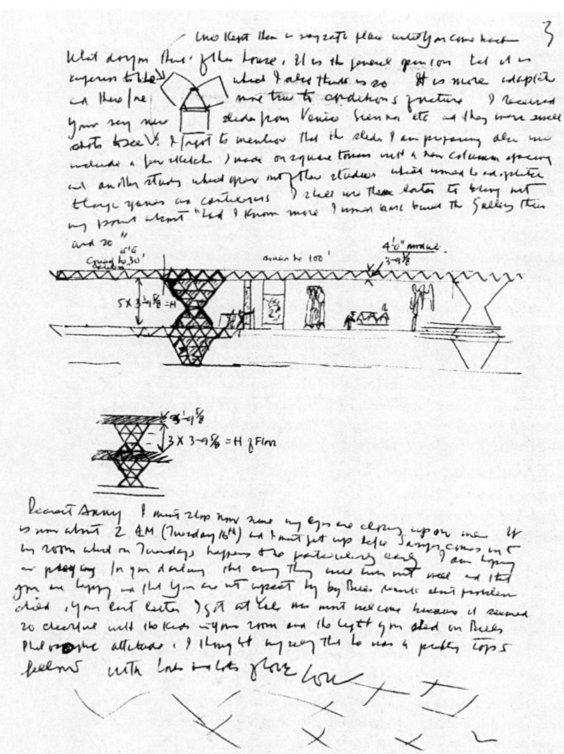
3
4'0" module
5 X 3-9 3/8 = H
3 X 3-9 3/8 = H of Floor
with love and lots of kisses Lou

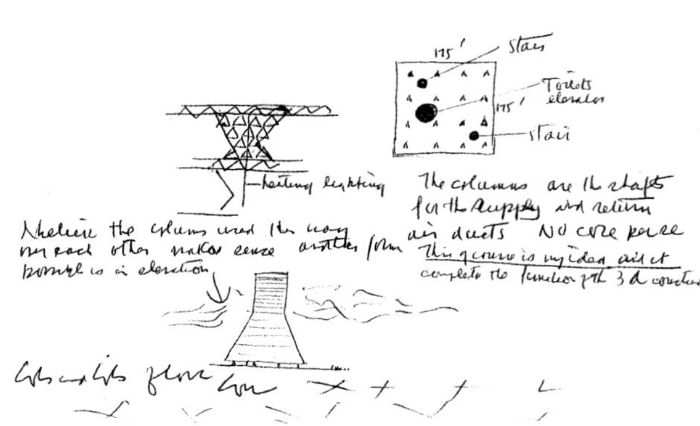
175'
stairs
Toilets elevators
stair
heating lighting
The columns are the shafts for the supply and return air ducts NO CORE HERE
Lots and lots of love Lou

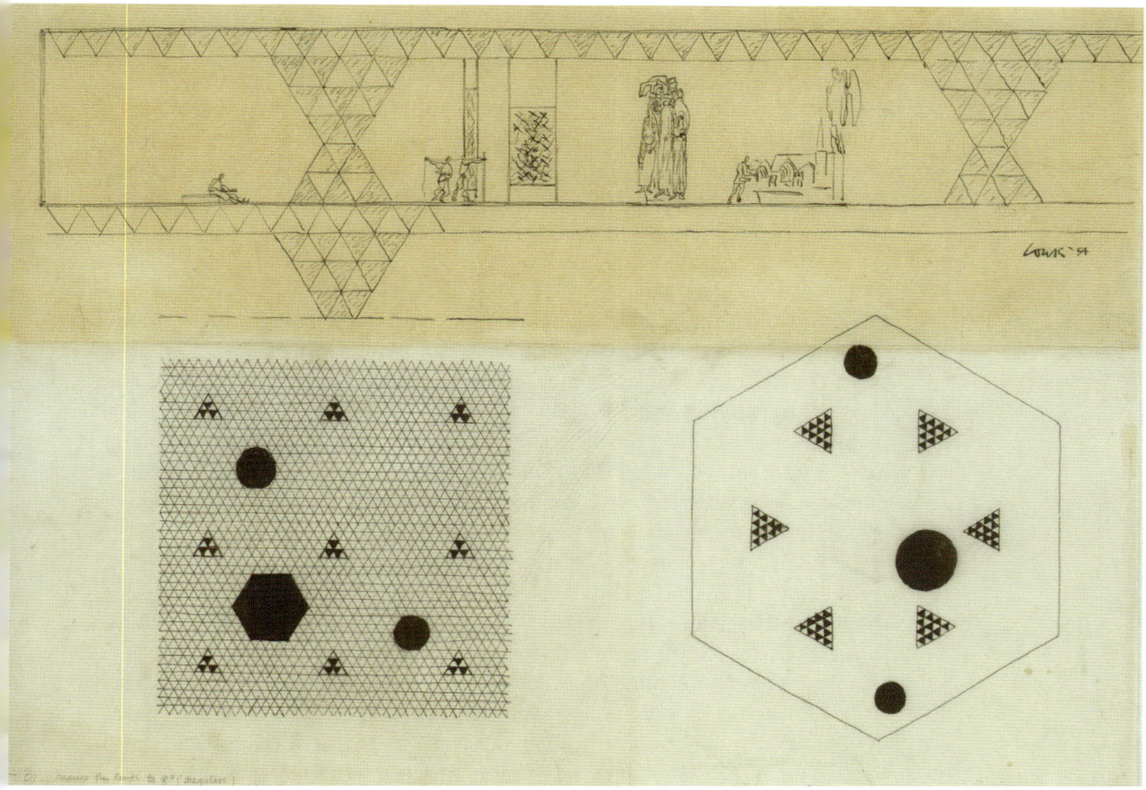

102-103. L. I. Kahn, sketches attached to a letter addressed to A. G. Tyng, 7 March 1954 (A. G. Tyng, ed., *Louis Kahn to Anne Tyng, The Rome Letters 1953-1954*, New York, Rizzoli International Publications, 1997, pp. 108-09).

104. L. I. Kahn, sketch attached to a letter addressed to A. G. Tyng, 16 March 1954 (A. G. Tyng, ed., *Louis Kahn to Anne Tyng, The Rome Letters 1953-1954*, New York, Rizzoli International Publications, 1997, p. 113).

105. L. I. Kahn, sketch attached to a letter addressed to A. G. Tyng, 19 March 1954 (LIK).

106. Yale University Art Gallery and Design Center, New Haven, Connecticut, 1951-53. Evaluation of structural system. Plan and section (LIK, 030.I.A.370.7).

107. Project for City Hall Building, Second Scheme, Philadelphia, Pennsylvania, 1954.
Section and plans (LIK, 030.V.B.13.14).

System. An initial sketch of this column form appears in March 1954 and shows how the whole column has become a combination of parts, starting with the original tetrahedron of the floor system. Depending on the height between the floors, the column can have a lower portion that widens towards the base, and therefore be composed of three or five drums in the form of tetrahedra. Precisely in response to the Space Frame of Mies van der Rohe, without overhanging sectors, Kahn's new system calls for 30-feet cantilevers – the span from column to column is 100 feet.

The sketches that illustrate Kahn's system show the section of a building created as a critical variation of the Yale Art Gallery, as is also demonstrated by the presence of artworks on the main level.[57] In a lecture at the Baltimore Museum of Art, Kahn illustrates the new structure: "how I would plan the Gallery now," he says.[58] Precisely to define the structure, he asks Venturi to build a "model of several floors of construction."[59] "[...] the columns are the shafts for the supply and return air ducts core per se," he tells Tyng. "This of course is my idea and it completes the function of 3D construction. I believe the columns used this way over each other make sense; another possible form is in elevation."[60]

Kahn's studies on tetrahedra are also urged on by the considerations expressed by Samuely to the students of Yale in May 1954 and commented on by Kahn: "He believes that the tetrahedron is basic in all structures but does not call a tetrahedron of Fuller's dimensions the criteria. He believes that tetrahedra of any combination of angles become a structural basis of form."[61]

Only when the column is no longer a post with a capital and becomes a complex combination of hollow tetrahedra in exposed reinforced concrete, capable of containing inside it, as in the Tetrahedron Floor System, the functional organs and vertical conduits that branch out in the floors, does Kahn introduce in his language another definition that not coincidentally comes from that of the "hollow stones" with which he envisioned prefabrication of the tetrahedra: "hollow columns."[62]

The design process that leads Kahn and Tyng to the definition of the tetrahedral structure of the City Hall Building focuses on amplification of the potential of the geometric system discovered with the floor system of the Yale Art Gallery. By now Kahn states, regarding the Yale Art Gallery, after the experimentation with different column forms, that the "order could have

[57] "I shall use them later to bring out my point about 'had I known more I would [have] built the Gallery thus and so,'" Kahn explains regarding the sketches (Louis I. Kahn, letter to Anne G. Tyng, 16 March 1954, *ibid.*, pp. 116-117, p. 117).
[58] Louis I. Kahn, letter to Anne G. Tyng, 19 March 1954, *ibid.*, (pp. 118-119), p. 118.
[59] *Ibid.*, p. 119.
[60] *Ibid.*
[61] Louis I. Kahn, letter to Anne G. Tyng, 2 May 1954, *ibid.*, (pp. 127-129), p. 128. In the margin, an article by Samuel in "The Architect's Journal" is mentioned.
[62] Kahn arrives at a new definition of the "hollow columns" which he also documents with slides (Louis I. Kahn, letter to Anne G. Tyng, 19 May 1954, in, Tyng, *op. cit.*, pp. 134-137, p. 134). In June 1954 he again changes the configuration of the tower on R Plaza – now the plan is triangular and the plaza is said to be "in the style of Pisa" (Louis I. Kahn, letter to Anne G. Tyng, 13 June 1954, *ibid.*, pp. 140-146, p. 145).
[63] Louis I. Kahn, letter to the Foreign Buildings Office of the State Departement, 30 June 1954 circa, "Louis I. Kahn State Department," box 34 (Goldhagen, *op. cit.*, p. 238, note 60). See also [Anne G. Tyng, Louis I. Kahn], *Spatial triangulation. City Hall Philadelphia, Pennsylvania*, in, "Progressive Architecture," June 1954, special issue, *Toward new structural concepts*, pp. 102-103.
[64] Louis I. Kahn, letter to the State Department in Washington, box 34 (Goldhagen, *op. cit.*, p. 241, note 34).

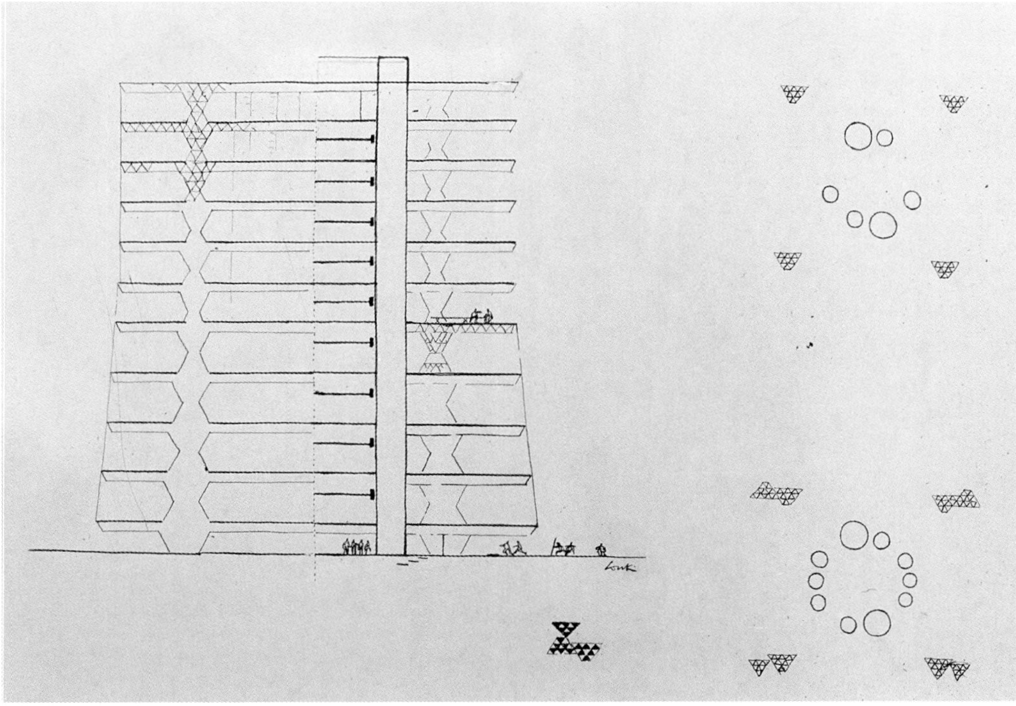

been more developed to include a vertical control system integrated with the columnar system."[63]

"The [Yale Art] Gallery is finished," he writes between April and July 1954, "[and] I find that the order could have been more developed to include a vertical control system integrated with the column system. The cantilever possibilities of the deep construction combined with the wider shear head of a hollow column would create a system of fewer columns. Each column would carry the air supply and return and other utilities thereby distributing service at the column points and giving better zone control. I am now applying this idea to a religious building in Philadelphia."[64]

The Adath Jeshurun Synagogue and the metaphorical values of structure

Studies and reflections following the opening of the Yale Art Gallery indicate that Kahn is capable of imagining a structure in reinforced concrete beyond any possible variation of the Domino skeleton, rich with a complexity also based on the face-off with metal reticular structures and arising from the Tetrahedron Floor System, from which a series of fantastic columns are generated. The "hollow stones," "column clusters" or "hollow columns" are the ingredients of this original 3D structure Kahn hopes to finally be able to build, when in the summer of 1954 he is commissioned to design the Adath Jeshurun Synagogue and School Building, at 69th Avenue and Old York Road in Philadelphia.

As in other projects in progress at the time, Kahn decided to use the Tetrahedron Floor System and to identify the most appropriate column form for that geometry and for the space of the synagogue, as demonstrated by the designs of the first version, where the school wraps the synagogue which is inserted in a hexagonal layout, and those of the second version,

108, 110. Project for Adath Jeshurun Synagogue and School Building, Elkins Park, Pennsylvania, 1954-55.

108. Plan studies (H. Ronner, S. Jhaveri, *Louis I. Kahn Complete Work 1935-1974*, Basel, Birkhäuser, 1987, p. 74).

109. L. I. Kahn, sketch attached to a letter addressed to A. G. Tyng, 16 July 1954 (A. G. Tyng, ed., *Louis Kahn to Anne Tyng, The Rome Letters 1953-1954*, New York, Rizzoli International Publications, 1997, p. 155).

110. Section study (LIK, 030.I.A.410.17).

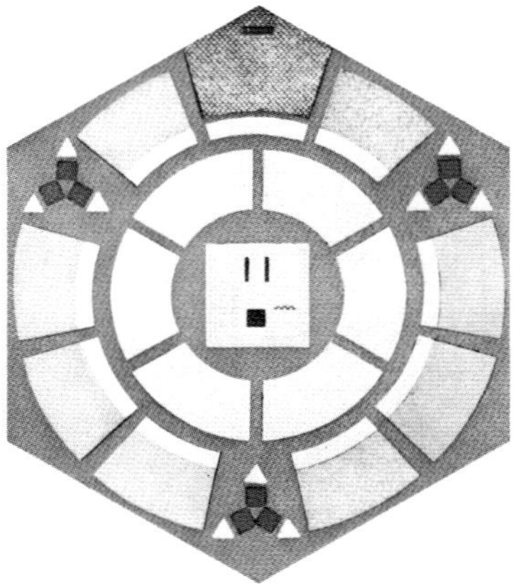

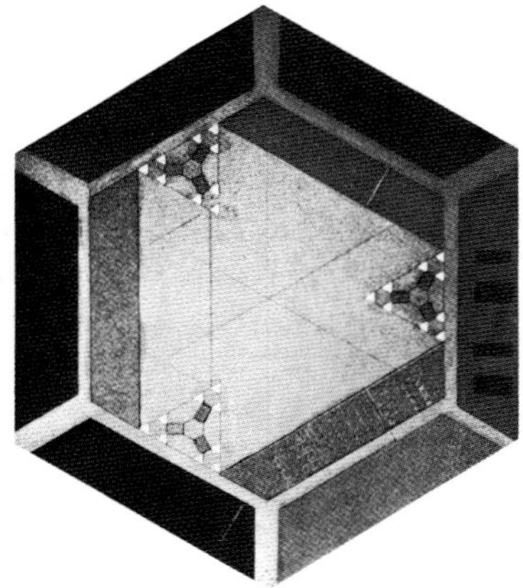

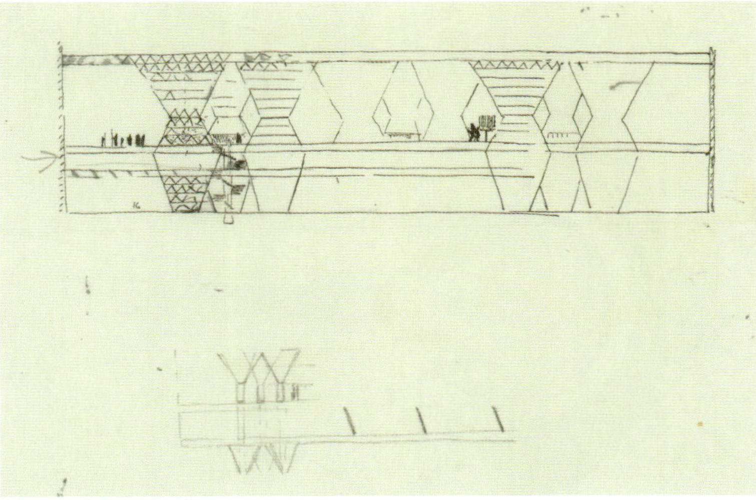

where the school is instead an oblong volume juxtaposed to the synagogue, which has a triangular plan inserted in a circle.[65]

For the floor system of the synagogue, Kahn experiments with the range of columns discovered during the course of his research after the Yale Art Gallery: a "cluster of four columns";[66] a "cluster" of three columns with a triangular section, arranged at the points of an equilateral triangle that contains a staircase at its center; and, finally, the clusters of tetrahedra triggered by the critical review of the structure of the Yale Art Gallery.[67] In the case of the "column clusters" the individual columns with a triangular section are hollow; so each of them belongs to the "hollow column" type, and together they form a support that borders another, larger void.[68]

The trees of Elkins Park depicted in Kahn's sketches are more than mere evocations of the landscape in which the building would have been immersed had it been built. The model of the definitive version, photographed without the roof to show the level of the hall with the altar and the pulpit, is emblematic of the symbolic value assigned to the three "column clusters," each composed of three hollow triangular columns, as if they were references to the trunks of the trees in the park, to remind the worshippers gathered in the synagogue of the atmosphere of the first outdoor sermons of the nomadic Jews.[69] In this case the intertwined rods of the Tetrahedron Floor System would seem like the branching of trees, confirming the metaphor applied by Kahn for the synagogue, "a place to meet under a tree or under many trees."[70] The glass envelopes would reinforce the

[65] Ronner, Jhaveri, Vasella, ed., *Louis I. Kahn: complete work 1935-74* cit., p. 80.
[66] Louis I. Kahn, letter to Anne G. Tyng, 5 July 1954, in, Tyng, *op. cit.*, pp. 156, 158.
[67] *The Louis I. Kahn Archive: Personal Drawings* cit., vol. I, pp. 350, 357, drawings 410.3, 410.17, 410.18.
[68] "This building," Kahn writes, "could be too important in proving the point of integration to a greater degree than that of the Art Gallery. I must not fail but really I am very confident that my recent discoveries about Architecture that is intrinsic (not only space frame) will see thru" (Louis I. Kahn, letter to Anne G. Tyng, 5 July 1954, in, Tyng, *op. cit.*, p. 158).
[69] One of the sources of inspiration is Paul Thiry, Richard Bennett, Henry Kamphoefner, *Churches and Temples*, New York, Progressive Architecture Library, Van Nostrand Reinhold, 1953 (Goldhagen, *op. cit.*, pp. 94-95).
[70] Louis I. Kahn, *Synagogue Architecture in the U.S. by Rachel Wischnitzer. [...]. A Review by Louis I. Kahn*, n.d., [1955], LIK, 030.II.A.65.36.

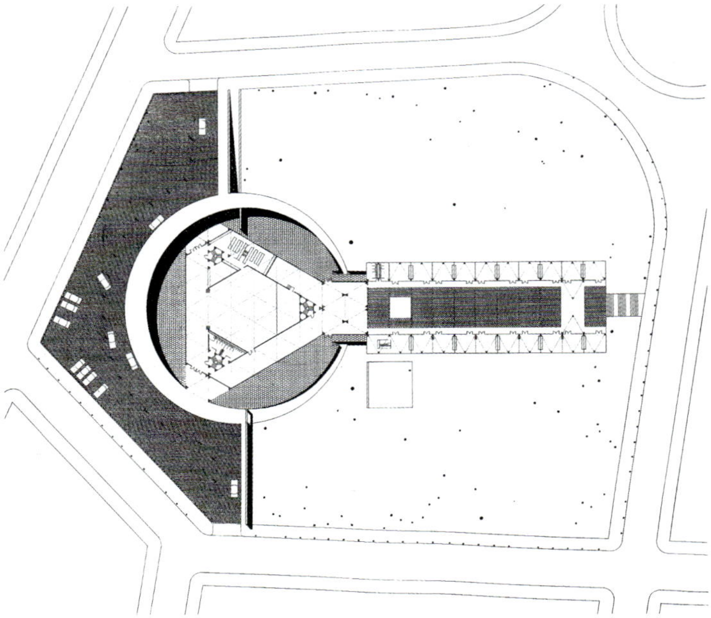

111-114. Project for Adath Jeshurun Synagogue and School Building, Elkins Park, Pennsylvania, 1954-55.

111. Ground level plan (LIK, 030.IV.B.410.1.1).

112. Second floor plan (LIK, 030.IV.A.410.1.4).

113. Elevation study (LIK, 030.I.A.410.13).

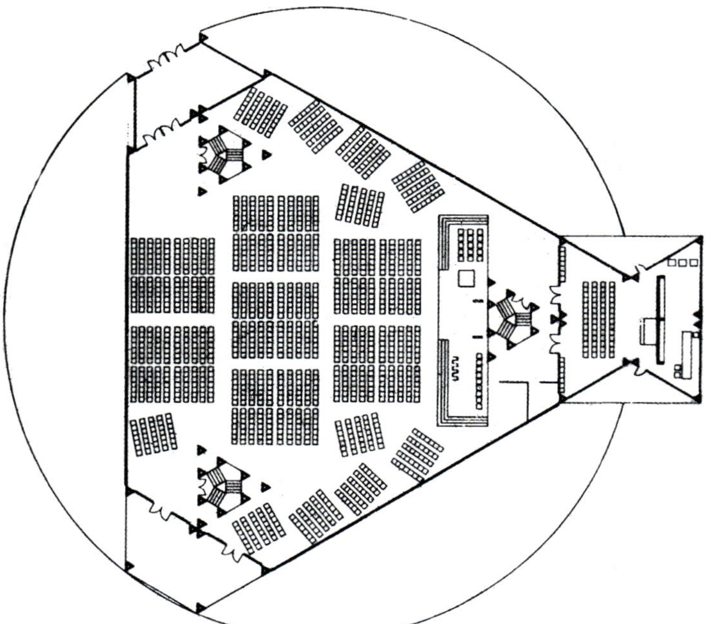

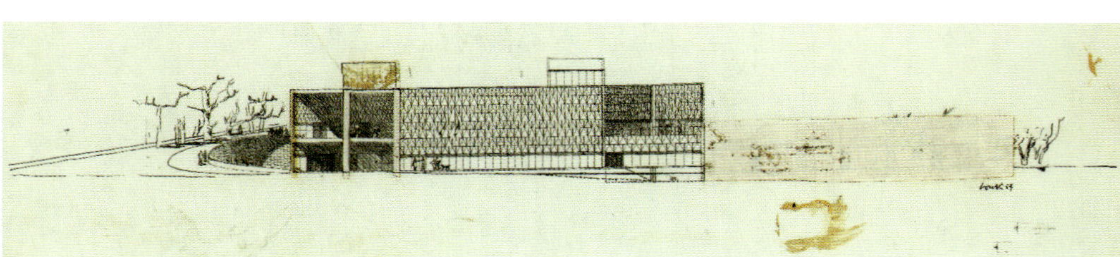

114. Model (LIK, 030.IV.A.410.1.3).

illusion of an outdoor space whose symbolic values are defined by the column-trees. The concrete envisioned for the Adath Jeshurun Synagogue and School Building would be exposed, for the floor system and the columns, in spite of the doubts expressed by one member of the committee charged with visiting the Yale Art Gallery and contacting the contractor. "That guy," Kahn reports, "may of course hate the exposed concrete and may also think Macomber too fancy for them. I have my finger crossed."[71]

This is Kahn's terse description of the final design: "Three major column clusters of 9 columns each in the form of an open triangle 26' on the side support a tetrahedral space slab. [...]. Each column cluster harbors a stairway as though captured in a great hollow trunk. The columns thus spread grip the floor and roof structure like outspread fingers."[72]

But in the design process the choice of the "column clusters" seems to be a result of the difficulties encountered by Kahn in developing a technical and functional solution for the columns with a spindle shape, generated by the combination of perforated tetrahedra, forming magnificent transparent sculptures in perforated concrete in the space. It is precisely during the project for the Adath Jeshurun Synagogue that Kahn defines as "angled columns" this sort of support directly derived from the Tetrahedron Floor System.[73] These "angled columns" would have contained the staircases, creating a theatrical ascent, as Kahn himself explains: "The stair is in the center of the 3-D column supports (hollow to take the services) and the effect of going up the stair from the coat room and lounges below should be good with light from above lighting the cluster of column supports."[74] But the insertion of the stairs in this type of column of the Space Frame is anything but easy, and Kahn admits that he has not yet found "the answer but will find it because it is right."[75]

In the pursuit of technical solutions for "the more possible type of construction"[76] of the "angled columns" Kahn calls in the engineers of Keast & Hood: Gianopulos, already on the team for the Yale Art Gallery, and a

[71] Louis I. Kahn, letter to Anne G. Tyng, 18 August 1954, in, Tyng, *op. cit.*, (pp. 168-171), p. 171.
[72] Louis I. Kahn, *A Synagogue (Adath Jeshurun of Philadelphia)*, in, "Perspecta," n.d., [1955], No. 3, p. 62.
[73] Louis I. Kahn, letter to Anne G. Tyng, 3 September 1954, in, Tyng, *op. cit.*, (pp. 174-177), p. 174.
[74] Louis I. Kahn, letter to Anne G. Tyng, 18 August 1954, *ibid.*, (pp. 168-171), p. 171.
[75] Louis I. Kahn, letter to Anne G. Tyng, 18 August 1954, *ibid.*, (pp. 168-171), p. 171.
[76] Louis I. Kahn, letter to Anne G. Tyng, 3 September 1954, *ibid.*, (pp. 174-177), p. 174.

young engineer, Charles Etter, who had just completed his PhD at the Massachusetts Institute of Technology in Boston, on the question of the "3-D" structure.[77] Kahn's discussions with Etter focus on ways of constructing the tetrahedra. Kahn goes back to thinking of the tetrahedra in terms of prefabricated reinforced concrete parts, as in the case of the floors system for the project of the City Hall Building, to avoid the technical difficulties encountered in the construction of the floor system of the Yale Art Gallery. The account written by Tyng explains the reasons behind Kahn's renunciation of the "angled columns." This is the context in which he introduces a definition of concrete that expresses his vision of the nature of the material: "molten masonry."

"I learned that I must talk to them with less enthusiasm and more probing in their minds and training. I found that a star frame means literally a frame of members – the members having direction held together where three of the members come together. No membrane is involved. I did not reach them because I used the term 'space frame' when I should have used 'space slab' (a word I invented for the concrete construction at Yale). As soon as I did use space slab he said 'oh that's different!'. Space frame – space slab, there is really a difference. How to form such slab is still a problem in spite of Yale because it is by experience and one should stop until the answer is humanly complete. I began to think (this all happened in the presence of Etter who just asked me repeatedly how do you do this instead of venturing how himself) and came up with the general attitude. Concrete is masonry with reinforcing for tension built in. This molten masonry is marvelous but it is heavy but should do wonders with the tension ability added and much lighter than. [...]. Now as masonry it may/can be formed by making each [tetrahedron] = a stone (hollow stone) – or we could assume with the absolute purity of form for a likely way equally valid from my point of view. [...]. The continuous direction of the rows of tetrahedrons are thought of as joists of concrete with the reinforcing following the [tetrahedron]. Pattern and opening provided throughout for the servicing of the slab. Actually that picture is wrong or inaccurate because you would probably get forms like the Yale (of a series of hollow containers dropped in place of a self-spacing design). I must lick the cost problem but the [tetrahedron] can be beat! for looks. I intended to make each about 4' (like now at Yale). The only reason I don't want to as yet try something new for the synagogue is my present knowledge of the [tetrahedron]. And because I get no stimulus from the engineers."[78]

When the Jewish community tells him of its intention to abandon the project, Kahn draws, in 1955, a section of the synagogue reduced to a pure structural web of lines, in which the geometry of the tetrahedra of the floor system generates columns with angled rods that emerge from the roof, while the floors extend into the void beyond the supports – an indirect response to the Space Frame of Mies van der Rohe. The whole structure stands out against the backdrop of the landscape of the park, with the trunks and branches of the trees. This is Kahn's last major attempt to

[77] Louis I. Kahn, letter to Anne G. Tyng, 18 August 1954, *ibid.*, (pp. 168-171), p. 168.
[78] Louis I. Kahn, letter to Anne G. Tyng, 3 September 1954, *ibid.*, (pp. 174-177), pp. 174-177.
[79] Louis I. Kahn, *Synagogue Architecture in the U.S. by Rachel Wischnitzer. [...]. A Review by Louis I. Kahn*, n.d., [1955], LIK, 030.II.A.65.36.

build another Tetrahedron Floor System, after that of the Yale Art Gallery, resting on a new type of column.

In the same year in which the project for the Adath Jeshurun Synagogue comes to an end, with the drawing of its ideal section, Kahn writes a review of the book by Rachel Wischnitzer, *Synagogue Architecture in the United States, History and Interpretation*, where after having criticized the Gothic Revival for its distance from the "order of construction" – "it can be explained psychologically" – he describes the potential of contemporary structure, the one he dreamt of building for his synagogue, returning to the concepts of "hollow stones" and "hollow columns":

"Today our growing knowledge in the structural nature of the materials tends to minimize the miracle of span. The columns are thin and the space can be great. Stability, once expressed by solid walls, piers and vaults of masonry, is now expressed by columns, cantilevers, slabs and space frames, in steel and concrete. We can build strong, open structures of grace and little weight. In a sense we build of 'hollow stones'. Such structures suggest their exposure and the harboring within the structure of the service needs. Such mechanical elements as air conditioning, plumbing, heating, electrical, together with elevators, stairs, closets, anterooms, washrooms and hallways may be considered the 'servants' of the larger spaces. This suggests a concept of space-order in which these minor spaces may have a consistent architectural relationship to the larger spaces they serve. It is conceivable that these servant-rooms are hollow columns themselves supporting the greater spaces. Thus Architectural order implies integration of space, structure and mechanics. Most buildings today do not show evidence of having achieved this integration. This may be considered as a stage of awkwardness in our growth to reach architectural maturity. The maturity of this architecture will depend on motivations of the community and the strength of the will to create what a space wants to be. [...]. It may also be expected that the first evidence of this integration may be 'archaic.'"[79]

After the scheme of the tetrahedral structure also extended to the columns, born of the critical review of the Yale Art Gallery, and after the attempt to apply that scheme to the Adath Jeshurun Synagogue, it is not surprising to read, in an article published in "Perspecta" in 1955 on the work in New Haven, Kahn's own critical reflections on the Tetrahedron Floor System supported by traditional pillars:

"Mr. Kahn acknowledges that, while formulating his design concept, he was not fully aware of the potentialities of multi-planar truss systems. If he were building the Gallery today, he most probably would have followed the geometry of the ceiling through in the spacing of the columns, and thus

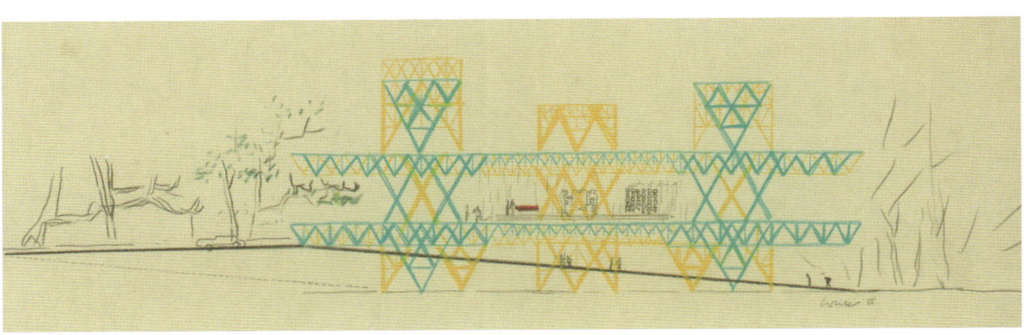

115. Project for Adath Jeshurun Synagogue and School Building, Elkins Park, Pennsylvania, 1954-55. Section study (LIK, 030.I.A.410.16).

come closer to the translation of forces. A hexagonal module, for instance, would have enabled him to transmit direct loads in all six directions, instead of only in two, and thus would make the system a genuine 'space frame'. Such a system would make possible cantilevers along the edges of the building, which would eliminate some awkward intersections and the basic inconsistency of a triangular order inscribed into a rectangular one; the number of columns would be reduced. A non-rectangular column module could also have stimulated a rich expression of the inside on the outside."[80]

"Square space idea" and masonry order

Pillars, beams and floor framework, with or without place for conduits, are the primordial elements utilized by Kahn to determine an elementary unit of structure with a square plan, corresponding to the measurements of a room, to repeat in order to build a house according to logics of grouping set case by case by the landscape, the slope of the land, the type of layout and theoretical and artistic principles. The idea of finding an elementary structural unit dates back, in Kahn's work, to the project for the Parasol House. After the construction of the Yale Art Gallery, Kahn transforms that structure, first conceived in metal, into one to be made with masonry and carpentry, which would influence several successive variants in concrete alone. While at the Yale Art Gallery Kahn invents what he calls the "space slab," now for the houses he invents the "square space idea."[81]

The potential of the square-plan structural unit is tested in the projects in 1954-55 for the house of Francis Heed Adler, on Davidson Road in Philadelphia, and for the house of Weber DeVore, on Montgomery Avenue in Springfield Township, Montgomery County, Pennsylvania. "Each house is a cluster of square areas 26 feet on a side. They grow out of the same order. The designs are different," Kahn explains in a note.[82]

In both houses the materials of pillars, beams and infilling walls are selected to make the various load-bearing roles of the structural organism visible through their textures. The order of the masonry composition becomes a fundamental design question. In the DeVore House the bricks that form the infills with the interspace are placed "on edge"[83] to make it clear that they do not bear loads. "Brick of cavity walls laid in non supporting manner to distinguish from the brick supporting piers," Kahn indicates.[84] While in the Adler House the structure of the roof is made with wooden beams,[85] the choice of the material of the beams and joists of the roof of the DeVore House remains enigmatic.

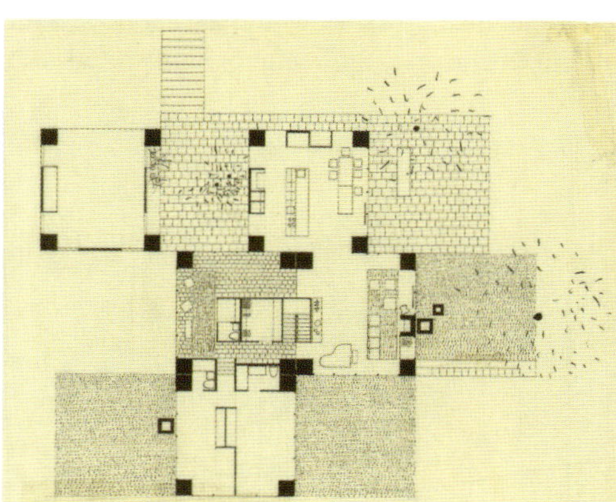

[80] *Order and Form* cit., p. 51. Kahn explains the structure of the Mill Creek towers in these terms: "the interior room of vertical circulation became the column of this tower, and the exterior was the column for the wind bracing of this tower, plus the support of themselves" (cit. in, Latour, *Louis I. Kahn: Writings, Lectures, Interviews* cit., p. 65).

[81] Louis I. Kahn, letter to Anne G. Tyng, 3 September 1954, in, Tyng, *op. cit.*, (pp. 174-177), p. 177.

[82] *Two houses*, in, "Perspecta," n.d., [1955], No. 3, pp. 60-61. On the Kahn houses see George H. Marcus, William Whitaker, *The Houses of Louis Kahn*, New Haven, Yale University Press, 2013.

[83] See the details represented in the drawing "House for Mr. Mrs. Weber De Vore," n.d., LIK, 030.I.C.425.001.

[84] *Two houses*, in, "Perspecta," n.d., [1955], No. 3, (pp. 60-61), p. 61.

[85] *Two houses* cit., p. 61.

[86] Louis I. Kahn, letter to Anne G. Tyng, 3 September 1954, in, Tyng, *op. cit.*, (pp. 174-177), p. 177.

[87] Louis I. Kahn, letter to Anne G. Tyng, 24 July 1954, *ibid.*, (pp. 165-167), p. 165.

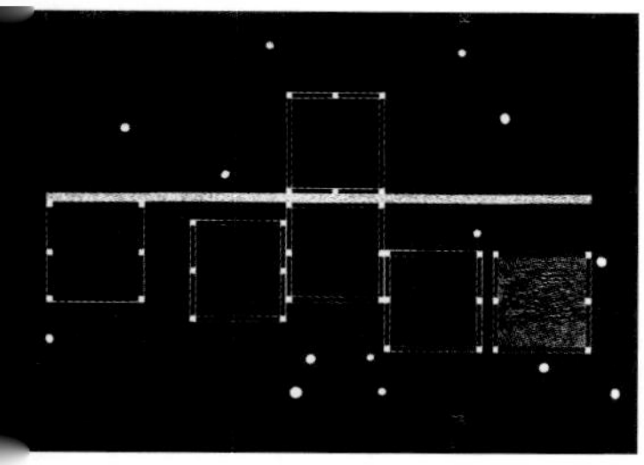

Thanks to the sloped assembly of the joists above the beams, "worked with flat nailing,"[86] Kahn recreates in the DeVore House the cavities that in the Yale Art Gallery were obtained by means of the pouring of the concrete into special pyramid-shaped formwork. Thus the structure of the Tetrahedron Floor System generates, in the design of the DeVore House, a similar system for the passage of physical plant systems, but extremely simplified in its construction.

The idea for the joists of the DeVore House comes in part from the research of Samuely, according to what Kahn himself states in a passage that is also significant for their eventual making in prefabricated concrete: "This ∧ triangular beam idea (Samuely) is immediately applicable (very good in concrete) (fair in wood) (good in steel) is therefore within the killing time demands of the client. On the Adler House I believe we could benefit from time a little more."[87]

It may be no coincidence that in the official text of presentation of the two houses, Kahn indicates only the material of the roof beams of the Adler House. The fact that in other projects during the same year of 1954 he is utilizing prefabricated reinforced concrete beams with a form identical to those of the DeVore House would lead us to believe that this is also the material selected for the joists of the roof of the house.

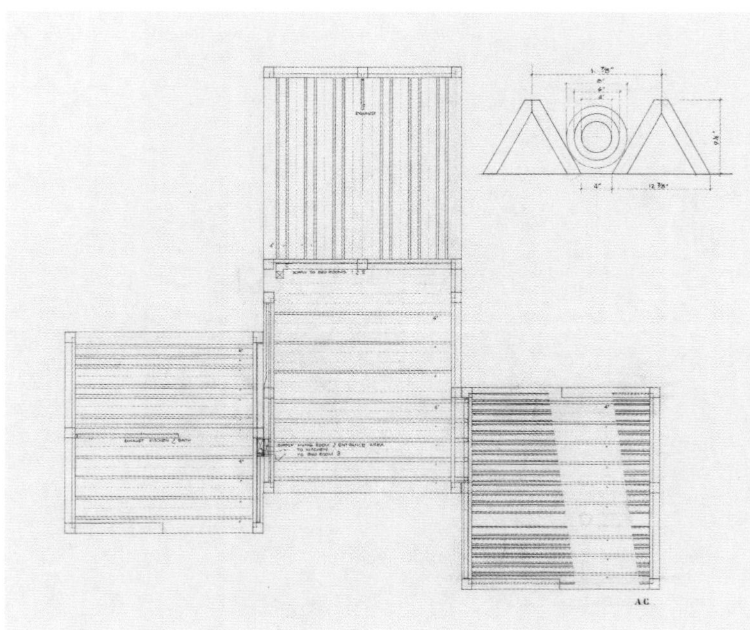

116. Dr. and Mrs. Francis H. Adler House, Philadelphia, Pennsylvania, 1954-55. Plan (LIK, 030.I.A.415.9).

117-119. Mr. and Mrs. Weber De Vore House, Springfield, Pennsylvania, 1954-55.

117. Schematic plan (LIK, 030.IV.B.425.1.1).

118. Elevation ("Perspecta," [1955], No. 3, p. 6).

119. Reflected ceiling plan (LIK, 030.I.C.425.1).

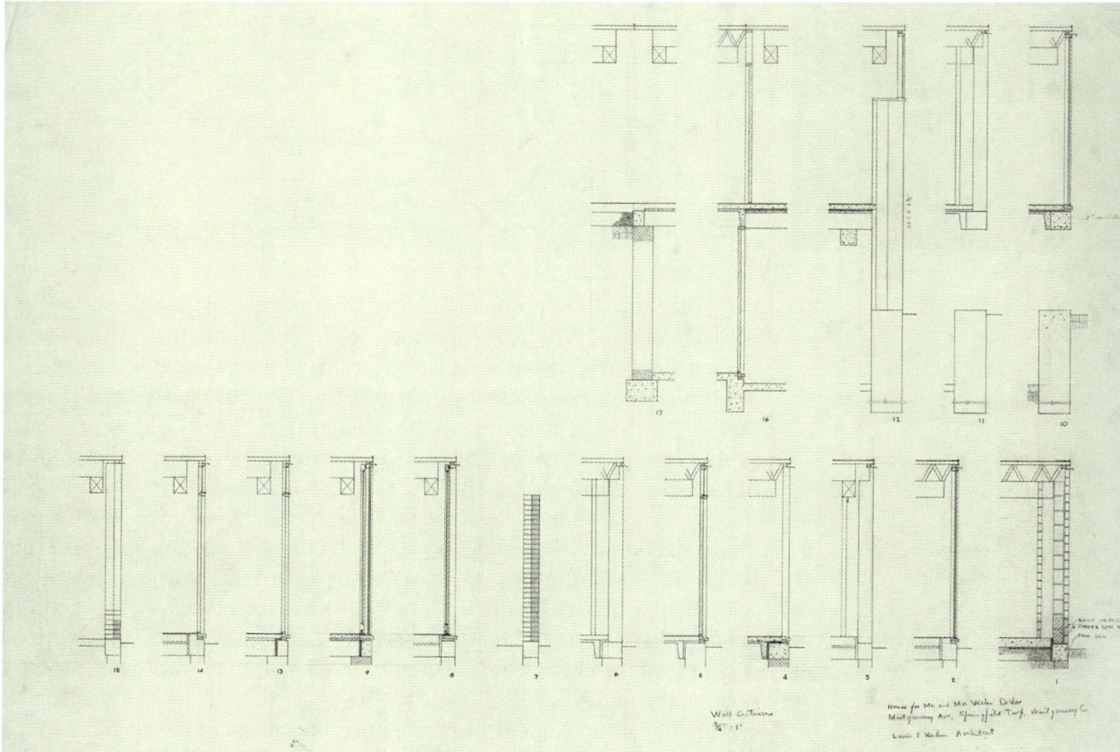

120. Mr. and Mrs. Weber De Vore House, 1954-55. Wall sections (LIK, 030.I.C.425.1).

121-122. L. I. Kahn, sketches attached to a letter addressed to A. G. Tyng, 24 July 1954 (A. G. Tyng, ed., *Louis Kahn to Anne Tyng, The Rome Letters 1953-1954*, New York, Rizzoli International Publications, 1997, pp. 166-67).

123. L. I. Kahn, sketch attached to a letter addressed to A. G. Tyng, 29 September 1954 (A. G. Tyng, ed., *Louis Kahn to Anne Tyng, The Rome Letters 1953-1954*, New York, Rizzoli International Publications, 1997, p. 179).

124. L. I. Kahn, sketch attached to a letter addressed to A. G. Tyng, 25 October 1954 (LIK).

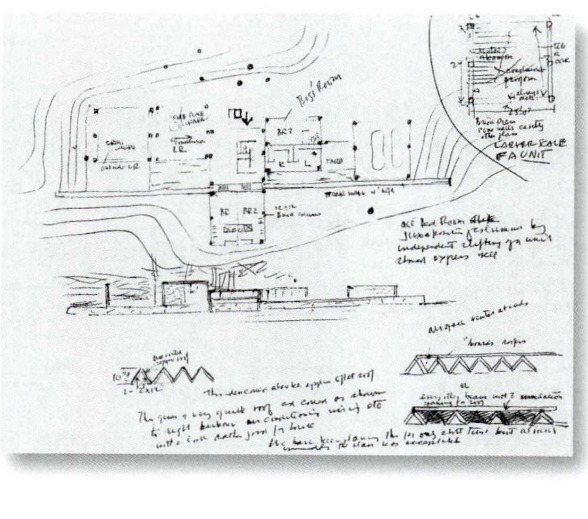

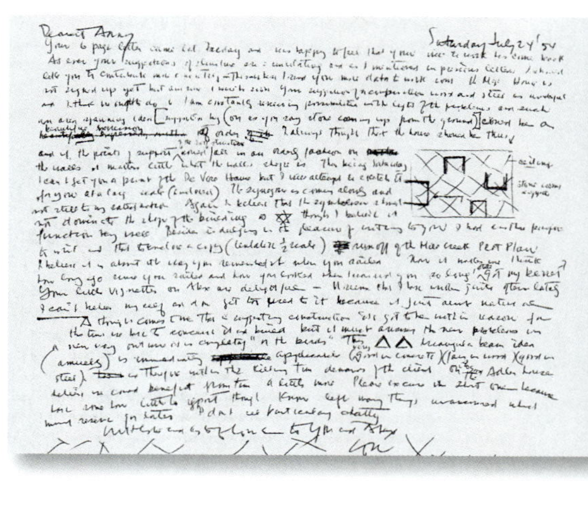
Dearest Anne
Saturday July 24 '54

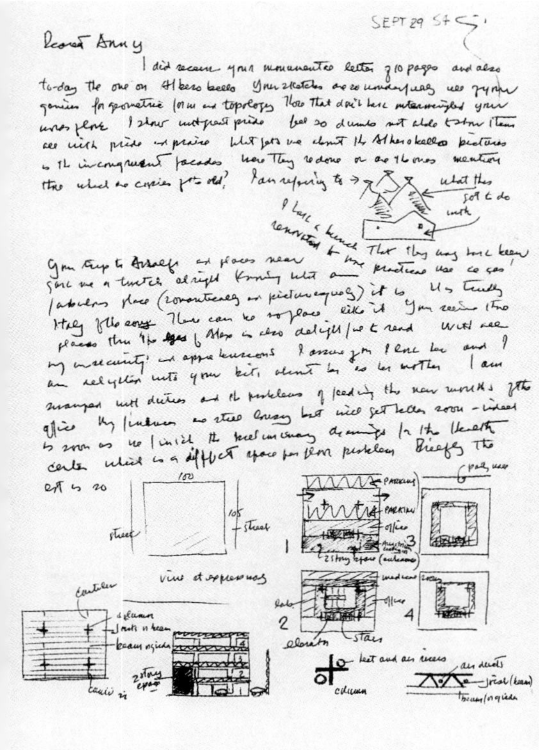

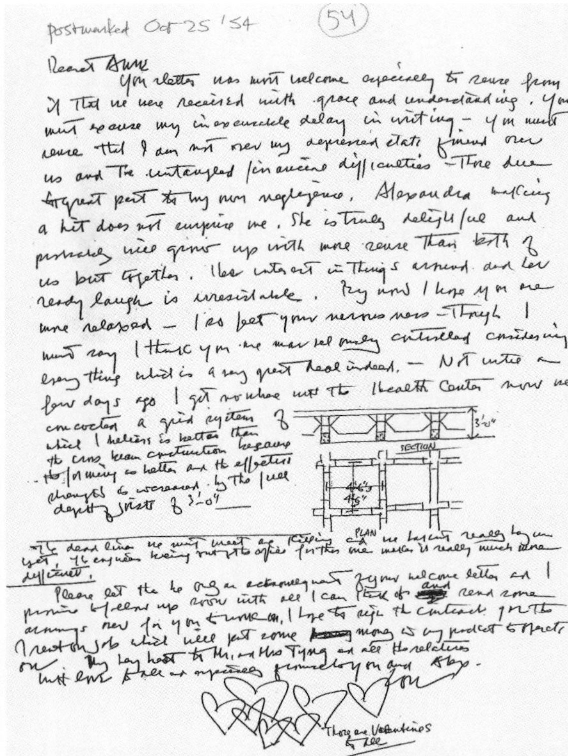

Vierendeel beams with floor system of the AFL Medical Services Building

The V-shaped system of beams and joists, for Kahn, emerges as a possible alternative to the Space Frame derived from the Tetrahedron Floor System in the summer and fall of 1954. The design of the DeVore House is but the first step in the development of a structure that – starting in September – Kahn imagines applying on a large scale, in the American Federation of Labor Medical Services Building on Vine Street in Philadelphia (designed and built in 1954-57, demolished in 1973).

In the preliminary phases of the project, at the end of September, a few days after having signed the contract, Kahn is thinking about a floor system with a geometry similar to that of the DeVore House, but to be made with prefabricated parts to assemble sloped over the beams, to obtain the cavities for the passage of physical plant conduits.[88] The columns have a cruciform section, with extended flanges, to create other cavities in which to place the vertical conduits that then branch out along the voids of the V-shaped joists.[89] Simplification of the structure leads to a device still marked by the presence of cavities and still to be made with reinforced concrete, but now freed of the geometry derived from the Tetrahedron Floor System. The research on floor system forms in reinforced concrete with physical plant

[88] See the sketch attached to Louis I. Kahn, letter to Anne G. Tyng, 29 September 1954, in, Tyng, *op. cit.*, (pp. 178-181), p. 179 (the contract for the project is dated 14 September 1954; the first contacts with the AFL date back to May 1954).

[89] Louis I. Kahn, letter to Anne G. Tyng, 29 September 1954, *ibid.*, (pp. 178-181), p. 178, p. 179. See also Louis I. Kahn, letter to Anne G. Tyng, postmarked 25 October 1954, *ibid.*, p. 182.

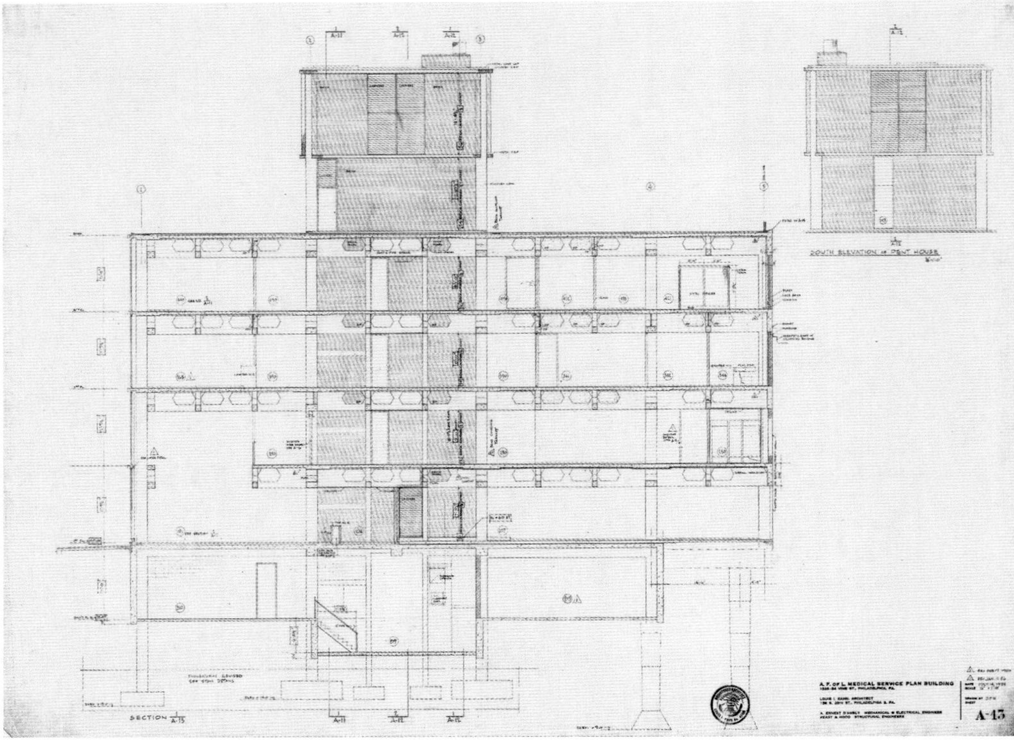
SOUTH ELEVATION of PENT HOUSE
SECTION
A. F. OF L. MEDICAL SERVICE PLAN BUILDING
LOUIS I. KAHN, ARCHITECT
A. ERNEST D'AMBLY MECHANICAL & ELECTRICAL ENGINEER
KEAST & HOOD STRUCTURAL ENGINEERS
A-13

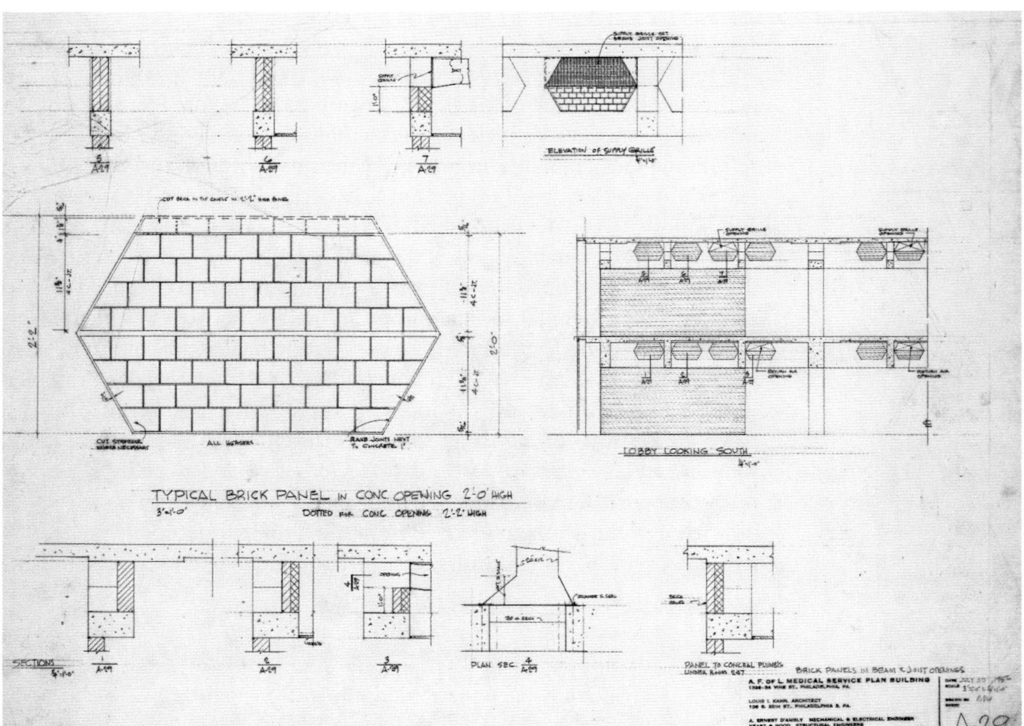
ELEVATION OF SUPPLY GRILLE
ALL HEADERS
LOBBY LOOKING SOUTH
TYPICAL BRICK PANEL IN CONC. OPENING 2'-0" HIGH
DOTTED FOR CONC. OPENING 2'-2" HIGH
SECTIONS
PLAN SEC. 4
PANEL TO CONCEAL PLUMB'G UNDER ROOM 247
BRICK PANELS IN BEAM & JOIST OPENINGS
A. F. OF L. MEDICAL SERVICE PLAN BUILDING
LOUIS I. KAHN, ARCHITECT
A. ERNEST D'AMBLY MECHANICAL & ELECTRICAL ENGINEER
KEAST & HOOD STRUCTURAL ENGINEERS
A-29

125-127. American Federation of Labour Medical Services Building, Philadelphia, Pennsylvania, 1954-57.

125. Section, 14 July 1955 (LIK, 030.I.C.420.1.A-13).

126. "Brick Panels in Beam & Joist Openings," 30 July 1956 (LIK, 030.I.C.420.1.A-29).

127. View (LIK, 030.IV.A.420.1.5).

cavities could not help but consider the use of one of the most interesting beams in the evolution of 20th-century structures, due to its large openings: the Vierendeel beam, which Kahn and Stonorov had already applied in steel for the roof of the Parasol House. The sole trace of the geometry of the earlier structures is the hexagonal configuration of the cavities in the Vierendeel beams for the AFL Medical Service Building.

Kahn creates a main framework with a large square grid of Vierendeel beams, putting the pillars at the intersections – some of them with the cross form of the initial studies. Inside the large squares he places a secondary framework, again with Vierendeel beams, to form nine square compartments.

The Vierendeel beams have an unconventional configuration. Precisely because the concrete is poured in place, their upper bar is replaced by a continuous slab, again in reinforced concrete, that binds them in a system. This is a detail that indicates the affinities in the construction process for the slabs between the Yale Art Gallery and the AFL Medical Service Building.

The making of the structure happens by means of formwork with wooden boards that leave visible imprints after stripping. In the Vierendeel beams the boards have the length of the segment between the intersections of the beams, to avoid imprints of vertical joints that would break up the image of the beam. In the assembly of the boards of the top slab, Kahn is influenced by the pattern used by Le Corbusier on the platform of the Unité d'Habitation in Marseille: each square compartment formed by the beams features different orientation of the assembly of boards, producing a decorative effect of alternating imprints.

The cinder blocks of the walls, laid without cladding, make a decisive contribution to the image of a structure of primeval brutality. The envelope of the different levels is an alternating succession of colossal panels of gray polished granite and panes of glass.

In Kahn's research on the expressive values and meanings of concrete, the AFL Medical Services Building marks another fundamental phase; it

128-129. American Federation of Labor Medical Services Building, Philadelphia, 1954-57.

128. View (M. D. Meyers Archive, AAUP, 136.115.1).

129. Detail of the floor structure (LIK, 030.IV.A.420.1.11).

demonstrates the refusal to make use of the forms of concrete poured in panels, so as not to reduce the structural potential of the material to the fabrication of slender enclosures. Also in his later works, the panel is always made with sheets of stone, marble or steel, never in concrete. So from the Yale Art Gallery to the AFL Medical Services Building we can already see two fundamental forms of Kahn's concrete as they emerge: prefabricated in blocks or bricks, and poured in place in special formwork, but always used to create load-bearing or self-supporting structures, never a mere question of cladding.

The blocks for the Bath House, or the art of American concrete

Halfway through the 1950s Kahn already outlines an idea for contemporary concrete that continues to evolve in keeping with the principles indicated in *Monumentality*, towards a skeleton with technical and constructive characteristics specified against the backdrop of an ongoing comparison to steel structures and the complex geometrics of the Space Frame. Precisely because Kahn constantly monitors the evolutions of the Space Frame, his reinforced concrete becomes first of all the pursuit of the configuration of a sculptural structure with cavities for a "space slab" conceived as an organ capable of performing multiple functions. Kahn demonstrates that his vision of concrete is focused, at least from the early 1960s, on those "gigantic sculptural forms of the skeleton frame" glimpsed at the time of *Monumentality*. So it is not surprising that with Tyng he shares an attraction to structures like those of Nervi. For Kahn the wall remains a construction to be done with bricks or concrete blocks while the panel – apart from the glazing of the windows – is made with sheets of black slate or gray granite. It is evident that the chromatic range of the materials for claddings and walls has a direct relationship with the concrete of the skeleton, starting with the clay brick that will continue to be a part of the future works of Kahn. To understand what is taking form as Kahn's poetics of materials, it is worth thinking about the hint made in *Monumentality* of materials for completion of the "giant major skeleton of the structure" without cladding: those materials, according to Kahn, should "feel at ease in its presence."[90] It is precisely the search for agreement of colors, textures and finishes that guides Kahn by now in the choice of materials to combine with the exposed concrete of the skeleton.

In a lecture in 1955 at Tulane University, in New Orleans,[91] Kahn makes statements that offer a clear picture of his thoughts on projects like the Yale Art Gallery or the DeVore House, in which he has had to come to grips with the question of the wall and its materials, "brick or cinder block."[92] The works to date demonstrate that the blocks play a particular role in his vision of concrete. But it is not until the start of 1955 that Kahn begins to sense the possibility not only of leaving them exposed in interiors, but even of using those ordinary blocks for entire works with walls without cladding. This is a breakthrough in the 20th-century history of prefabricated concrete blocks, because those used by Kahn are not specially designed with artistic forms like the "textile blocks" of Wright.

[90] Kahn, *Monumentality* cit., p. 583.
[91] Louis I. Kahn, *This Business of Architecture*, lecture at Tulane University, New Orleans, 1955, in, Latour, *Louis I Kahn, Writings, Lectures, Interviews* cit., pp. 62-66.
[92] "To build a wall requires the order of the header and stretcher system of laying a brick wall; but whether you use Flemish bond or English bond is a matter of design" (*ibid.*, p. 63).

The opportunity to build the first work only with cinder blocks, the prototype of a series of future works of Kahn destined to also influence the evolution of European Brutalism, comes with the projects the members of the Trenton Community Center commission for the Jewish Community Center, to host creative activities and meetings, a gymnasium with dressing rooms, a daycare center and offices, and for the Bath House, an outdoor swimming pool with dressing rooms and showers. Both buildings are to be constructed on land in Ewing Township, owned by the local Jewish community, immersed in greenery at the edge of the woods and close to the residential zone.[93]

The projects for the Jewish Community Center and the Bath House begin at the start of 1955, but Kahn concentrates on the development of the Bath House, for which construction is slated to start in the spring – this would be the only work of the original project to actually be built.

The geometrics of the concrete construction studied in the first designs for the Bath House reveal the artistic potential Kahn seeks in the deployment of materials. The plan of the project in February is a painting with an abstract composition, with the pavements and two swimming pools, a sort of "homage to the square" in the manner of Albers. The fact that the figures of the pools are part of the same invisible square is not secondary for an understanding of the criteria of Kahn's artistic composition of concrete elements. The texture of the blocks extends to the enclosure bordering the dressing rooms up against one side of the square.[94]

There is little doubt that Kahn intends to repeat the brash use of the most ordinary concrete, which precisely in this project demonstrates its potential for graphic abstraction, for an art that is a precursor not just of the taste for gray of Johns, but also of the artistic use by Sol LeWitt of normal concrete blocks.

The abstract painting in concrete of the Bath House contains the generating principle of the successive variations, all the way to the completed project. When the dressing rooms are detached from the edge of the square to occupy the terrain closer to the Jewish Community Center, their enclosure in any case defines a figure with a square base, with its center on the axis of the invisible square that includes the two swimming pools. The concrete blocks continue to characterize the enclosure of the dressing rooms.

In the design prepared from 13 March to 4 April the dressing room is generated by two concentric squares, of which the central one is subdivided into rooms and covered with a pitched roof.[95] Kahn calls for a texture of blocks related to the overall design, so that the square that generates the plan is also evident in the detail of the deployment of the concrete. This is why the measurements of the concrete block form a cube, 8 x 8 x 8 inches. The coping of the walls is formed by a row of slender slate sheets, based on the chromatic tuning utilized for the Samuel Radbill Building.[96]

The roof of the central square is in exposed reinforced concrete. The pitches are designed not to extend beyond the coping, so that the walls are the only visible elements, confirming the image of the enclosure. Kahn's

[93] See Solomon, *op. cit.*
[94] The plan has no indication of the material of the walls, but it is probable that Kahn envisions them, from this project phase, in concrete blocks.
[95] In the plan of the project of 13 March 1955 we read "design building so that later entire building can be roofed" (*The Louis I. Kahn Archive: Personal Drawings* cit., vol. I, p. 371, drawing 430.2)
[96] "Dressing & filter building for swimming pool," 4 April 1955, LIK, 030.I.C.430.003.

130-131. Bath House, Jewish Community Center, Trenton, New Jersey, 1954-59.

130. Plan, 7 February 1955 (LIK, 030.I.C.430.3).

131. "Dressing & Filter Building for Swimming Pool," 4 April 1955 (LIK, 030.I.C.430.3.13).

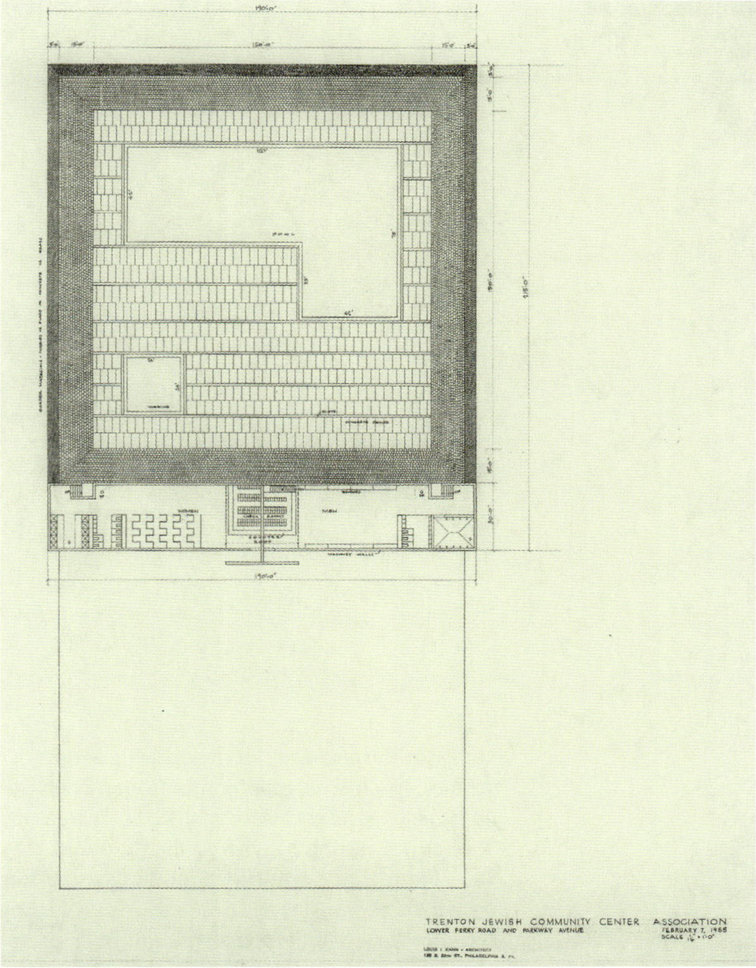

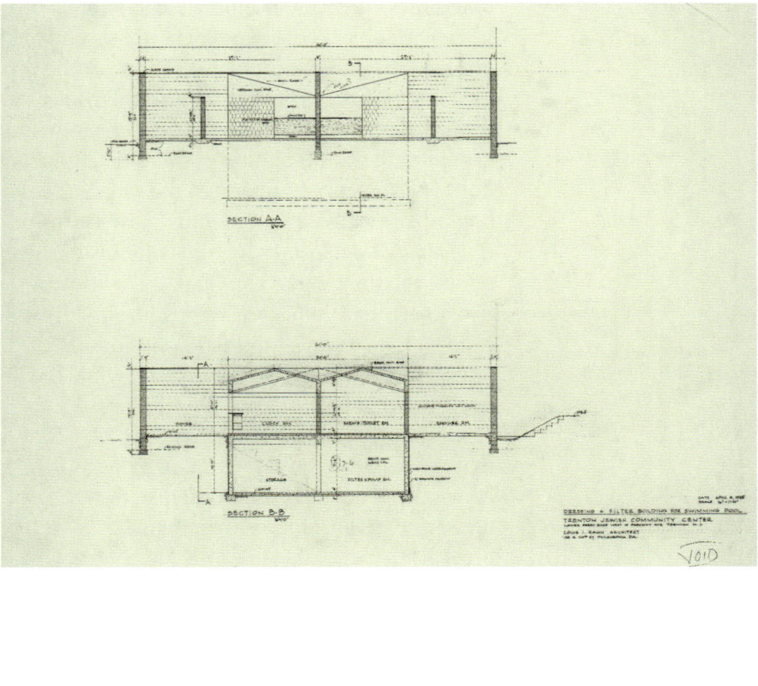

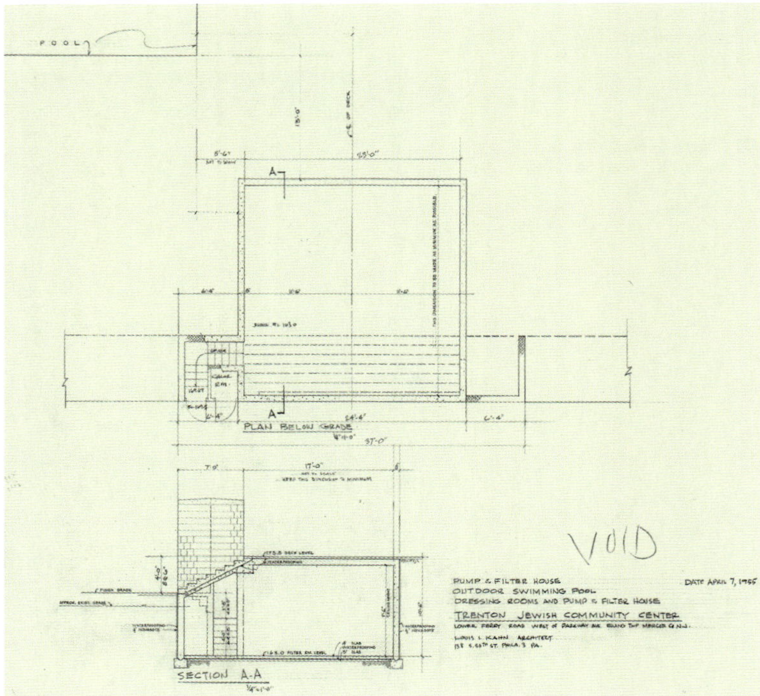

132-133. Bath House, Jewish Community Center, Trenton, New Jersey, 1954-59.

132. "Pump & Filter House," 7 April 1955 (LIK, 030.I.C.430.3.12).

study for the enclosure of the dressing rooms decorated by painting on the surface, like hieroglyphs, comes from this phase of the Bath House project – one side of the enclosure is braced by two triangular buttresses, of the type Kahn drew during his visit to the Acropolis of Athens.[97]

From 7 to 11 April, Kahn designs the steps for access to the pools, framed by two prisms with a square plan containing a room and access to the technical space under the steps. The walls of the prisms are still in cubical concrete blocks and the roof is a slab of reinforced concrete.[98] When Kahn draws these prisms the entire geometry of the dressing rooms changes.

The cubical concrete blocks are assembled to create four square units grouped around a center, also square, in the project documented by drawings on 14 April. The structure of the roof becomes a truss of metal rods resting on the squares at their ends. Precisely the transformation of the roof generates the configuration of the walls of the units, each composed of four hollow elements of support of the rods, or "columns," and of the enclosure sectors, or "walls." The "columns" have the form of the hollow prisms. "8' x 8' ∏ shaped enclosures are the columns of the roof structure," we see written on the drawing of the plan.[99] The idea of these "∏ shaped enclosures" dates back to Kahn's reflections, in July 1954, on the supports of a roof called "stone room supports."[100] It is another form of the "hollow columns" but with an opposite value, because while the first "hollow columns" come from the geometry of the Tetrahedron Floor System, those of the Bath House are generated by the concrete blocks and thus have a mural nature that makes it possible to contain true "rooms" in the void. It is no coincidence that the metal truss designed for the roof of the Bath House has by now become an object utterly extraneous to the masonry construction in blocks.

Kahn's theorem on masonry construction with concrete blocks is destined to remain in the state of an imperfect formulation. The unit of masonry texture and the differentiation of the structural roles between columns and walls, which Kahn seeks to achieve in the development of the details, do not manage to resolve into the coherent expression reached in the DeVore House, precisely due to the choice of using concrete blocks of the same size, but full for the columns and hollow for the walls. For the non-load-bearing walls, in any case, he is forced to introduce ribbing in the cavities. Kahn makes the different static functioning of the columns and walls visible, separating them by means of a slender crevice or "expansion strip" along which he has to place special connection "ties."[101] Once transformed into a segment disconnected from the columns, each wall is stiffened by two vertical bars that cross the hollow blocks and attach to the foundation and the course at the top. In this new configuration of the concrete block wall the previous slate coping gives way to a "solid top course."[102] But even the course is not visible, given the fact that Kahn has it built with blocks of the same size as the others, but full ones. Thus the vertical metal bars become the true non-visible columns of a skeleton formed by the foundation and the upper course, which are also invisible. What seems like a continuous wall of cubical blocks is actually a system of crossbeams and columns. The fact that in a sketch of the plan of the wall Kahn introduces protrusions at the position of the vertical bars is a more than significant clue of the conflict between the desire to reinforce the continuity of the block wall and its further articulation with ribbing that indicates the position of the reinforcements.[103] Kahn finds himself faced by a similar question of expression of reinforcement in the case of the brick construction for Ahmedabad.

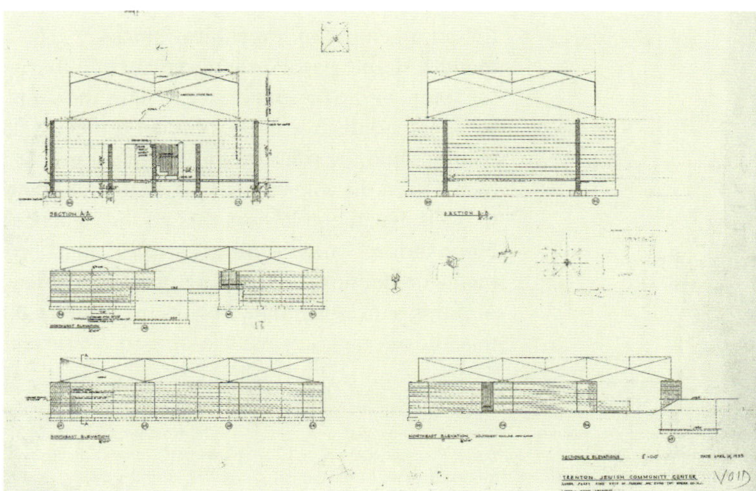

133. "Sections & Elevations," 14 April 1955 (LIK, 030.I.C.430.3.14).

[97] *The Louis I. Kahn Archive: Personal Drawings* cit., vol. I, p. 371, drawing 430.2.
[98] "Pump & Filter House Outdoor Swimming Pool Dressing Rooms and Pump & Filter House," 7 April 1955, LIK, 030.I.C.430.003; "Pump & Filter House Outdoor Swimming Pool Dressing Rooms and Pump & Filter House," 11 April 1955, LIK, 030.I.C.430.003 (in the latter drawing the building is shown without a roof).
[99] *The Louis I. Kahn Archive: Personal Drawings* cit., vol. I, p. 378, drawing 430.11, transcriptions, p. LXVI.
[100] Louis I. Kahn, letter to Anne G. Tyng, 24 July 1954, in, Tyng, *op. cit.*, p. 165.
[101] See the drawing "Sections & elevations," 14 April 1955, LIK, 030.I.C.430.001.
[102] *Ibid.*
[103] *Ibid.*

In May 1955, Kahn prepares further details of the project and the specifications of the "Bath House and Apparatus Room for outdoor swimming pool" of the Jewish Community Center.[104]

For the construction of the concrete block walls he indicates a range of technical recommendations. Each block should be free of manufacturing defects, such as "cracks, chips, sprawls or other defects that may impair strength or show in the finished work."[105] This unusual type of indication is fundamental to achieve a masonry composition with blocks that will be left exposed. Kahn calls for "load-bearing blocks" for the columns, to form solid supports for the roofing, and "non-load-bearing blocks" for the "walls." For the construction of the base and summit courses of the walls, and their free ends, "solid blocks" are used. In the specifications Kahn also advises preparation of the holes in the blocks to reinforce with steel bars ("voids in blocks shall be slushed full where tie-rods run through, at frames, etc."[106]). Alternating installation of special metal reinforcement, of the Perma-wall or Dur-o-wal type, is foreseen along the courses to consolidate the arrangement of the cubical blocks.[107] The mortar for the masonry is made with one part Portland cement and five parts "clean sharp sand." All the joints of the masonry have to have the same thickness. The mortar has to be spread flush with the blocks ("joints cut flush"[108]). Only between the columns and the walls do the specifications call for a joint recessed by one inch. Given the convention of covering the cement of floors and the blocks of walls, Kahn emphasizes that all the cement surfaces, both the "concrete block surfaces" and the "floor surfaces," must be left exposed, without treatments or claddings. "Surfaces not to be painted," is the title of the chapter of the specifications that dictates the absence of claddings. It is significant that the warning against cladding in *Monumentality* – "It need no longer

[104] "Bath House and Apparatus Room for outdoor swimming pool, Trenton, Jewish Community Center," specifications, 11 May 1955, LIK, 030.II.A.108.62.
[105] *Ibid.*
[106] *Ibid.*
[107] At the moment of the drawing up of the specifications, the blocks are still cubical, as shown in the drawing "Revisions to apparatus room for outdoor swimming pool," 17 May 1955, LIK, 030.I.C.430.003.
[108] The fact that the "mortar was deliberately smeared rather than carefully flush jointed" cannot be attributed to Kahn, according to the specifications and, more in general, in keeping with what he has always indicated on his worksites for the laying of bricks or blocks (Neil Levine, *Kahn's Edge: The Provocative Historicism of the Trenton Jewish Community Center*, in, Kries, Eisenbrand, Von Moos, *op.cit.*, pp. 101-114, p. 105). For a "brutalist" interpretation of the work see Denis Scott Brown, *A Worm's Eye View of Recent Architectural History*, in, "Architectural Record," vol., CLXXII, February 1984, (pp. 69-81), pp. 71-73.
[109] "Bath House and Apparatus Room for outdoor swimming pool, Trenton, Jewish Community Center" cit.
[110] See the back of the photograph, LIK, 030.IV.A.430.003.
[111] See the drawing "Roof structure & roofing bath house and apparatus room for outdoor swimming pool," 20 May 1955, LIK, 030.I.C.430.003
[112] *Order of spaces integrated with order of construction*, in, "Perspecta," n.d., [1957], No. 4, p. 59.

134-135. Bath House, Jewish Community Center, Trenton, New Jersey, 1954-59.

134. View (M. D. Meyers Archive, AAUP, 136.116.2).

135. View (LIK, 030.IV.A.430.1.4).

be clothed for eye appeal" – now guides the preparation of worksite documents. The covering with "one coat of Alkatite Cement and Stucco Paint Primer" whose color has to be indicated by Kahn can be used only for certain sectors of the Bath House, where it is necessary to protect the blocks from liquids: the "column areas" where there are "toilets and urinals" and the "shower walls." A final treatment is indicated for all the walls with exposed blocks: "Clean down walls at completion and leave free of stains, marking and other defacement."

For the parts of the Bath House to be made in concrete poured in place, from the walls of the technical spaces to the stairs, the type of formwork and implementation varies, depending on whether the concrete has to remain directly visible or not (the walls of the technical spaces are to be made with formwork of unfinished boards).[109]

In the definitive project the fundamental passage is the elimination of the square blocks, replaced by rectangular blocks in order to construct more solid segments of walls and columns. In the new terminology of the two basic constituent parts of the Bath House, the columns and walls become, respectively, "column areas" and "enclosing walls."[110] The slab on top of the columns is called the "concrete column cap."[111] The "cap" is a reinforced concrete slab poured in place in plywood formwork. Kahn is not concerned with obtaining a regular geometric imprint pattern in its making. The roofs of the single units take on the form of a truncated pyramid, with an "oculus" at the top,[112] to create a profile similar to the roofs envisioned for the Jewish Community Center. The surfaces of the blocks are left exposed, without any finishing. In keeping with this choice that favors the gray of the concrete, the marble sheets originally planned for the partitions in the bathrooms and showers are replaced by sheets of slate.

The quality of the rectangular blocks is indicated in the documents. "All block work below grade shall be built of concrete block, above grade it shall

be Waylite block," we read in the addition to the specifications.[113] The construction of the walls and columns is done with rectangular blocks of Waylite,[114] made with light concrete with a pale gray color, composed of New Jersey sand and containing small fragments of stones from the Delaware River, in a pale shade of orange, reddish blue, blue-gray and white.[115]

The truncated pyramid roofs are added in the fall of 1955 and made with exposed wood carpentry and an economical dark gray layer of "asphalt shingles," left rough like the Waylite blocks of the walls and columns, the concrete of the floors and the slate sheets.

All the materials, from the concrete of the blocks to that of the slabs and floorings, to the mortar of the joints, the wood of the roofs and the layer of asphalt shingles, are deployed in a rudimentary way and left exposed. The shades of gray of the concrete, the black slate and dark wood form an essential chromatic range that later becomes a constant in other works by Kahn. It is no coincidence that the image of the cinder block as a humble material is evoked by Kahn in thoughts that begin with the brusque statement that "architecture need not necessarily be beautiful," in the lecture at Tulane University.[116] That "archaic" about which he wrote regarding the synagogue becomes, precisely in the Bath House, the "degree zero" Kahn would often cite, associating it with the concept of "beginning."[117] All of Kahn's research on concrete happens in terms of the discovery of the degree zero of the material. It was not possible to get beyond the degree of archaism achieved in this work in the use of concrete.

The fact that Kahn builds the Bath House as a sort of philosophical inquiry into the question of the materials of architecture seems to be confirmed by certain short remarks in his notebook under the word "wonder": "Marble and stone, brick and block, marble facing and rubble backing. The Romans introduced brick used dramatically with facing of marble. The attitude toward blocks is the same. How can block be a noble material in itself, as brick is though not used as a base material. Brick though hidden by stuc or marbles was by the Romans used with impressive integrity. Now that the religious buildings built by the church during the Renaissance [illegible] the marble quarry – the brick integrity of the Romans was revealed."[118]

The mural of blocks: the transparency of the "rough"

On the wall of the hollow column near the entrance to the Bath House Kahn, together with the architect Marie Kwo who works in his office, has a thin layer of white paint placed that allows the flaws of the blocks and joints to show through, so that the isodomic masonry defines the regulating pattern of the mural.[119] The surfaces of the blocks become units of measure of design in black and orange traced on the white background. In certain blocks the diagonal of the rectangle divides the face into two sectors, one of which is painted black, or two diagonal lines are traced. In both cases it obtains the effect typical of the representation, in graffito or paint, of the masonry bonding on the plaster facing, widespread in European me-

[113] "Contract amendments to the documents," n.d., LIK, 030.II.A.35.3.
[114] The measurements are 19.5 x 19.5 x 40 cm (7.67 x 7.67 x 15.74 inches).
[115] Restoration done by Michael Mills, of Farewell Mills Gatsch Architects, LLC 200 Forrestal Road Princeton, New Jersey.
[116] Kahn, *This Business of Architecture* cit.
[117] In 1953 the book *Le degré zero de l'écriture* is published by Roland Barthes, which poses the crucial question of the existence of a neutral writing, beyond style (it is translated into English in 1967, *Writing Degree Zero*, London, J. Cape, 1967). Kahn likes to use the degree zero image to indicate, in his conversations, a return to the origins. Kahn's use of this expression is recalled, among others, by Michael Graves

136-137. Bath House, Jewish Community Center, Trenton, New Jersey, 1954-59.

136. Study for a mural, 1956 (J. Hochstim, *The Paintings and Sketches of Louis I. Kahn*, New York, Rizzoli International Publications, 1991, p. 304, No. 430).

137. Mural (photograph by the author, 2010).

dieval and Renaissance houses Kahn may have seen in Rome. But in the case of the decoration of the Bath House the blocks of concrete, not their representation, generate the design, as if the truth of the construction had incorporated the cladding. The various graphic motifs designed starting from the circle create lines and abstract fields in which we can glimpse waves, clouds, birds, the horizon line (located at the fifth course from the top) and reflections, all figures that introduce the world of the outdoor swimming pool.

and Jules D. Prown (*Michael Graves, Revised version of the interview with Kazumi Kawasaki, A + U, Architecture and Urbanism, November 1983, Extra Edition*, in, Alessandra Latour, ed., *Louis I. Kahn, l'uomo, il maestro*, cit., (pp. 163-167), p. 165; and *Jules D. Prown, Interview with Alessandra Latour, New Haven, Connecticut, June 23, 1982, ibid.*, pp. 133-143, p. 138).

[118] Louis I. Kahn, notebook, LIK, 030.VII.1.

[119] Louis I. Kahn, study for a mural for the Trenton Bath House, 1956, collection of Sue Ann Kahn, published in, Jan Hochstim, *The Paintings and Sketches of Louis I. Kahn*, Rizzoli, New York, 1991, p. 304.

In the experimentation on concrete beyond the forms of *béton brut*, the fresco on the hollow column of the Bath House, which allows the roughness of the blocks to show through, corresponds to the *sculptures moulées* made in the wall of the Unité d'Habitation in Marseille, as Kahn has chosen a form of concrete that cannot be shaped on the worksite, namely that of prefabricated blocks.

The Bath House becomes the opportunity to define the concept of "compartment spaces," as described in remarks Kahn jots down in a notebook. In the same notebook, Kahn writes a page entitled "Mural – Trenton" where he discusses the mural of the Bath House.[120] The fact that he mentions the imperfect surfaces of the concrete blocks demonstrates the extent to which that background is the very essence of the mural, as if the paint and the colors were simply a veil of water through which one could get a vibrant glimpse of the rugged bed. "Rough" is the adjective that emerges from the phrases written by Kahn, and it adds a particular value to his considerations on the mural of Trenton, in a period in which international criticism is beginning to discuss and interpret the concept of *brut*. That concrete block wall, exposed or visible through the coat of paint conceived "as a construction problem," is Kahn's response to the *béton brut* grained with the imprints of formwork made with rough wooden boards.

"The enlargement for scaled drawing to full scale may be understood better by actual photographic enlargement of the small scale. The contours from accidentals are brought out as though not to be imulated [emulated] steel suggests the breaking up into smaller parts *on the construction* of the line. The single stroke at small scale becomes a constructed movement at full size. – Doing the mural at the Community Center at Trenton […] the scale drawing was only directional and that the lines and fields had to be built up from the limits of the block the medium and the surface and treated as a construction problem. The original idea of the discipline act by the existing 8 X 16 blocks still […] hold however. The block actually is rough the scale drawing is smooth. The actual joints between blocks are not uniform [,] the joints in the scale drawing are uniform. The transformation from drawing to an actual vertical wall – which cannot be tilted or laid flat as a drawing, requires the use of a ladder and the jumping from it to view the work and back again. Many new experiences enter which are not present in working of drawing. The actual mural must reflect the kind of labor different from that of making a drawing or painting. The struggle must be evident – felt by the observer. One must stop to consider unsatisfying the effect on our reuse of a painting let us say 100' long painted to imitate the studio sketch 100" long. It is a question of scale. Scale would be defined as related to the labor and the instrument of labor: Is it the hand? The spray? The bulldozer or crane? Also it is related to the readability of how it is done. We cannot expect the spray to simulate the action of the hand nor the hand to imitate the texture or unit or unity of the spray. Nor can the hand get acting [sic] paint a single line 10 feet long without rewetting the brush several times and probably modifies the color in so doing. The home painter prides in disguising his brush marks and in the matching of colors uniformly. The artist painter

[120] Louis I. Kahn, notebook, LIK, 030.VII.4.
[121] *Ibid.*
[122] José Luis Sert, *Complicated and Delicate*, in *Opinions of the Yale Art Gallery and Design Center*, in, "Progressive Architecture," May 1954, (pp. 15-16, 22, 24), p. 22.
[123] Boris Pushkarev, *Order and Form. Yale Art Gallery and Design Center, designed by Louis I. Kahn*, in, "Perspecta," n.d., [1955], No. 3, (pp. 47-58), p. 51.
[124] *Ibid.*, p. 55.

shows his brush strokes and considers each new brush stroke as a new color (even though the same color is used). Scale is the brush stroke. If the hand gets to be as big as a crane then the scale is equally as justifiable as the scale produced by the hand. These two scale[s] must not resemble each other. Leger's immense Mural (U.N.) from a 10" x 10" sketch suffers from lack of scale."[121]

New Brutalism at New Haven?

While the Bath House is being built, the architecture reviws are manifesting increasingly widespread interest in Kahn's architecture, after the success of the Yale Art Gallery. The circulation of articles on the Yale Art Gallery brings the work to the attention of international critics. The phrases written by Sert in 1954 reveal an awareness of the continuity of experiences on exposed concrete between Le Corbusier and Kahn: "the Yale University Art Gallery, with its remarkable ceiling pattern making use of rough concrete and its walls of unpainted cement blocks, follows a new trend toward a more sculptural expression in architecture which possibly started with Le Corbusier's apartments in Marseilles."[122]

The most significant and well-documented presentation of the Yale Art Gallery is in the review of the students of the Yale School of Architecture "Perspecta" which offers it ample space, with many photographs, in issue No. 3, published in 1955. The text on the work written by Boris Pushkarev and introduced by the two Kahnian terms *Order and form* contains no reference to the question of *béton brut* introduced by Sert. Otherwise, we can notice that in the series of photographs published in the same issue under the title *The evolution of the craftsman?* there are three close-ups of the concrete surfaces of the Unité d'Habitation in Marseille, as well as a work by Nervi mentioned by Kahn in *Order is*, published in the same issue. Without ever calling *béton brut* or Le Corbusier into play Pushkarev, in any case, taking the words from Kahn, describes the novelty of the quality of the concrete surfaces of the Yale Art Gallery:

"Mr. Kahn loves the construction joints, the form imprint; he hates patch-up work. Thus he left the tetrahedral rib structure raw, without cleaning or painting, to keep the freshness of surface which derived from its forms. The day the first forms were stripped and the tetrahedrons appeared in daylight was an event anticipated weeks in advance."[123]

Precisely because one of the characteristics of the work is the making of walls left exposed, without cladding, Pushkarev includes information on the care the workmen put into the process, setting this worksite apart from others in New Haven for the choice of the type of finishing, which is almost done by hand in spite of the exceptional technical character of the structure. It is precisely this aspect that links the article on the Yale Art Gallery to *The evolution of the craftsman?*

"The workmen." Pushkarev writes, "realizing that materials will be left exposed and that whatever is done in the beginning will remain visible, showed conscientious care. Several left conventional jobs which were under way in the city at the same time, to work on a project that stimulated their interest and pride. Notable results of their efforts are: the high quality of wood formwork on the concrete columns; the precision of brick-work on the South façade and of the exposed interior concrete-block walls. Kenneth Froberg, the superintendent on the job, should be credited with maintaining high standards of workmanship throughout."[124]

The article that brings renown to the term brutalism, following the definition coined by Le Corbusier of *béton brut*, is *The New Brutalism*, published in December 1955 by "The Architectural Review" and written by Reyner Banham, who also has images of the Yale Art Gallery close at hand. The fact that he discusses the "composition" can to some extent be seen as a response to certain theoretical formulations of Kahn on "order," "design" and "growth" contained in "Perspecta." An important part of Banham's article focuses on the comparison between the Yale Art Gallery and the Hunstanton Secondary Modern School in Norfolk by Peter and Alison Smithson, designed in 1949 and opened in 1954. Both of the works are indicated by Banham as paradigms of the New Brutalism. The caption for the small images of the Yale Art Gallery is eloquent in this sense: "formality, expressed structure, exposed materials; the most truly Brutalist building in the New World."[125] Thus the work by Kahn enters the ranks of the masterpieces that represent the new current that focuses on the accentuated expressive potential of all materials, including concrete. But the Hunstanton School and the Yale Art Gallery also represent two different takes on the New Brutalism, according to Banham, and one of the fundamental differences lies precisely in the care Kahn puts into constructive details:

"[...] 1, Formal legibility of plan; 2, clear exhibition of structure, and 3, valuation of materials for their inherent qualities 'as found.' This summary can be used to answer the question: Are there other New Brutalist buildings besides Hunstanton? [...]. Only one other building conspicuously carries these qualities in the way that Hunstanton does, and that is Louis Kahn's Yale Art Centre. Here is a building which is uncompromisingly frank about its materials, which is inconceivable apart from its boldly exhibited structural method which – being a concrete space-frame – is as revolutionary and unconventional as the use of the Plastic Theory in stressing Hunstanton's steel H-frames. Furthermore, the plan is very formal in the disposition of its main elements, and makes a kind of symmetry about two clearly defined axes at right angles to one another. And this is a building which some Brutalists can apparently accept as a constituent New Brutalist structure. But, with all due diffidence, the present author submits that it still does not quite answer to the standard set by Hunstanton. For one thing, the Smithsons' work is characterized by an abstemious under-designing of the details, and much of the impact of the building comes from the ineloquence, but absolute consistency, of such components as the stairs and handrails. By comparison, Kahn's detailing is arty, and the stair-rail and balustrading (if that is the word for stainless netting) is jarringly out of key with the rough-shuttered concrete of the main structure. This may be 'only a matter of detailing' but there is another short-fall about Yale Art Centre which could not be brushed off so easily. Every Smithson design has been, obviously or subtly, a coherent and apprehensible visual entity, but this Louis Kahn's design narrowly fails to be. The internal spaces will be clut-

[125] Reyner Banham, *The New Brutalism*, in, "The Architectural Review," vol. CXVIII, December 1955, No. 708, (pp. 355-361), p. 360.
[126] Colin Rowe, letter to Louis I. Kahn, 7 February 1956, LIK, 030.II.A.65.22.
[127] *Ibid*. Also as a result of Banham's article, a group of students of the Carnegie Institute of Technology in Pittsburgh sends Kahn a questionnaire on the principles of the "new brutalism." See Marian Z. Augustyniak and Frederick T. Entwistle Jr., students of the Carnegie Institute of Technology of Pittsburgh,

tered with display screens which, in the nature of his programme and his solution of it, must be susceptible of being moved, so that formal clarity is always threatened. But beyond this the relation of interior to exterior fails to validate the axes which govern the plan. Available viewpoints, the placing of the entrances, the handling of the exterior walls – all tend to lose or play down the presence of planning axes. No doubt there are excellent functional reasons for the doors being where they are, and excellent structural reasons for the walls being treated in the way they are – but if these reasons were so compelling, why bother with an axial plan anyhow? [...]. The definition of a New Brutalist building derived from Hunstanton and Yale Art Centre, above, must be modified so as to exclude formality as a basic quality if it is to cover future developments and should more properly read: 1, Memorability as an Image; 2, Clear exhibition of Structure; and 3, Valuation of Materials 'as found.'"

We know nothing of Kahn's reaction to Banham's article, except through some indirect references made by Colin Rowe who expresses his own perplexity regarding the inclusion of the Yale Art Gallery in the new current. It is in these terms that Rowe informs Kahn, on 7 February 1956, of the publication of the article by Banham: "Did you see in the *Architectural Review* a deplorable article on 'the New Brutalism' – very chauvinistic and patronising – suggesting first of all that you were a 'new brutalist,' which as far as I know you would never claim to be, and then turning around and damning you because you didn't fulfill the N.[ew] B.[rutalism] canon."[126]

Also because Banham in his article criticizes *Architectural Principles of the Age of Humanism* and the Neo-Palladians, with which Rowe identifies, Rowe sends Kahn a copy of the book by Rudolf Wittkower, where he believes Kahn can find some questions they have discussed together ("because of matters arising out of our conversation"). Then he addresses the distinction and opposition between composition and growth as he had discussed it with Kahn. The fact that Banham too tackles the question of composition is not extraneous to what Rowe writes. The conception defended by Kahn of "Growth" would characterize his entire future idea of construction using exposed concrete and the meaning of every technical detail visible on its surfaces, to the point where precisely the technique of making concrete he utilizes will represent the most lucid and strong response to the concept of "Composition" defended by Rowe.

"I remember that something of our conversation in Philadelphia revolved around principles which we designated as GROWTH and COMPOSITION," Rowe writes to Kahn. "Principles which are opposed. And I took the side of Composition, and then had great difficulty in explaining what I meant by it. You deplored Composition because it appeared to be no more than a manipulation of forms for the sake of effect. You wanted to Grow a building, and I, I think, suggested that I wanted to Compose it. Or at least I was very emphatic about the PARTS."[127]

letter dated 14 April 1957, LIK, box 72. Besides to Kahn, the questionnaire was also sent to Patrick Abercrombie, Lawrence Anderson, Marcel Breuer, Gordon Bunshaft, William Dudock, Walter Gropius, Wallace Harrison, Philip Johnson, William Lescaze, John Marrill, Richard Neutra, Nathaniel Owins, Eero Saarinen, Louis Skidmore, Alison & Peter Smithson, Frank Lloyd Wright.

chapter three
precast and poured in place

The Washington University Library, or the "implicit shape"

The question of prefabrication of the parts of a reinforced concrete skeleton begins to take form in Kahn's work with the studies for housing in Israel and the project for the City Hall Building. But it is not until 1956 and 1957 that Kahn has a chance to study prefabrication in greater detail, thanks to the launch of his first important construction in which he uses this type of exposed concrete. Applying prefabrication, for Kahn, does not mean limiting oneself to one construction system rather than another for a given structural form, as already demonstrated in the projects for the houses to be built with the Precast Vacuum Concrete Method. The whole design of the elements is defined on the basis of the specific technical characteristics of the fabrication process. Prior to the start of the worksite that ushers in Kahn's technical experimentation on the prefabrication of reinforced concrete, one project points to his intentions regarding the characteristics the structure should assume in keeping with processes of making and implementation.

In February 1956, Kahn is invited to take part in the competition for the Washington University Library in Saint Louis – he submits his project in May. It may seem paradoxical that in this project too, as in the Yale Art Gallery, the City Hall Building, the Adath Jeshurun Synagogue, the AFL Medical Services Building or the Jewish Community Center, Kahn is looking for a structural system in reinforced concrete with a high degree of originality in the design and assembly of the members, given the fact that the structure he creates for the library seems so banal that it might appear to be an elementary exercise of skeletal structure with square-plan spans. It is the design of the connections between the parts that reveals the goal of

138, 140-141. Project for Washington University Library Competition, St. Louis, Missouri, 1956.

138. Perspective sketch (LIK, 030.I.A.465.10).

139. B. F. Voichysonk, *Implicit shapes*, Basic Drawing Course, Yale University (F. A. Horowitz, B. Danilowitz, *Joseph Albers: to Open Eyes, The Bauhaus, Black Mountain College, and Yale*, London, Phaidon, 2009, p. 164).

140. Detail section and ceiling plan (H. Ronner, S. Jhaveri, eds., *Louis I. Kahn Complete Work 1935-1974*, Basel, Birkhäuser, 1987, p. 106).

141. Section (LIK, 030.I.C.465.1).

finding a special solution with respect to the traditional juxtaposition of pillars and beams.

Above each pillar Kahn creates protruding segments of overhanging beams connected together by thinner beams. In certain sketches the connections between the beam segments are not even depicted, and the span seems to dissolve in vertical posts with cut arms.[1] In the study sketches for the Adath Jeshurun Synagogue Kahn had already imagined a similar design for the reinforced concrete structure.[2]

The fragmentation of the image of the span allows Kahn to obtain a structure with a design similar to the geometric figures, evoked only in certain portions, that Albers assigns to his students at Yale University and defines as "implicit shapes."[3] Precisely because they have been transformed into "implicit shapes" offered to the imagination, the spans of the reinforced concrete structure with the fragmented beams end up resembling the trees with nude branches drawn by Kahn in the views of the Washington University Library. After all, does the separation from the traditional image of the span not transform a structural member into a sculpture, a forerunner of the beam-sculptures of Robert Morris? The sculpture of the three irregular trunks erected on the podium in front of the entrance to the Washington University Library also alludes to the trunks of the trees (perhaps the sculpture, too, is to be made of concrete?).

If we consider the fact that Kahn is constantly looking for a structural system that can give each of his buildings a character of "monumentality," it seems probable that the form of the parts of the structure for the Washington University Library can indicate an integration, in the design of the pillars and beams, of the worksite assembly process of the prefabricated pieces with the crane. If this is the case, then the configuration of the pillar

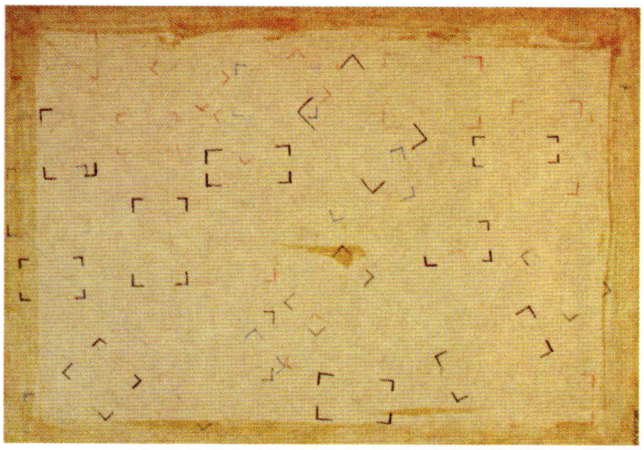

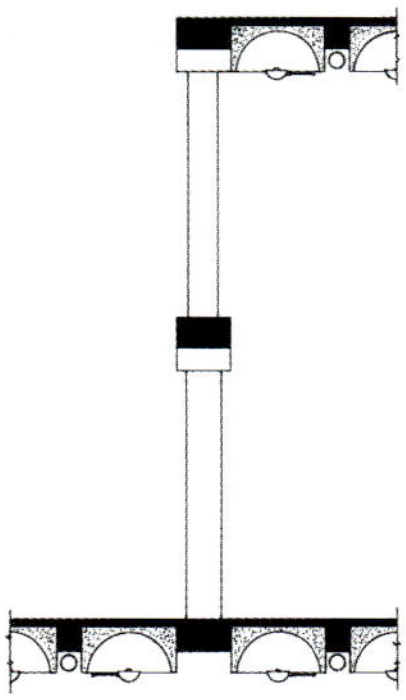

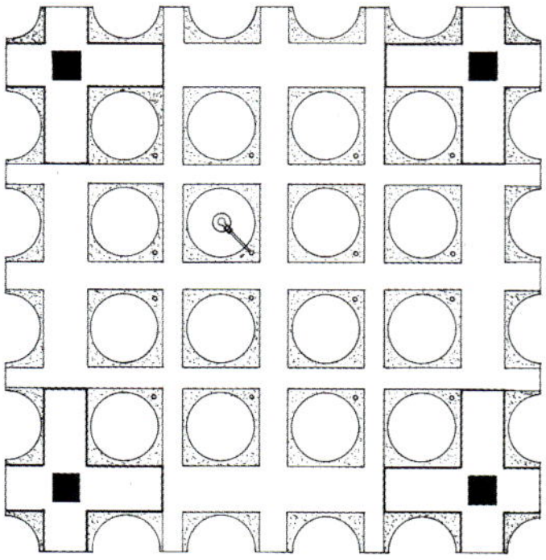

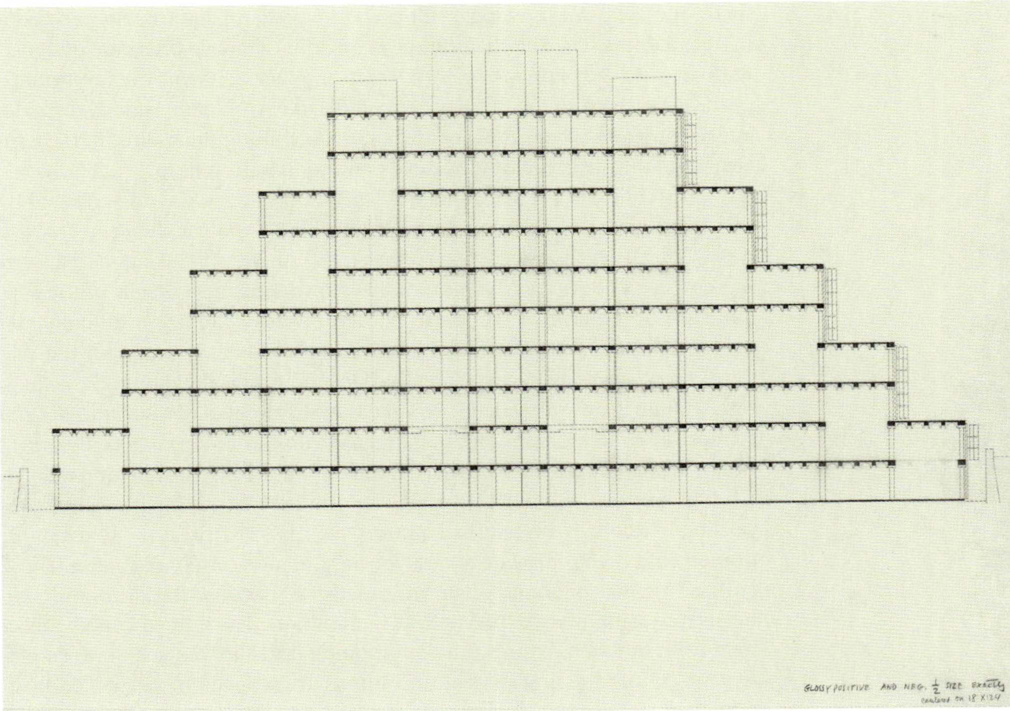

and the beam as fundamental parts of the skeleton, and their composition in the span, may have been redefined by Kahn precisely in terms of the process of their construction: the fundamental element of the skeleton becomes a pillar with two corbel-like branches at the top, taking the form of an elementary T structure conceived for easy raising by a crane. The combination of such units reproduces the geometry of the traditional span. The joint between the T-shaped elements takes on a force of image that goes well beyond the junction to be sealed with the plastic material.

The idea that the skeleton is the result of a process, and therefore that its members and their articulations can be redefined case by case in keeping with their construction, reveals the way Kahn's thinking is transforming at their root certain central figures of 20th-century architecture, against the backdrop of the typical processes of prefabrication, reaching the point of reflecting the contemporary American artistic currents that will converge in Process Art.

Again in the Washington University Library, as in the series of projects that begin with the Yale Art Gallery, the floor system is more than a simple load-bearing structure. Kahn seeks another type of floor in exposed reinforced concrete, still capable of incorporating the lighting and air conditioning systems. To this end he develops a system conceived as a variation on the Waffle Floor – the patented slab. Instead of utilizing the usual square prefabricated gaps of this type of slab, he envisions the arrange-

[1] *The Louis I. Kahn Archive: Personal Drawings* cit., vol. I, p. 458, drawing 465.10.

[2] *Ibid.*, p. 352, drawing 410.9.

[3] *Cf.* Horowitz, Danilowitz, *op. cit.*, p. 164.

ment of a series of openings, still prefabricated and square, but each containing a dome. The fact that these special dome-shaped coffers are designed with sound-absorbing material to be left in place confirms the importance of the experience of the floor system of the Yale Art Gallery, and Kahn's desire to integrate every provisional element, including the molds, in the construction of a structure that becomes an organ with multiple functions.

The choice of the semi-spherical surface of the molds is determined by the lighting system: a fixture arm projects light against the center of the dome, which acts as a reflecting surface (each square span contains 16 lighting domes). One cannot help but notice, in the technical details of the construction of the slab of the Washington University Library, a fundamental evolution of the system with the small metal vaults envisioned for the Yale Art Gallery, precisely to create ideal surfaces for artificial lighting: the ceiling, or cladding, is now directly incorporated in the process of fabrication of the reinforced concrete slab. The prefabricated domes are arranged to form a network of channels in a square grid to contain the physical plant conduits.

"The structure," Kahn explains, "is composed of concrete columns supporting a two-way system of square bays made of precast, lightweight, domed coffers that provide spaces for lighting and air conditioning. These coffers, formed of sound-absorbing material, become the deck and side forms for cast in place concrete beams and floors. The domed coffers reflect light from centered fixtures. Air conditioning ducts are housed between coffers and under the beam system. Columns are reduced in size as loads diminish but all perimeter columns are designed to take one additional story. Future expansion is thus possible in small units of space."[4]

"Tower of triangular concrete frames" for the Universal Atlas Cement Company

The question of prefabrication is by now a central focus for Kahn, and the design of the Washington University Library is but one among others that share formal qualities that depend on this mode of deployment of concrete. By now the material is starting to appear nude in every part of the building. Hollow stones, hollow columns, walls of blocks and skeletons of pillars and beams form the foundations of the vocabulary of structures in exposed concrete of the complex of monumental buildings that in 1956-57 Kahn imagines scattering through the center of Philadelphia to give form to the Civic Center, in place of the earlier projects based on the tetrahedral grid of the Yale Art Gallery.

With the sketches for the Civic Center, Kahn confirms all the civic potential of his concrete: it is the stone of the contemporary civilization, capable of restoring the appearance of values of medieval monuments, from cathedrals to fortresses; it is also the material capable of generating imposing, powerful spaces, such as those of the monuments of the Roman Empire, for a civic center that is an alternative to the values expressed by the steel and glass towers of American financial capitalism.

In the Civic Center, Kahn calls for large buildings with a circular plan, with 19 floors above ground level, veritable "tower gateways"[5] capable of absorbing the flow of vehicle circulation and protecting the center from the

[4] Louis I. Kahn, *Space Form Use: A Library*, in, "The Pennsylvania Triangle," vol. XLIII, December 1956, No. 2, pp. 43-44, now in, Latour, *Louis I. Kahn: Writings, Lectures, Interviews* cit., p. 70.

[5] Louis I. Kahn, *Order of movement and renewal of the city*, in, "Perspecta," n.d., [1957], No. 4, p. 61.

penetration of automobiles. Thus Kahn's citadel erected in the heart of Philadelphia would be defended, like Carcassonne, by a belt of unusual grand circular towers, each containing a garage for 1500 automobiles, to be stored in levels, of which four above ground, which would be topped by a platform, or plaza, with a roof garden, in turn surrounded by a thick fourteen-story strip for offices and hotels. Great vertical openings would allow light to enter the raised plaza, giving the Tower Gateways the connotation of ruined buildings. From the plaza these breaches would create views of the city, as at the Capitol in Rome. The measurements and the aura of the Tower Gateways, after all, are also Roman.

The skeleton of the Tower Gateways is configured on a colossal Michelangelesque order: the square-section columns are three stories high and feature at the two extremities the same prismatic capital-base that serves for mechanical connection in their stacked assembly (probably secured by means of bolts). After the invention of the T beams for the Washington University Library, these members for the Tower Gateways represent other column forms specially configured to be made in prefabricated reinforced concrete and inserted on the worksite. The position of the columns sets an alternating rhythm of wide and narrow spans, so as to also include the elevators and service spaces in the pattern. At ground level the columns form a continuous portico with a height of just one story.

Beside the City Hall Building, Kahn depicts other towers with an exceptional profile, variations on the tetrahedral Space Frame – the tower with

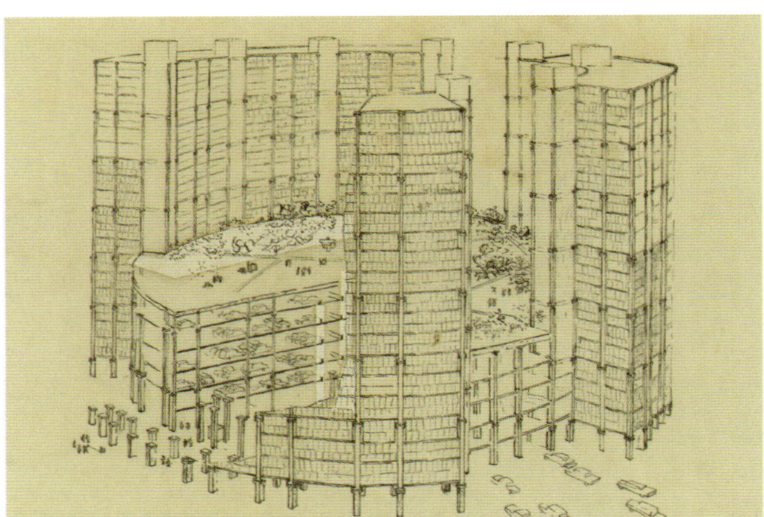

142-143. Project for Civic Center, Philadelphia,1956-57. Perspective sketches (LIK, 030.IV.A.620.1.6; LIK, 030.I.A.455.24).

144-147. Project for City Tower, 1957.

144-145. Sections and plan (LIK, 030.IV.B.385.1.1; LIK, 030.IV.B.385.1.2).

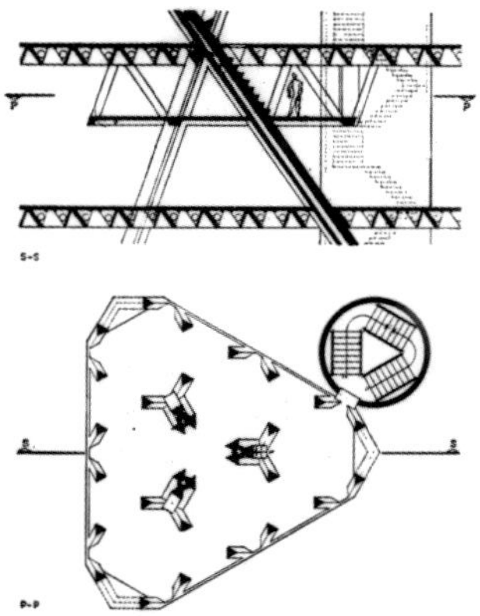

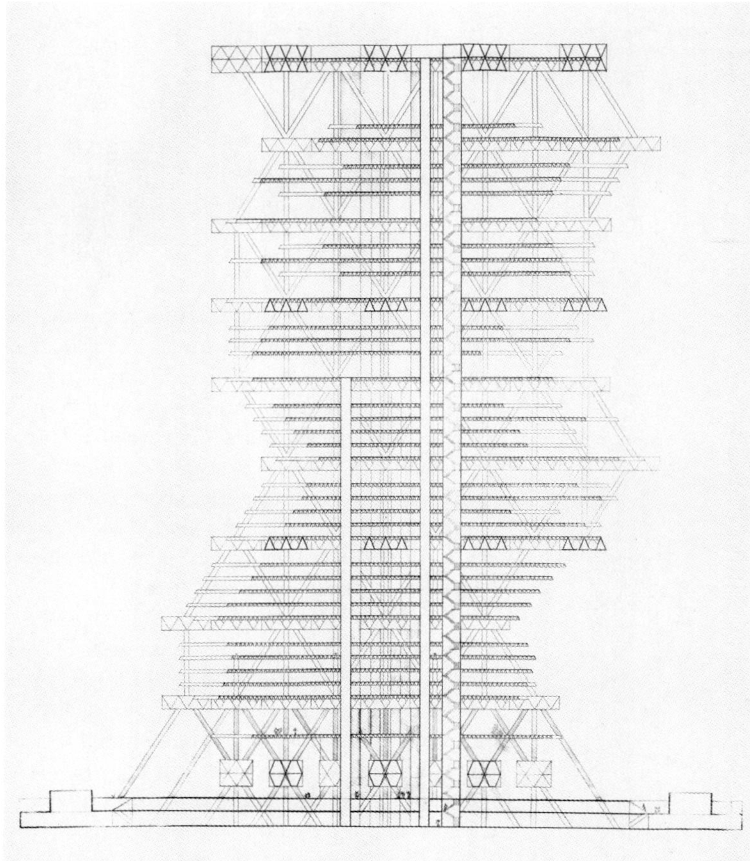

a pyramid base is envisioned with a triangular grid structure, as shown in a working sketch.[6]

The evolution of the City Hall Building project meets with a final, definitive chapter precisely in the context of the new version for the Civic Center of Philadelphia. The building is more than doubled in height, taking on the appearance of a tower, and both the grand "capitals" described in previous years and the platform at the base, or "plaza," with four access ramps to the underground garage, make their appearance. The new version of the City Hall Building seen in the sketches for the Civic Center corresponds to the building Kahn develops for the Universal Atlas Cement Company in 1957.[7]

Specializing in concrete constructions of all kinds, and engaged since the 1930s in spreading the word in the United States on the possible innovative uses of concrete, the Universal Atlas Cement Company, with head-

[6] *The Louis I. Kahn Archive: Personal Drawings* cit., vol. I, p. 331, drawing 385.86.

[7] In 1971 Roland E. Engelke, of Universal Atlas Cement, informs Kahn that he has received "numerous requests for the pamphlets and ad reprints of our futuristic series featured in 1957" (Roland E. Engelke, letter to Louis I. Kahn, 18 November 1971, LIK, 030.II.A.45.23). The photograph of the model sponsored by Universal Atlas Cement is published in "Architectural Forum" and "Architectural Record" (advertisement of Universal Atlas Cement in, "Architectural Forum," vol. CVI, May 1957, No. 5, n.p.; advertisement of Universal Atlas Cement, in, "Architectural Record," vol. CXXI, May 1957, No. 246, n.p.).

146. Model (LIK, 030.IV.A.385.3.13).

147. L. I. Kahn, *A City Tower*, booklet made by the Universal Atlas Cement Company, [1957] (LIK, 030.II.A.45.58).

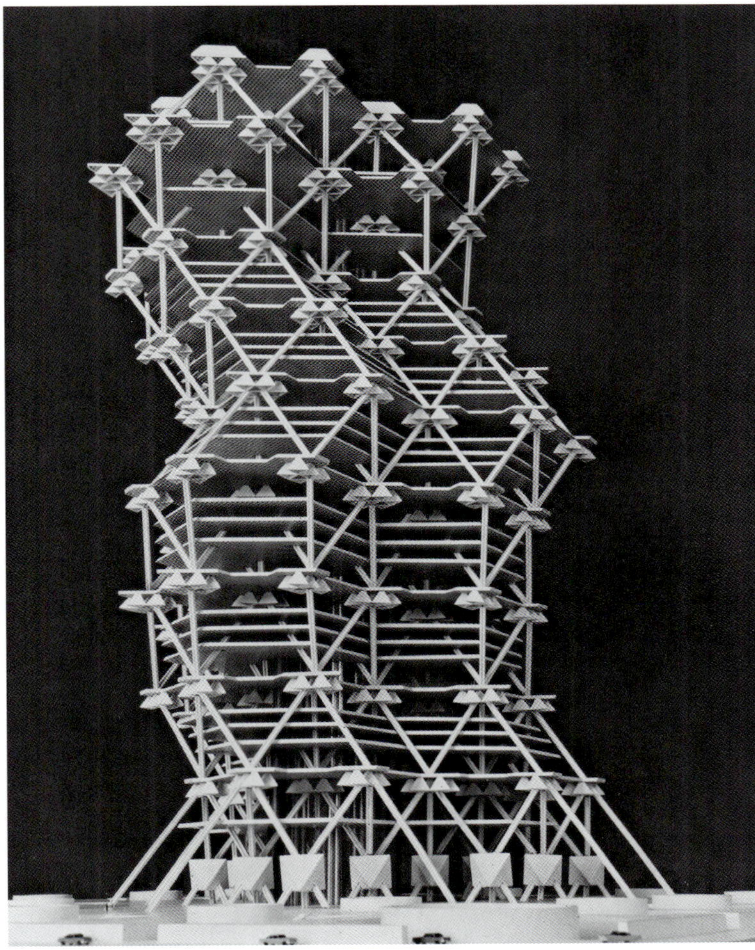

quarters in New York at 100 Park Avenue, headed by Charles B. Baker and equipped with a research staff of chemists, engineers, physicists and technicians,[8] turns to Kahn for consulting on innovative applications of reinforced concrete after the company's technicians see the Yale Art Gallery and the developments of the tetrahedral floor system in the design for the City Hall Building. "Atlas Cement commissioned me to design a new idea for concrete," Kahn recalls, in relation to the version of the City Hall Building of 1957.[9]

The development of the project by Kahn in collaboration with Tyng is for the construction of a model of the tower with the plaza, in order to publish a booklet and advertising brochure for the Universal Atlas Cement Company.[10] The model, made by Robert Damora in the early months of 1957, serves to demonstrate the potential of the tetrahedral Space Frame applied to the skyscraper.[11] We should not rule out the idea that Kahn hopes the model can take on a value for the definition of a future structure similar to that assigned by Le Corbusier to the Domino skeleton.

The version of the project publicized by the Universal Atlas Cement Company is known as the "city tower," "a tower of spaces for public areas and city departments" no longer designed only for the center of Philadelphia;[12] it is also called "Tomorrow's City Hall: high-rising tower of triangular concrete frames."[13] In a caption, the work is described as "an imaginative tower conception by Kahn indicating the design results of thinking in terms of structural strength in triangulation for tall buildings."[14] In the advertising brochure, the City Tower is distributed in "Architectural Record" and "Architectural Forum" in May 1957 with the stated aim of the Universal Atlas Cement Company "to promote interest in architectural contributions for a greater America through the medium of concrete."[15]

All the elements of the structure of the City Tower are in prefabricated concrete, prestressed and left exposed. The "structure" is described as "a vac-

[8] *Cf.* Earl J. Hadley, *The Magic Powder. History of the Universal Atlas Cement Company and the Cement Industry*, G.P. Putnam's Sons, 1945.

[9] Louis I. Kahn, letter to Universal Atlas Cement, 29 October 1971, LIK, 030.II.A.45.58. "Our futuristic series featured in 1957" is the definition of Universal Atlas, written in response to Kahn's letter (Roland E. Engelke, letter to Louis I. Kahn, 18 November 1971, LIK, 030.II.A.45.58). Confirming the interest in Kahn's project, Universal Atlas states that it has "received numerous requests for the pamphlet" of 1957 (*ibid.*).

[10] The booklet is printed at the end of 1957, as shown by J. A. Sullivan, letter to Louis I. Kahn, 30 December 1957, LIK, 030.II.A.34.20. "The 4-pages on the 'City Hall Building' has been published," says Sullivan, Director of Advertising & Public Relations of Universal Atlas Cement Company.

[11] *Cf.* Robert Damora, letter to Louis I. Kahn, 10 April 1957, LIK, 030.II.A.34.20. "When the model costs began to get out of hand, I decided to go along with this rather than sacrifice your aims because of costs. [...]. Universal Atlas has been a good client and is assuming its responsibilities in helping the architectural aims of the program as I am trying to develop it. Your model has been sent to Cornell University at Atlas' expense for use in your exhibit at the White Museum."

[12] [Louis I. Kahn], *A City Tower*, booklet published by the Universal Atlas Cement Company, n.d., [1957], LIK, 030.II.A.45.58.

[13] Universal Atlas Cement Company, project presentation flyer, n.d., LIK, 030.II.A.34.20.

[14] [Louis I. Kahn], notes, n.d., LIK, 030.II.A.34.20. Also in 1957, the City Tower and the plaza are published in "Perspecta" (Louis I. Kahn, *Order of movement and renewal of the city* cit).

[15] "Tomorrow's City Hall: high-rising tower of triangular concrete frames," advertisement of Universal Atlas Cement, 1957, LIK, 030.II.A.45.58.

[16] [Kahn], *A City Tower* cit.

[17] *Ibid.*

[18] Louis I. Kahn, Anne G. Tyng, advertising flyer of Universal Atlas Cement Company, n.d., [1957], LIK, 030.II.A.45.58.

[19] Ronner, Jhaveri, Vasella, *op. cit.*, p. 73.

[20] S.P.F. Humphreys-Owen, *Physical Principles Underlying Inorganic Form*, in, Lancelot Law Whyte, ed., *Aspects of Form. A Symposium on Form in Nature and Art*, London, Lund Humphries, 1951, edition of 1968, (pp. 8-22), p. 14, fig. 2. The same book provides the slide in Kahn's collection of the Aulonia Hexagona and the bone of "a vulture's wing." The contribution in which the images photographed by Kahn are included is that of Conrad Hal Waddington, *The Character of Biological Form*, *ibid.*, pp. 43-56 (for the figures, p. 53); the images come from D'Arcy Wentworth Thompson, *On Growth and Form* cit., pp. 708 and 981. Waddington asserts that "man-made mechanical forms" are usually "designed to serve one single function, or at most two or three," while for living things the functions are multiple and only in things perfected over centuries do we find something similar to the "natural forms of life" (Waddington, *op. cit.*, pp. 45-46).

[21] [Kahn], *A City Tower* cit. See also, *Louis Kahn [...]*, in, "The Architectural Review," vol. CXXII, May 1957, No. 724, (pp. 344-345), p. 345.

148. Project for City Tower, 1957. Study drawing (LIK, 030.IV.A.385.1.8).

149. L. L. Whyte, ed., *Aspect of Form. A Symposium on Form in Nature and Art*, London, L. Humphries, 1968, p. 14.

uum precast, prestressed concrete triangulated strut frame, integrally braced by the cross framing and intersecting of the column system at every 66-foot level."[16] The struts and columns stand out for the differing role of their cavities: the "struts" contain the aeration system, while the columns "are hollow for the use of conduits and pipes." The "capitals" contain "storage, toilets and sub-stations for mechanical services," reached by means of service staircases shaped on one of the inclined rods of the structure. Only three of the capitals, those at the center of the plan, can be reached from the main stairs (these capitals contain the restrooms). The capitals have a variable number of tetrahedral parts and rods in keeping with their position in the plan.

The tetrahedra of the floor system, specified in "precast light-weight concrete," as Kahn had already proposed for the City Hall Building, have cavities to contain "air conditioning, lighting, wiring and all piping."

Six groups of three cylinders each are positioned at the center of the plan of the City Tower to contain the elevators, while three larger hollow cylinders each contain a triangular staircase based on the model of the stairs of the Yale Art Gallery.

The envelope of glazed panels and sunscreens in aluminium is seen as an "intermediary element between the building and outside forces," and as such it is developed as a "beginning of structural reaction against these forces"[17] – sun, temperature, wind. On the lower levels of the City Tower the structure forms an even more monumental portico than the one planned for the City Hall Building, while on the upper levels the number of floors diminishes to offer a view from below of the impressive grid of tetrahedra.[18]

It is worth noting that certain graphic schemes of the City Tower[19] are directly derived from the three-dimensional representations of a "kind of terrace" of the molecules of a sphere.[20]

It is at this point in the evolution of the City Hall Building project into the City Tower that Kahn introduces the question of the dynamic forces generated by wind thrust against a skyscraper, precisely to justify the complex structure of inclined rods: "This form must express: the buttress to wind forces […]."[21]

Introduced in the Yale Art Gallery as a structure with an exceptional geometry, the tetrahedral floor system generates a new structural system, the "triangular concrete frames" that take on civic values in the studies for the City Hall Building, returning – in Kahn's statements on the City Tower – to a question of structure, of wind resistance, revealing the creative process that makes Kahn's works in exposed reinforced concrete into or-

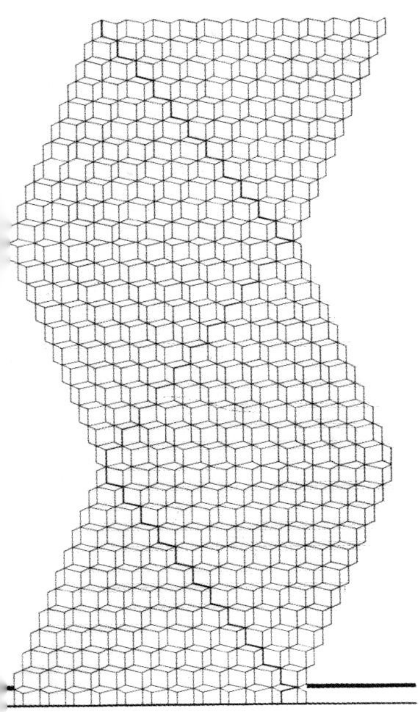

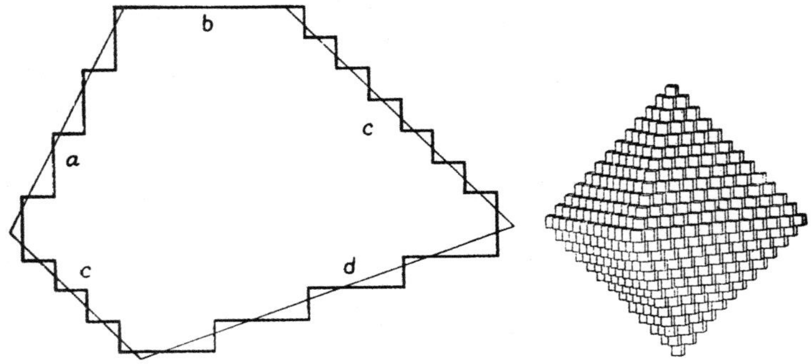

ganic systems with a symbolic purpose. At this point, Kahn's skyscraper seems like it is being shaken by a tremor.

These are the terms Kahn and Tyng use to sum up the meaning of the City Tower designed for the Universal Atlas Cement Company:

"Utilizing one of nature's strongest, simplest shapes – the triangle – these triangular frames of precast, prestressed concrete buttress this high-rising structure. It grows upwards in a succession of pyramid forms to provide maximum areas of easily divided space, with a variety of ceiling heights. Through the hollow concrete sections run arteries which carry air conditioning, heating and other service lines, while a central core consolidates elevators and stairways. Thus, a meaningful form evolves out of a municipal structure's needs – and is expressed in concrete, the versatile building material."[22]

Thanks to the City Tower of the Universal Atlas Cement Company, the Space Frames and geometrics of Buckminster Fuller, Wachsman or Le Ricolais transfigure what seemed like the unavoidable model of structure for contemporary architecture, the Domino skeleton, also opening up new possibilities for the integration of technological organs in the structural system. Precisely concrete, with its possibility of being shaped in forms with large cavities, permits this new conception of a skeleton traced inside by vital fluids. The example of the Unité d'Habitation in Marseille, with its pilotis, ventilation chimneys and platforms for ducts and air circulation, continues to be of fundamental importance for Kahn's vision of a hollow structure in exposed concrete.

With the City Tower, the passage from the hypothesis of a monumental structure of steel tubes formulated in *Monumentality* to a hollow structure in exposed concrete can be said to have been completed. The fact that for Kahn concrete has taken on the value of a civic material, an alternative to the meaning attributed to steel, can be seen explicitly in one of his statements from 1957: "concrete columns are better" than those of steel, he says. "They're real columns," he continues. "Of course, steel is a marvelous material. You can do wonderful things with it, build great machines, but in architecture you're not building airplanes after all, are you?"[23]

The publications of the Universal Atlas Cement Company bring Kahn's dream of being able to make the most fantastic structure ever conceived in reinforced concrete for a skyscraper in the 20th century to an end. The booklet and the advertising brochure rely on the forceful imagery of the photograph of the model as a last chance to find financing for the construction of a City Tower.

Metaphorical concrete for the pavilions of the Day Camp

The construction of the four pavilions of the Day Camp, in Trenton, play facilities for children near the Bath House of the Jewish Community Center, for which Kahn continues to develop new design variations, becomes an opportunity to experiment with the deployment of concrete in another form, again to create a narrative written with the words of construction and materials. For this work Kahn collaborates with John M. Hirsh and Stanley R.

[22] Kahn, Tyng, advertising flyer, cit.
[23] Cit. in, Walter McQuade, *Architect Louis I. Kahn and his Strong-Boned Structures*, in, "Architectural Forum," vol. CVII, October 1957, No. 4, (pp. 135-143), p. 135.

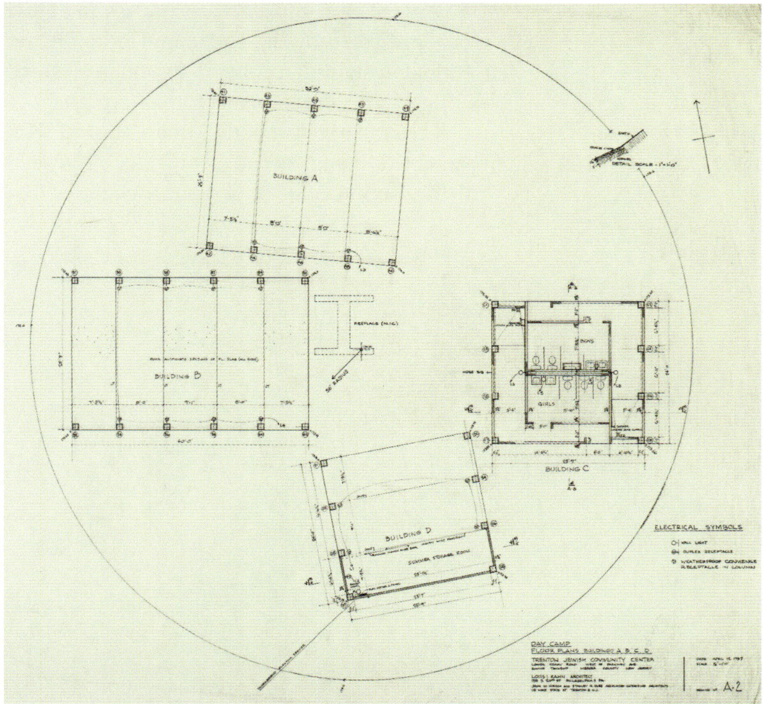

150. Day Camp, Jewish Community Center, Trenton, New Jersey, 1954-59. "Floor Plans Building A, B, C, D," 12 April 1957 (LIK, 030.I.C.430.2).

Dube, architects based in Trenton. The project and the construction happen from April to August in 1957. As in the case of the Richards Laboratories, which he is developing precisely in the first half of 1957, for the pavilions of the Day Camp Kahn turns to the two basic forms of concrete, prefabricated and poured in place. The different configuration of the formwork for the various parts of the pavilions is essential for the symbolic character of the reinforced concrete structure. All the pavilions are built on the same elementary structure of pillars, beams and roof slabs. But Kahn creates a true compositional exercise, starting with the same rectangular span.

The pavilions have different plans. Two of them form squares with three spans each, while the other two form rectangles of different lengths (one with four, the other with five spans). The rectangular pavilions are left completely open, while the square pavilions have spans closed with exposed brick walls, to contain the bathrooms and a space with a counter for a stovetop and sink.

All the pavilions are placed inside a circular enclosure indicated on the ground only by a layer of "shale chips" to be placed over a layer of gravel, and they are shifted according to angles calculated to generate unexpected overlaps of the rows of pillars, in a dynamic vision of the complex. The circular enclosure and the pavilions define the open area, the shelters for protection from rain and sunlight, and the services for children at play or engaged in creative activities in the park. The position at the edge of the park is decisive for the configuration of the constructive elements of the pavilions and for the choice of the modes of deployment of the reinforced concrete.

Kahn's pursuit of a structural system capable of staging metaphors simply with materials and construction processes can be seen in the various ways of pouring concrete used for the parts of the pavilions.

151-152. Day Camp, Jewish Community Center, Trenton, New Jersey, 1954-59.

151. Photograph of the construction site (M. D. Meyers Archive, AAUP, 136.30.11).

152. View (LIK, 030.IV.A.430.1.19).

For the roof slabs Kahn chooses prefabricated panels, the Flexicore panels produced by the Flexicore Company, in which reinforced concrete is poured in smooth formwork to obtain clean surfaces with a workshop finish. Along the edges, "board forms" are used.

The construction of the square-section reinforced concrete pillars happens on site, using special formwork composed of terracotta tiles, arranged with large joints and left in place (flue tile of 13 x 13 / 6-24 inches length).

Kahn indicates that the joints of the tiles should be without mortar: "formwork – for columns use 13" x 13" terra cotta flue tile stacked without mortar joints and properly aligned and braced."[24] The placement of the concrete inside these formwork is planned in two or three pourings per day, of 24 inches each. The height of the tiles influences that of the pillar, so much so that the constituent unit of the formwork system becomes the proportional module of the trunk. The height of the trunk as far as the foundation plinth corresponds to six tiles, of which 4.5 are above ground.

In the pavilions with a square plan the laying of the bricks is specified in an ordinary way, "with no particular care," to underline the rustic character of the construction. The masonry is anchored to the columns with metal brackets. In the specifications, Kahn indicates that the brick should be painted, also outside, with one coat of "white Portland cement paint." But the bricks are left exposed, without paint finish.

Precisely the permanent formwork of the pillars made with tiles become essential for the definition of the metaphorical concrete of the Day Camp. Thanks to those formwork the pillars become abstract trunks with bark, like the trees that form the backdrop of the pavilions. In the dynamic viewing of the always different sequences of pillars of the various pavilions, that terracotta bark creates an artificial park without imitation, through a rigorous and archaic expression of the fabrication system of the concrete, which is thus transformed into poetry – the term "poetry" is written by Kahn on the page of a notebook on "The Joint."[25] The distance between the two rows of pillars of each pavilion, set based on the load-bearing capacity of the roof slabs, generates the large size of the voids, also to contrast the dense pace of the pillars in the rows. The result is a particular way of framing the landscape, which from the part facing the park cuts away the sky and concentrates the gaze on the trees. One lingers under those concrete slabs as under the foliage of leaves, in a place that evokes the myth of the origins of the trilithic system in wooden construction, and the idea of the tree as the generator of a space for the gathering of human beings. The entire reduction of the architecture of the Day Camp to its most elementary and essential structure in reinforced concrete is Kahn's lyrical tribute to the education of children for constructive imagination (and not just children):

"Large piece of land. Now if you can't afford to build anything, the act of making the recognition that there must be a large piece of land is a better act than fitting it out. The fitting it out will come. Not only that, but you will be able to sense what you cannot sense as an appropriate space today. Maybe the first act on the piece of land is to build a kind of shelter, not unlike that of a stoa, which is partitioned, and it's there as kind of shelter which is secure in shelter because the trees won't shelter you during the rain; so this is built as a kind of first thing. You wait, you go under this thing when it's raining, you wait for the rain to be over, and you go back to play again, or you go home, or wait for a bus. But this gives you a sense of the vitality of play, and this is what you're supporting."[26]

[24] "Specifications for Constructions of Buildings and Pavilions for Day Camp," 16 April 1957, LIK, 030.I.C.430.002.

[25] Louis I. Kahn, notebook, LIK, 030.VII.I.

[26] Louis I. Kahn, *Clearing*, in, "Via," 1973, 159, now in, Solomon, *op. cit.*, p. 125.

Projects for the Jewish Community Building: the shadow of concrete

The Bath House and the Day Camp remain the only constructed works designed by Kahn for the Jewish Community Center complex Trenton. Their design and construction happen against the background of the preparation of the project for the main building of the complex, the Jewish Community Building, which begins in 1955.

In the variations developed in 1957 one of the crucial questions for the character of the building has to do with the form of the roofing of the various parts. The different solutions rotate around the invention of a reinforced concrete roof to be made with flat vault segments, in order to use simple formwork, but in a geometric arrangement like that of the structural cells of the versions in the project of 1955-56. The fact that the roofing structure defines the character of a space can be gleaned from a number of statements by Kahn on domes or vaults. Kahn's goal is to invent complex roofs in reinforced concrete, avoiding the logic of shells with variously bent surfaces that are in fashion in contemporary engineering. It is meaningful that Kahn calls his roofs "triangular domes."[27]

In the version at the start of 1957 the pattern of pillars is already arranged, as in the subsequent versions that same year, in a square grid that starts with a group of four pillars, at the points of a square, to form a modular sequence of three spaces of different sizes, each with a particular function.

The pillars are envisioned in brick with three sections, all generated by the figure of the square: the large square; one L section with flanges corresponding to two sides of the large square; finally, the small square, corresponding to one fourth of the large square. If we consider the fact that the three sections are complementary figures of the same square, the symbolic meaning of the design of the two swimming pools of the Bath House becomes clear – an ideogram of the structure of the Jewish Community Center. The L pillars are placed in the "small spaces" at the points of a square, alluding to a single colossal structure. Precisely for this characteristic they correspond, like the spans of the Washington University Library, to the "implicit shapes" of Albers.

The arrangement of the beams contributes to the character of the two main spaces. While in the gymnasium the beams are set parallel, in the

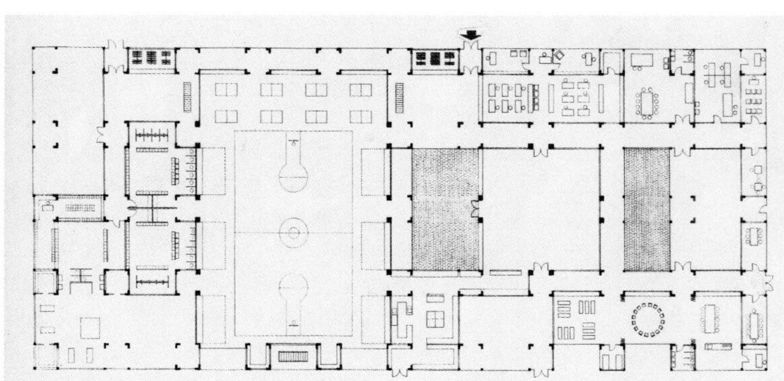

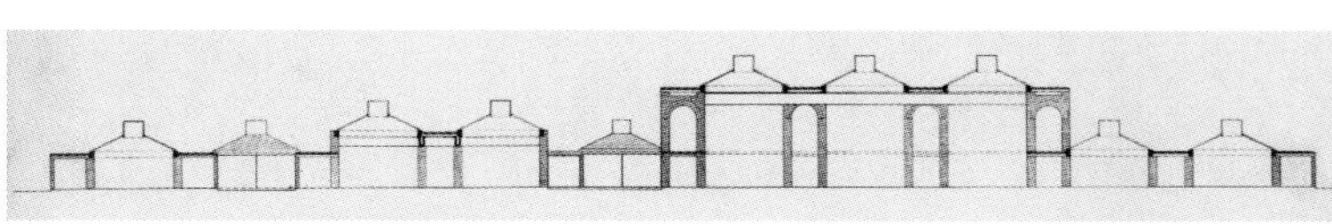

153-156. Community Building, Jewish Community Center, Trenton, New Jersey, 1954-59.

153-154. First floor plan and southwest-northeast section (LIK, 030.IV.A.430.2.6; LIK, 030.IV.A.430.2.5).

155. Plan (LIK, 030.IV.A.430.2.7).

156. Model (LIK, 030.IV.A.430.2.19).

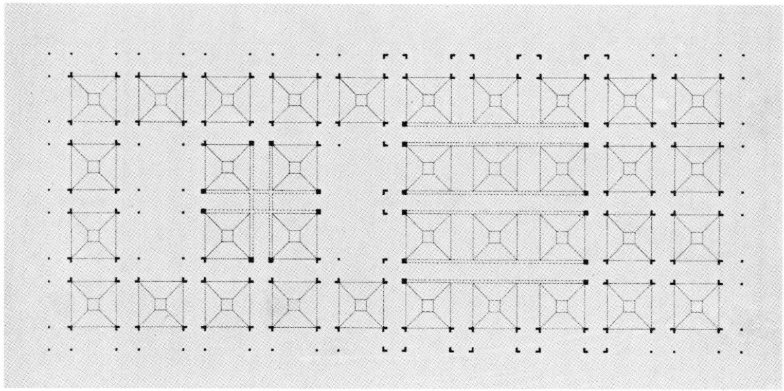

Social Hall their crossed arrangement helps to create a centrality that expresses the place's multiple functions – "a meeting room, a dance class, the dance, the party, the teenage shindig, the reception room, the exhibition hall, and numerous other activities."[28]

In the "Description of the spaces and functions of the community building," dated 21 May 1957, Kahn explains the characteristics of the project and its construction systems in detail. The structure of the roof of the gymnasium is defined as "undulating concrete beam slabs," with skylights for daylight and artificial lighting. Certain studies document a type of beam with inclined flanges for the penetration of light, which could correspond to the "undulating concrete beam slabs." The Social Hall is also covered with a "ceiling of 4 high coffers or triangular domes."[29]

Kahn subdivides the construction of the building into "structure" and "enclosure."[30] The indications on the fabrication of the structural elements define a concrete construction system that is being developed, at the same time, for the Richards Laboratories. All the parts of the "precast concrete roof system" were to be produced by the Lakewood Prestressing Company, located in Lakewood, New Jersey, about 38 miles from Trenton – the same company is also called in for the construction of the Richards Laboratories.

[27] "Description of the spaces and functions of the community building," 21 May 1957, LIK, 030.II.A.108.58.
[28] *Ibid.*
[29] *Ibid.*
[30] *Ibid*

157-159. Community Building, Jewish Community Center, Trenton, New Jersey, 1954-59.

157. Perspective view (LIK. 030.IV.C.430.1a).

158. Perspective sketch, August 1957 (LIK, 030.I.A.430.24).

Instead, the slabs were to be the ones prefabricated by the Flexicore Company. The columns supporting the roof system of the gymnasium are prefabricated. All the other pillars, including the L-shaped members, are to be made in brick (as no material is specified, it is probable that Kahn was thinking of clay brick). Concrete poured in place is to be used for the base platform, the staircase and the balcony around the gymnasium.

In all the prefabricated parts the concrete is to be left exposed. With the indication "to be left exposed as interior finish," Kahn underlines the value of the prefabricated concrete as the fundamental surface for a new form of wall decoration without other ornament. The exception to the rule of leaving the concrete exposed is represented by the beams of the gymnasium roof. Their importance in the design of the gymnasium is accentuated by the unusual decision of intense coloring of the material: "These beams will be excitingly painted."[31] Probably Kahn intends to use the color to control the play of shadows with intense chromatic effects, erasing its potential with a color that lights up the shade. This can be deduced from "Abstract of planes and forms" of 1948-50[32] or the drawings of Egyptian capitals from 1951,[33] which confirm the luminous and chromatic potential of shadows in Kahn's vision, now applied to the gray range of concrete. At the same time, many travel drawings and paintings demonstrate that Kahn is attracted by the chromatic potential of shadow, often conveyed with tones that might correspond to the "excitingly painted" effects described for the beams of the gymnasium. "It is non anticipated that concrete will have to be painted," we read in the same description. Precisely in the choices of colors of the shadow on the surface of the concrete is it possible to understand that the material and its gray surface are becoming the ideal reflector for a pure light in Kahn's thinking.

For the "enclosure" – a term that covers all the skins of the building, from the glass of the skylights to the doors, to the asphalt of the roof – where it takes the form of the walls of the main spaces Kahn makes use of brick, defining the system with a term typical of cladding: "brick veneer masonry."[34] Where the walls form secondary rooms, Kahn specifies masonry made with cinder blocks of the Waylite type, to be left exposed. These blocks are to be used in the washrooms, showers, the health club, locker rooms and kitchen. All the internal faces of the concrete block walls are indicated as painted (no color is specified). In other cases, wood paneling would be applied.

The views of certain areas of the Community Center document spaces defined by the various textures of the materials left exposed: stone or con-

159. Perspective view, August 1957 (LIK, 030.IV.C.430.1.4).

crete slabs for the floors, the bricks of the pillars with certain borders in marble, and the concrete of the pyramids of the roof, with skylight.[35]

In the version of the project presented to the committee of the Community Center, Kahn adds brick arches connecting the pillars of the structural units that form the highest spans of the gymnasium.[36]

Already in May and until the versions presented in October 1957, Kahn organizes the various parts thanks to the differentiation of the roofs.[37] It is in this phase that he writes of "small servant spaces" and "spaces of significance."[38] The presence of reinforced concrete becomes predominant in some cases, until the project in July, when the structures of the two main spaces of the gymnasium and the Social Hall undergo radical modifications. The reinforced concrete roof beams are combined with contiguous "triangular domes," standing out from the system that generated them and continues to mark the other spaces of the Community Building.

In the Social Hall the reinforced concrete beams slope upward to meet at the top, branching into two pitches for each square compartment, and then open to create the skylights.[39] By now the beams have become structural organisms capable of shaping light. All the premises are here for the system that will be used in the hall of the church in Rochester. Kahn himself explains that a large space becomes space "of significance" and "demands the large span."[40] In the same drawing Kahn depicts, with quick, essential strokes, the visual effect of the sequence of imprints of the boards on the concrete of the roof along the sloping pitches and vertical sides of the skylights. In the four compartments, rather than following the geometry of the triangular pitches, Kahn favors the original square, while on the vertical faces of the beams he arranges the boards following the long sides, to accentuate the dimension of the structure. On the small vertical face of the beams he designs a motif over the imprints, alluding to paint to be applied directly on the concrete surface, as in the mural of the Bath House.

[31] *Ibid.*
[32] Hochstim, *op. cit.*, p. 238.
[33] *Ibid.*, p. 293.
[34] "Description of the spaces and functions of the community building" cit.
[35] See the perspective of the men's locker room of the Jewish Community Center, 1957, published in, Goldhagen, *op. cit.*, p. 133.
[36] This version of the project is published in, "Architectural Review," May 1957, No. 724, p. 344.
[37] This version of the project is published in, "Architectural Forum," vol. CVII, October 1957, No. 4, p. 142.
[38] *The Louis I. Kahn Archive: Personal Drawings* cit., vol. I, p. 385, drawing 430.24, transcriptions, p. LXVI.
[39] *Ibid.*
[40] *Ibid.*

160-162. Community Building, Jewish Community Center, Trenton, New Jersey, 1954-59.

160. Plan of the structure (LIK, 030.IV.A.430.2.27).

161. Perspective sketch of gymnasium with the indication of the colors, 1957 (LIK, 030.I.A.430.25).

162. Perspective view (LIK, 030.IV.A.430.2.23).

In the gymnasium Kahn refrains from making the roof rest on the groups of four pillars, each connected to the arches. The system of groups of brick pillars generates a narrow flat roof that runs continuously around the gymnasium. Instead, to support the large roof, he introduces a new pattern of pillars, no longer in brick but in reinforced concrete, organized at the same rhythm as the previous arrangement, but with slightly different spans to avoid the overlapping of the two orders of pillars, in brick and concrete. The parallel roof beams are reconfigured as a roof with just two long pitches, while the "triangular domes" are transformed into squares with just two pitches and a skylight.[41] In the same gymnasium, Kahn introduces another reinforced concrete structure to make a rectangular Game Room at the first level, without a roof, reached by means of two staircases. The long walls of the Game Room, with oblong hexagonal openings, are two simple Vierendeel beams that make it possible to build the large load-bearing structure. The indications about "excitingly painted" beams of the gymnasium also apply to the concrete in the project in July, including that of the pillars and the new system with the Vierendeel beams.

The sole documentation of Kahn's color effects in the construction of the gymnasium is the perspective view drawn for the corner, starting from the Game Room. Pillars, walls and beams in concrete appear painted blue, while all the intrados surfaces in shadow are painted red (the inner surface of the slab of the Game Room, that of the steps and the corbel sectors of the wall on which the Vierendeel beams rest). The color exercises done by Kahn starting with a view of Egyptian columns with an architrave are important for an understanding of the reasons behind the choice of red, to counter the power of shadow.[42] Thanks to the color, the reinforced concrete of the gymnasium takes on a particular luminosity with respect to the concrete of the other spaces of the Jewish Community Center, against the prevalently gray and red background of the cinder block walls and brick pillars. This circumstantial use of color to apply to the concrete can only be seen as Kahn's personal interpretation of the polychromy invented by Le Corbusier for the *béton brut*, with the substantial difference that while Le Corbusier creates chromatic contrasts to compensate for the gray tones, Kahn sets out to give a particular aura to a main space, to define what this space "wants to be" and to adjust the intensity of light and shadow.

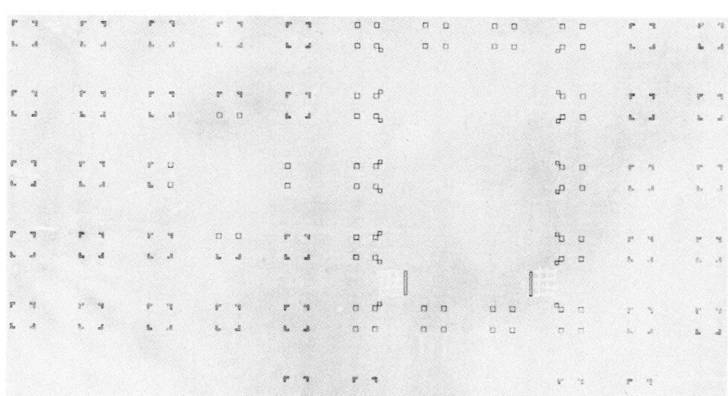

[41] See "Architectural Forum," vol. CVII, October 1957, No. 4, p. 142, and *The Louis I. Kahn Archive: Personal Drawings* cit., vol. I, p. 380, drawing 430.14.

[42] See the drawings published in, Hochstim, *op. cit.*, pp. 292-293, No. 405-407.

Jacobean Style in concrete and brick: the Richards Medical Research Laboratories

In the development of the model of the City Tower for the Universal Atlas Cement Company and on the worksite of the AFL Medical Services Building, Kahn interprets the potential of a use of reinforced concrete that differs from the systems that have emerged during the 20th century. All his works of the 1950s lay the groundwork for experimentation with a reinforced concrete skeleton equipped with functional cavities for physical plant, and articulated in a different way from that of the Yale Art Gallery.

The opportunity to invent and build a new hollow reinforced concrete structure comes to Kahn when the dean of the Department of Architecture of Philadelphia, G. Holmes Perkins, showed him the functional program prepared in October 1956 by the university committee chaired by Norman Topping for a new building of the University of Pennsylvania, to contain medical research laboratories and spaces for laboratory animals, and to form an extension along the Hamilton Walk of the building that is the symbol of the School of Medicine, the John Morgan Building. Starting in February 1957, Kahn works on what would become the Richards Medical Research Laboratories, named after Alfred Newton Richards, the pharmacologist and professor emeritus of the University of Pennsylvania.

From the earliest sketches, Kahn is looking for a device to graft onto the side of the John Morgan Building that is capable of expressing, with only its forms and materials, the powerful machinery with multiple conduits, the circulation of pure air and the control of contaminated discharges that are important factors for the scientific research conducted in the laboratories.

The metaphorical device for the laboratories takes the form of a structural tower unit with a square plan and colossal "columns" at the center of the sides. The columns have the dual function of bearing the load of the floor system and of containing the physical plant systems: the largest one contains the staircase and the machinery for the circulation and cooling of air; the others contain pipes and conduits for different fluids, gases, radioactive substances, electricity and air discharge. The physical plant network branches into the laboratories through special cavities in the slabs. The laboratories form a band of three towers, grouped together with the campus landscape in mind – "a central building around which the three major towers cluster," as Kahn puts it.[43] In the preliminary studies the towers are detached from the ground by means of columns that have the same landscape role, in terms of viewing and circulation, as the *pilotis* of Le Corbusier. Kahn's *pilotis* would offer a view, along the Hamilton Walk, of the vast Botanical Garden. A fourth tower, different from the others, is set aside for the laboratory animals. In the early studies this tower is placed up against the John Morgan Building. The functional program developed by Topping calls for isolation of the facility for the animals from the activities of the laboratories.

The red bricks of varying tones and the cream-color calcareous stone inserts typical of the Jacobean style of the John Morgan Building, the grand

[43] Louis I. Kahn quoted in, Richard Saul Wurman, *What Will Be Has Always Been: The Words of Louis I. Kahn*, New York, Access Press/Rizzoli, 1986.

[44] *The Louis I. Kahn Archive: Personal Drawings* cit., vol. I, p. 480, drawing 490.31, transcription p. LXVI. See also Matilda McQuaid, ed., *Envisioning Architecture: Drawings from The Museum of Modern Art*, New York, The Museum of Modern Art, 2002, pp. 116–117.

[45] The precast pieces, though conceived and implemented thanks to the collaboration of Komendant and Keast & Hood, are tested in their installation by the Jersey Testing Laboratory (Raymond A. Hood, letter to William H. Lindsay Jr., 24 October 1958, LIK, 030.II.A.25.19).

163-164. Alfred Newton Richards Medical Research Building, Philadelphia, Pennsylvania, 1957-60.

163. Model (LIK, 030.IV.A.490.15.3).

164. Perspective sketch (LIK, 030.I.A.490.31).

Quadrangle Dormitories with the turrets and the Zoological Laboratory (now the Leidy Laboratory of Biology) – all works by Walter Cope and John Stewardson built from 1895 to 1910 circa along the Hamilton Walk, precisely around the lot set aside for the new laboratories – represent the essential basis for the colors of the Richards Laboratories, which Kahn immediately indicates as those of the bricks and the exposed concrete. Without taking the grain, color and design of certain details, like the toothing of the stones at the corners of the buildings by Cope and Stewardson into account, it is impossible to grasp the full meaning of Kahn's concrete. Even the individual vertical conduits, so large and looming over the roof, beyond their functional logic, as to become a metaphor of the scientific activity conducted in the Richards Laboratories, and which in the preliminary studies have the profile of storybook chimneys, should be seen as other personal interpretations on Kahn's part of the Jacobean style – the chimneys and turrets rising over the roofs of the buildings of Cope and Stewardson next to Kahn's building are countless. Nothing is more distant from the towers of San Gimignano, in stone alone, than the chimneys and turrets in brick erected by Kahn for the University of Pennsylvania. No mistake can be made about what he states in a note on the drawing of the aerial view of the complex: "The stair towers between opened columns developing terminals for cooling and air handling. The exhaust system of air and fumes forming these towers present a silhoette [sic] complimenting the many chimneys and tower accents of the Jacobian [sic] dormitories nearby."[44]

When Kahn, in the summer of 1957, draws the silhouette with a Jacobean character of his new work of architecture for the campus, the Richards Laboratories have already become part of a larger complex of towers, still with a square plan, that also contains the rooms, library and laboratories of the Department of Biology of the University of Pennsylvania, directed by David Rockwell Goddard – it is approximately in this phase of the project that, among other things, the tower for the laboratory animals is separated from the John Morgan Building. From this point on the complex of the Richards Laboratories and the Department of Biology extends from the John Morgan Building to the Zoological Laboratory, completing the idyllic landscape composition envisioned by Cope and Stewardson around the Botanical Garden.

Towards the hollow floor structure with columns

Following approval of the preliminary design for the Richards Laboratories, Kahn calls in the engineer August E. Komendant, specialized in both prestressed and post-tensioned reinforced concrete, for the structural project, together with Gianopulos and Leidigh of Keast & Hood Company Structural Engineers,[45] the engineer Fred Dubin for the mechanical systems,

165-167. Alfred Newton Richards Medical Research Building, Philadelphia, Pennsylvania, 1957-60.

165. Elevation (LIK, 030.I.C.490.1.10).

166. Section (LIK, 030.I.C.490.1.8).

167. Plan (LIK, 030.I.A.490.2).

and the architect Ian McHarg for the design of the garden. He assigns the development of the drawings to three members of his office staff: David P. Wisdom, Thomas R. Vreeland and John E. MacAllister.

Kahn starts to develop a structural system in reinforced concrete with floors and columns capable of containing technical gear and ensuring uniform distribution of fluids in the laboratories.

Beams appear in the framework of the floor system with small overturned arches flush with the top slab for the passage of physical plant systems. Precisely due to this function, the arches become the characteristic motif of the exposed concrete edge beams of the facade – they are turned over and their profile is strategically shaped to transform them, at the point of the shifted corners, into semicurves suspended in the void. Between the edge beams brick infill is inserted, in keeping with the Jacobean style.

Conduits housed in the various columns feed the physical plant systems that run through the floors system. The colossal reinforced concrete column pierced by circles serves to anchor the band of chimneys, made with terracotta tubes, for the expulsion of air above the roof. The circles are arranged so that the voids straddling the floors form, at the same time, arched doors and upside-down arch openings to connect the conduits hung from the ceiling to the chimneys.

The vanishing of the pattern of circles marks the beginning of research on a framework of beams shaped with orthogonal lines, in visible hierarchies, designed to overlay the pattern of the conduits, which are also visible, thanks to their interwoven pattern.[46]

[46] Schematic, freehand, Elevation, n.d., [1957], LIK, 030.I.A.490.12, published in, Alexander Tzonis, ed., *The Louis I. Kahn Archive. Personal Drawings*, New York, London, Garland, 1987, p. 473.

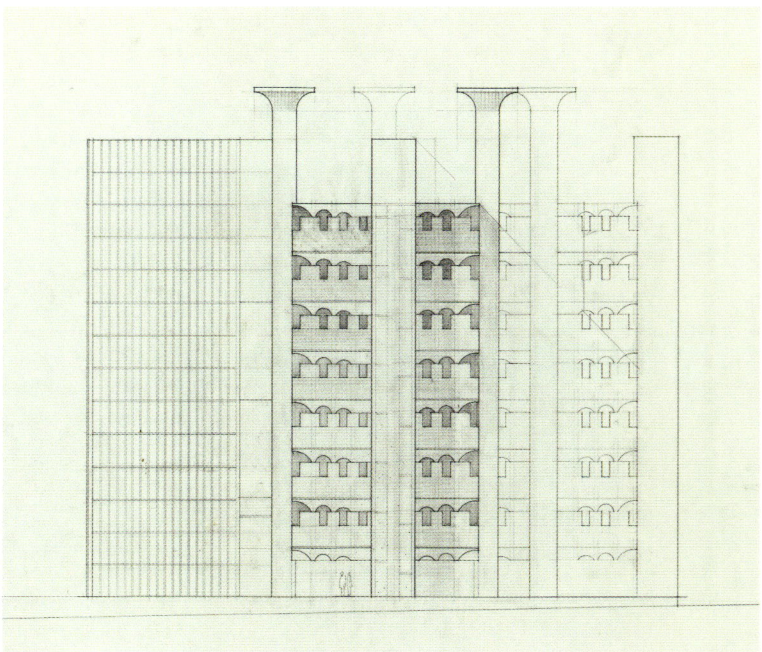

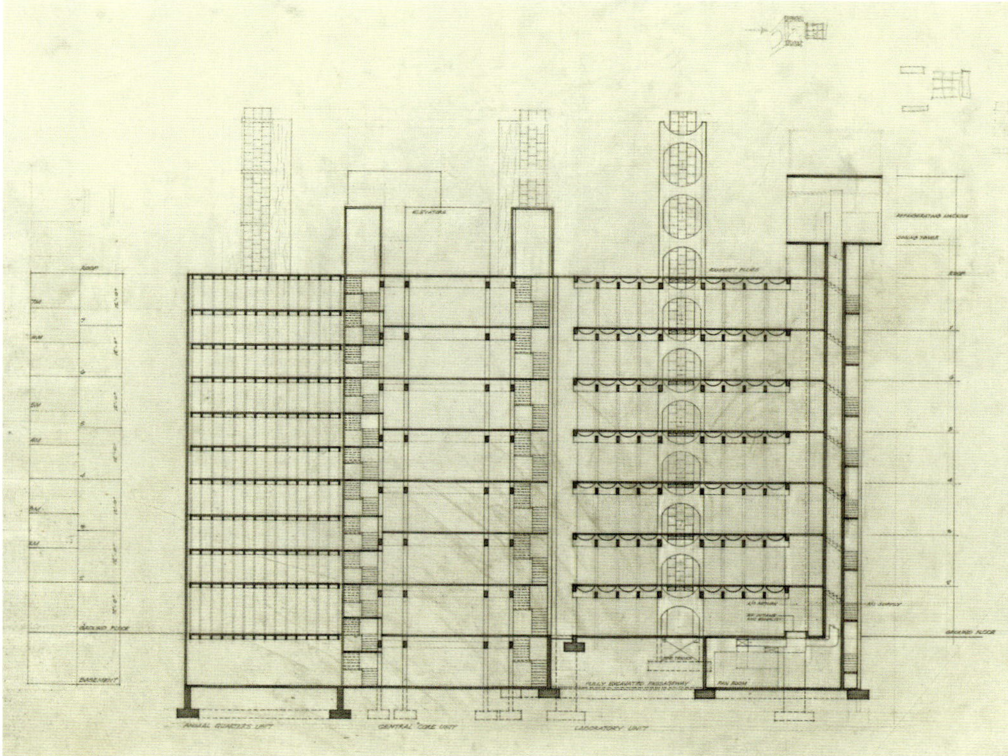
ANIMAL QUARTERS UNIT
CENTRAL CORE UNIT
LABORATORY UNIT
GROUND FLOOR
BASEMENT
FULLY EXCAVATED PASSAGEWAY
FAN ROOM

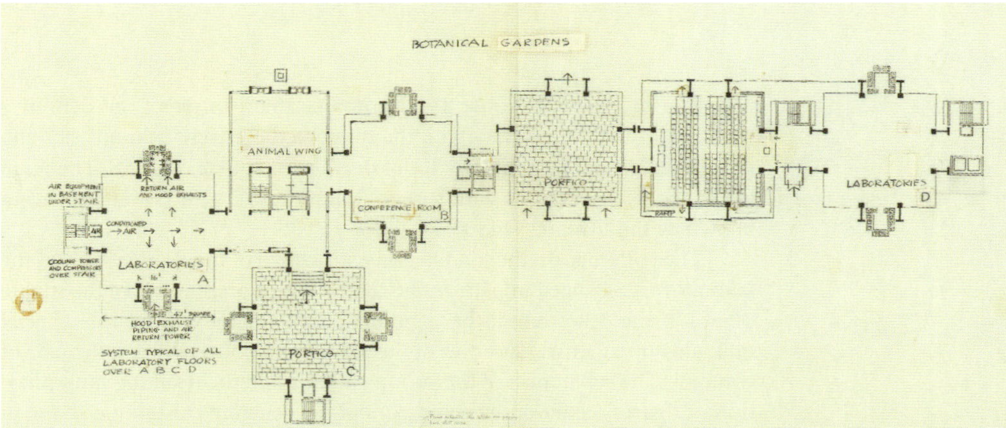
BOTANICAL GARDENS
ANIMAL WING
AIR EQUIPMENT IN BASEMENT UNDER STAIR
RETURN AIR AND HOOD EXHAUSTS
CONDITIONED AIR
AIR
COOLING TOWER AND COMPRESSORS OVER STAIR
LABORATORIES
A
16'
47' SQUARE
HOOD EXHAUST PIPING AND AIR RETURN TOWER
SYSTEM TYPICAL OF ALL LABORATORY FLOORS OVER A B C D
CONFERENCE ROOM
B
PORTICO
C
PORTICO
LABORATORIES
D

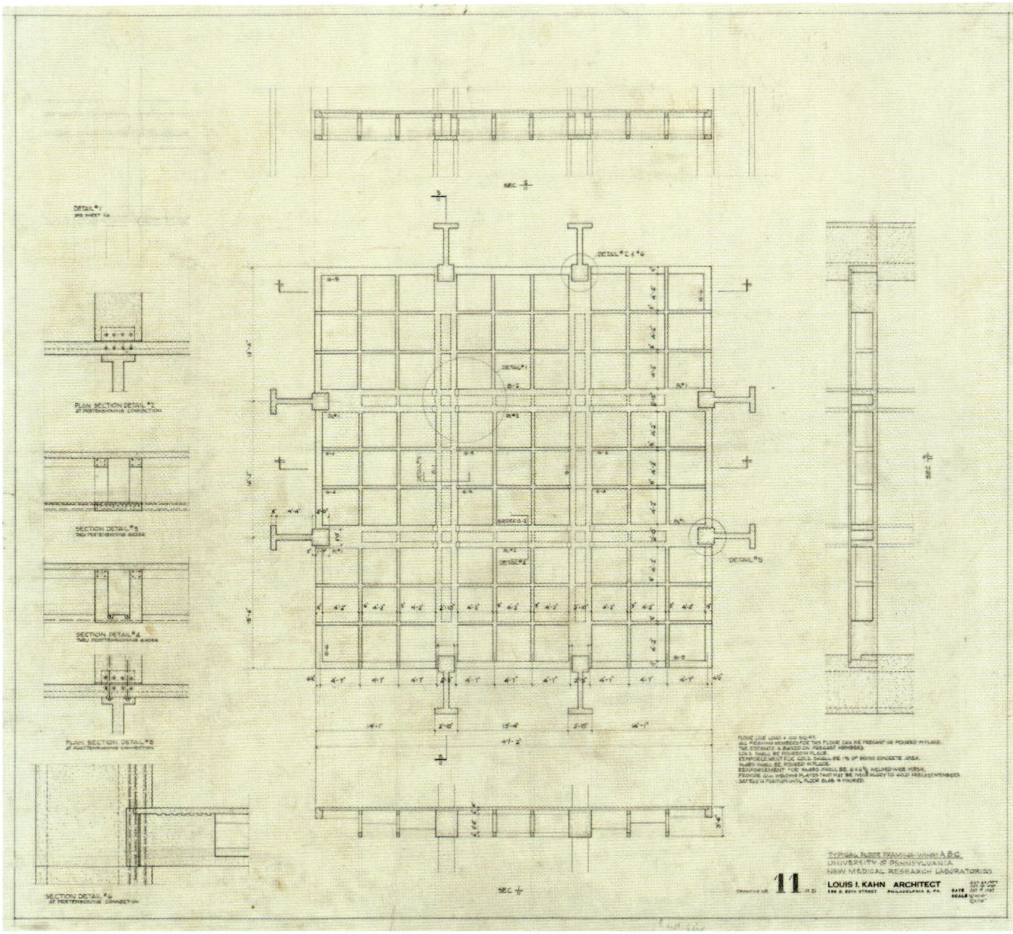

The project in October 1957 contains the features of the definitive structure of the floor system and the reinforced concrete columns, though certain details are modified by Kahn during the course of 1958. The naming of the towers is done with letters: A, B, C and D – the latter also being known as the "Animal Quarters."

The columns are the load-bearing structures of the floor system of the laboratories, arranged in pairs in the central sector on one side of the square plan. On the other hand, the conduits and stairs are placed in a special unit, also up against the center of the side of the floor, between the pair of columns. The separation of the various organs and the resulting multiplication of figures – column pair and conduits – generate the unique shape of the column section.

The large column is composed of two components, one large square and one slender rectangle, connected by a core. Nevertheless, its profile does not belong to the static genre of the double-T beam, because it is based on a different logic: only the squared piece forms the true pillar on which the beams of the floor system rest, while the boundless lengthening of the core serves to project the slender component well beyond the floor perimeter and beyond any static logic, transforming it into a figure that forcefully

168-169. Alfred Newton Richards Medical Research Building, Philadelphia, Pennsylvania, 1957-60.

168. "Typical Floor Framing. Wings A, B, C," 9 October 1957 (LIK, 030.I.C.490.3.19).

169. "Reflecded Ceiling Plan," 1 October 1957 (LIK, 030.I.C.490.1.8).

asserts, in the view of the complex, the pair of columns arranged beside what were the original columns containing the conduits and stairs.

The floor structure takes the form of a grid of main and secondary beams in reinforced concrete. Kahn gives this structure the generic definition of "typical floor framing."[47] The main framework forms nine square compartments that contain the secondary framework that determines nine other compartments, again square. All the beams are configured with large voids to facilitate passage of the conduits. Kahn looks for a particular design for those openings, a geometry that gets away from that of the beams of the AFL Medical Services Building, and one that can produce, in the combination of the various beams, a spatial graphic pattern of artistic value. The fact that Kahn sets out, in the design of the geometry of the floor system, to experiment with the artistic potential of the structure is also borne out by the choice of giving the same height to the main and secondary beams, going beyond the simple expression of the load-bearing essence of the system. The beams with the rectangular cavities produce a type comparable to the Vierendeel beam, but this is merely a result of the artistic process carried out by Kahn (this also happens in the configuration of the columns). It is no coincidence that in the documents related to the initial phase of the genesis of the floor system, we find no mention of the Vierendeel beam. The voids are so ample and the rods so slender that they create a variant of the Tetrahedron Floor System totally freed of the Space Frame applied for the Yale Art Gallery, and Kahn

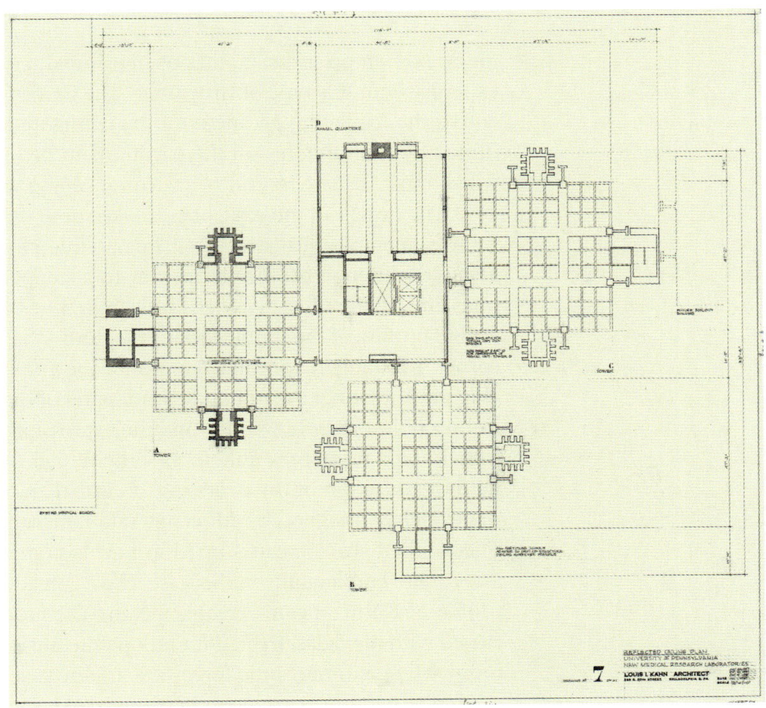

[47] See for example "Typical Floor Tower A, B, & C Erection Scheme," 21 May 1958, modified on 8 August, and 28 October 1958, LIK, 030.I.C.490.009. The same definition had already been given to the floor system in October 1957 ("Typical Floor Framing – Wings A, B, C," 9 October 1957, modified 10 October and 24 October 1957, LIK, 030.I.C.490.002).

170. Alfred Newton Richards Medical Research Building, Philadelphia, Pennsylvania, 1957-60. "South (Botanical Garden) Elevation," 10 October 1957 (LIK, 030.I.C.490.3.13).

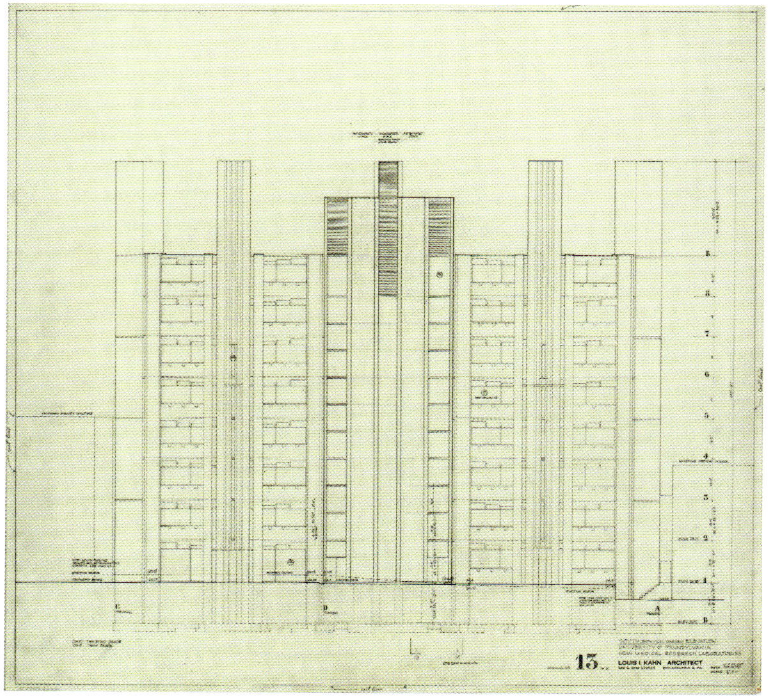

has to rely on prestressed and post-tensioned parts, thanks to Komendant's expertise with this type of structure.[48] The framework of the floor system becomes the regulating geometry of the configuration of the space, as is confirmed by several notes on the partitions to be inserted in the laboratories, which must "adhere to grid of structural ceiling wherever feasible."[49]

The design of the edge beams takes on particular value in Kahn's pursuit of a metaphorical structure. The previous shaping, obtained with small arches of different profiles, gives way to a stepped configuration of orthogonal lines, which in any case is not limited to the mere transcription, in a visible profile, of the framework of the floor system and the question of the corner overhang. It will suffice to notice the graphic proportions of the variable thicknesses of the full and empty parts of the beam to understand that Kahn's objective goes well beyond the expression of the floor system's static role. Kahn's mastery is able to give the design of the edge beams, starting with the question of the overhang, an artistic power comparable to that of the exercises assigned by Albers at Yale University in the course on color. The toothed strips created by Mon Levinson and called "vibrating boundaries"[50] are emblematic, in this case. If we consider the fact that the stepped profile of Kahn's beams creates rectangular modular compartments, then closed with plexiglas, treated for an opaque finish, the potential of the structure – where vibrant static forces are, in any case, at work – for the genesis of an abstract art in concrete becomes clear. One cannot help but notice, furthermore, that precisely the sophisticated profile of the edge beam introduces, in the sequence of the horizontal bands of concrete, brick and glass, a geometric motif of clamping of the large glass panes of the windows, like the toothing of the stones to the courses of bricks at the corners of the buildings of Cope and Stewardson.

The decision to create a floor system with beams perforated to contain the conduits in a weave with the clear character of an abstract art composition raises the question of the quality of the concrete to be left exposed and the fabrication of the individual rods. The erasure of the visibility of traces of the construction process emerges as a fundamental corollary to bring out the degree of artistic abstraction the framework possesses. This is the perspective in which we can grasp Kahn's preference, from the initial idea of the framework, for the prefabrication of the single pieces of the floor system.

A note reveals Kahn's intentions regarding the making of the reinforced concrete, on site or at the workshop, for the constituent parts of the "typical floor framing" and the columns. The possibility of prefabrication is mentioned only in the case of the beams, while the columns and top slab are to be poured in place. "All framing members," the note states, "for this floor can be precast or poured in place. The estimate is based on precast members. Col[umn]s shall be poured in place. Reinforcement for col[umn]s shall be 1% of gross concrete area. Slabs shall be poured in place. Reinforcement for slabs shall be 6 x 6 6/6 weld wire mesh. Provide all welding plates that may be necessary to hold precast members safely in position until floor slab is poured."[51]

Every concrete member of the floor system is to be left exposed, except in the portico where insulating panels applied between the compartments of the secondary framework would ensure insulation of the rooms above the slab.[52] In this phase of the project, Kahn does not indicate the formwork system for the pouring of the top slab; nor does he go into the making of the beams if it is decided that they should be poured in place.

For the columns, which are the only parts clearly to be poured in place, Kahn studies every constructive detail, from the base clad in granite sheet

[48] See the drawing "Typical floor framing – wings A, B, C," 9 October 1957, modified 10 and 24 October, LIK, 030.I.C.490.002.
[49] "Reflected Ceiling Plan," 1 October 1957, modified 9, 10 and 24 October 1957, LIK, 030.I.C.490.002.
[50] Horowitz, Danilowitz, *op. cit.*, p. 222. The exercises are produced in the "color course" taught by Albers at Yale University from 1950 to 1958.
[51] "Typical Floor Framing – Wings A, B, C," 9 October 1957, modified 10 October and 24 October 1957, LIK, 030.I.C.490.002. It should in any case be specified that certain drawings show the version of the floor system with all the parts poured in place (see the drawing "Typical wall sections," 10 October 1957, LIK, 030.I.C.490.002).
[52] "Typical Wall Sections," 10 October 1957, modified 24 October 1957, LIK, 030.I.C.490.002.

171. Detail of a building of the University of Pennsylvania, Philadelphia, Pennsylvania (photograph by the author, 2010).

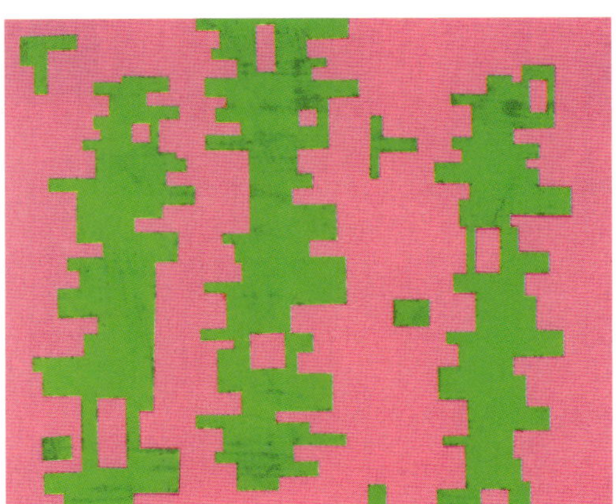

172. M. Levinson, *Vibrating boundaries*, Color Course, Yale University (F. A. Horowitz, B. Danilowitz, *Joseph Albers: to Open Eyes, The Bauhaus, Black Mountain College, and Yale*, London, Phaidon, 2009, p. 222).

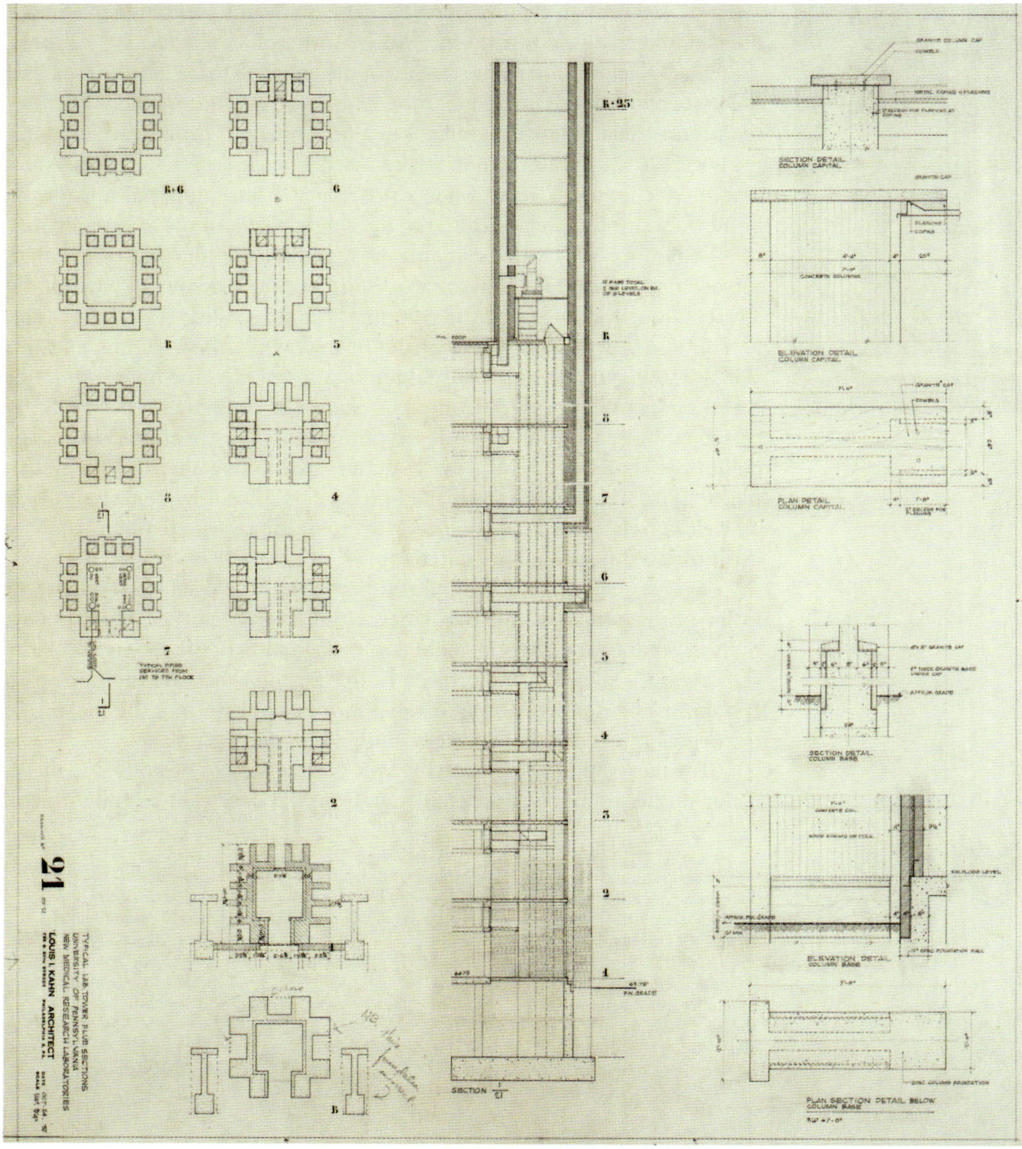

173. Alfred Newton Richards Medical Research Building, Philadelphia, Pennsylvania, 1957-60. "Typical Lab. Tower Flue Sections," 24 October 1957 (LIK, 030.I.C.490.3.21).

to protect the concrete from erosion, to the design of the imprints of the vertical boards, all of the same width for the core and the trunk, all the way to the "capital" which again, like the base, is formed by a rectangular sheet of granite attached to the top, likewise to protect the concrete from rain.[53] The choice of the boards for the formwork confirms that for Kahn concrete poured in place should continue to be treated according to the technical principles of *béton brut*.

A note clarifies the nature of the poured in place concrete: "Concrete: all 'in situ' concrete 3000 p.s.i. Reinforcing standard practice & governing codes. Forms – surfaces – common board forms to conform to architect's design and to be installed in strict accordance with form surface shop drawings. Design of form bracing is the responsibility of contractor within code limits. Rubbing and finishing of concrete surfaces after stripping of forms is not required and not wanted."[54]

The form of the volumes of the staircase and the conduit, and their systems of construction, undergo a number of different, significant variations. Brick is the fundamental material for the character of both types of volumes.

Though the organs have been separated, each in its own configuration, Kahn continues to keep the columns and towers, the vertical load-bearing structure and the conduits, together as if they still constituted a unit. The drawings entitled "Tower Flue Sections" is indicative, showing details of the construction of both the column and the "typical piping services."[55]

A more conventional reinforced concrete skeleton forms the exposed structure of the Tower of the Animal Quarters, with the cavities enclosed by dual masonry: bricks for the outside and cinder blocks for the inside (a system that nevertheless is changed at the time of construction).[56] On the back of the tower, with the aim of creating formal unity with the other laboratory towers marked by the volumes of conduits and staircases, Kahn shapes powerful, narrow solid vertical ribs, one central rib of brick that forms the chimney, or "incinerator stack," and two in exposed concrete to contain the "air exhaust stacks" – each of them contains two pillars of the main skeleton.[57]

Precast and prestressed concrete

During the course of 1958 the uncertainties that have emerged regarding the different reinforced concrete techniques for the Richards Laboratories find an initial resolution, focusing the dialectic pursued by Kahn between prefabricated and poured in place, exposed or brick-clad concrete. This does not mean that other hesitations having to do with the making of the concrete do not surface during the project development, revealing a never definitively answered question on the dual nature of the concrete of the Richards Laboratories.

Kahn and his collaborators do not reach a definitive design until they have first reconsidered the general dimensions and the number of levels of the towers, as a result of pressure from the client to reduce construction costs. It is significant that in the project in January, Kahn indicates just one

[53] "Typical Lab. Tower Flue Sections," 24 October 1957, LIK, 030.I.C.490.002.
[54] "First Floor Plan," 1 October 1957, modified 9,10 and 24 October 1957, LIK, 030.I.C.490.002.
[55] "Typical Lab. Tower Flue Sections," 24 October 1957, LIK, 030.I.C.490.002.
[56] "Reflected Ceiling Plan," 1 October 1957, modified 9, 10 and 24 October 1957, LIK, 030.I.C.490.002.
[57] "South (Botanical Garden) Elevation," 10 October 1957, modified 24 October 1957, LIK, 030.I.C.490.002.

174. Alfred Newton Richards Medical Research Building, Philadelphia, Pennsylvania, 1957-60. Plan, 10 January 1958 (LIK, 030.I.C.490.2).

conduit and one cylindrical staircase for each tower – the latter is a variation of the staircase of the Yale Art Gallery.

The few phrases in the minutes of the meeting with the clients, held on 5 February to examine the project, offer a glimpse of the lively discussion on Kahn's proposal to leave the concrete exposed. The main concern of the committee is to obtain such quality of the surfaces as to guarantee durability of the parts exposed to weather over time. Therefore the main question has to do not so much with the floor system, which since 1957 has been hypothesized in prefabricated concrete, as with the columns that Kahn proposes pouring in place, which are the most visible concrete parts, and those most vulnerable to corrosion. At this meeting the decision is made that will definitively orient the development of the project of the structure of the floors of the laboratories, including the columns, towards a single type of concrete.

"The advisability of using exposed concrete as proposed by the architect is a matter of extreme concern," the minutes of the meeting of 5 February report. "After considerable discussion it was decided that the columns should be precast and articulated by floors, that the concrete must be rigidly controlled, and that the resulting concrete must be of the greatest possible density so as to do all possible to achieve durability."[58]

Following the decisions made in February, all the documents clearly distinguish between the two types of concrete: "precast concrete" and "poured in place concrete." Kahn indicates the details of the two types through drawings and specifications. In the new project the stairs and conduits are contained in volumes with a rectangular plan – those of the staircases are slightly larger. Kahn adds a volume for the piping in the tower with the portico, which therefore becomes exceptional with respect to the others and is thus configured for its central position in the complex and its predominance in the view of the Richards Laboratories across the Hamilton Walk. Precisely the addition of a volume for the conduits (the staircase is shifted to the side along the street) permits Kahn to confirm the position

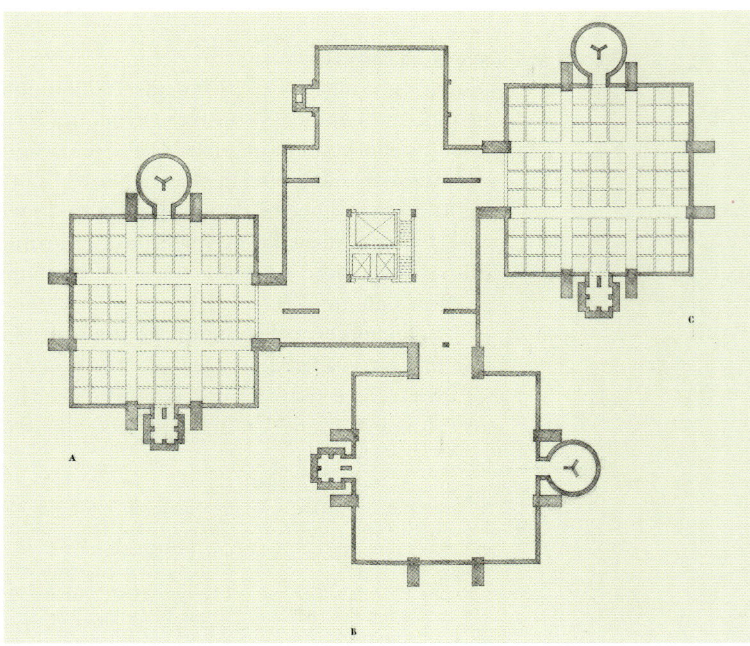

of the two diagonal flights of steps for access to the portico, which are fundamental in the sophisticated corner view that justifies, in the overall perception of the complex, the overhanging beams at the corners of the various floors. The portico thus encourages the passer-by to observe the device contained in the laboratories, and becomes the diorama of the view of the structure displayed like a sculpture, visible without the pipes for the fluids.

The precast concrete extends to all the skeleton elements of the three towers of the laboratories, from the columns to the main and secondary beams of the floor system – the "precast and prestressed concrete units" that form the "complete framing system."[59]

The structure of the floor system is similar to that of the design of October 1957, but each detail is indicated in terms of the prefabrication of the parts and their assembly at the worksite. The thicknesses of the beams increase at the connections, and the four beams of the main framework are to be cast in pairs: two continuous beams and two broken down into three pieces, to be reassembled on site.[60] It is in this phase of the study of the structure that the term Vierendeel is used for the secondary beams that form the pattern of nine small compartments enclosed inside those formed by the main beams.

The question of the formwork for the top slabs is central to the development of the secondary framework, and had not been addressed in the project of October 1957. The small compartments generated by the Vierendeel beams are closed by means of "Precast Concrete Slabs." The concrete topping is to be poured in place over these precast slabs, with a thickness of two inches, equipped with metal reinforcement – "topping" with Welded Wire Fabric (WWF).[61] Thus the system of secondary Vierendeel beams and precast concrete slabs seems to be conceived as a structure of formwork to be left in place, replacing the temporary scaffolding and formwork. The size of the pattern of the secondary compartments can support the precast slabs used as the formwork for the pouring. So in this version the construction process, in all its phases, becomes an integral part of the structure. In the Yale Art Gallery the special formwork for the top slabs also remained incorporated, taking on the function of acoustic paneling for the rooms. In the case of the Richards Laboratories, Kahn intends to show the observer a structure with surfaces made in the workshop, perfectly smooth and no longer bearing the imprints of traditional formwork, in an attempt to get beyond the *béton brut* utilized in the AFL Medical Services Building.[62]

[58] "Minutes of Meeting in Dr. Topping's Office, 9:00 A.M. February 5, 1958," LIK, 030.II.A.25.8.
[59] Kahn office, "Specifications for the Construction of New Medical Research Laboratories," 19 May 1958, LIK, 030.II.A.24.6.
[60] See the drawing "Details of Precast Units," 21 May 1958, modified 3 August and 28 October 1958, LIK, 030.I.C.490.009.
[61] Kahn office, "Specifications for the Construction of New Medical Research Laboratories," 19 May 1958, LIK, 030.II.A.24.6.
[62] See the drawing "Typical Floor Tower A, B, & C Erection Scheme," 21 May 1958, modified 8 August and 28 October, LIK, 030.I.C.490.009. The three phases of the assembly of the prefabricated parts ("Erection Phase 'A'," "Erection Phase 'B'," "Erection Phase 'C'") are indicated in a drawing with notes dated 21 May 1958. The first phase calls for the erection of the columns, held in place with an epoxy resin product of the Permagile type, followed by post-tensioning and grouting of the joints. In the next phase the main structures of the floor system are assembled, first the two continuous beams ("Prestressed girders MK.1") and then the two beams each composed of three parts, for post-tensioning after positioning ("Post-tensioned girders MK.2"). Once the main structures have been installed, the program moves on to the welding of the MK.2 girders with the MK.1 girders, grouting and post-tensioning where required. The third phase calls for assembly of the edge beams ("Precast wall beams MK.3, MK.3', MK. 4"), their connection, the deployment of the parts of the secondary structure of the floor system ("Precast Vierendeel trusses MK5.6" and "members MK.7") and of the "Precast Concrete Slabs," and the pouring on site of the upper part of the top slab ("Topping"), with a thickness of two inches, reinforced with "welded wire fabric" (WWF). After the connection, a further grouting of the joints between the MK.2 members is planned between the edge beams, followed by post-tensioning.

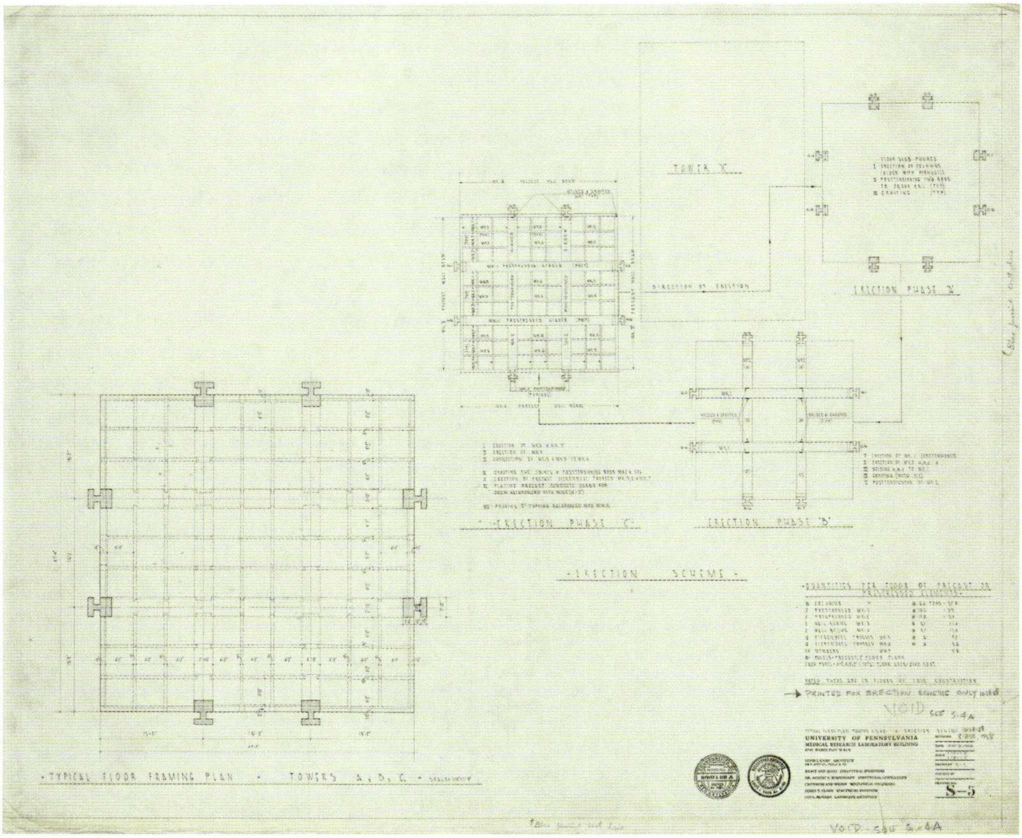

175-176. Alfred Newton Richards Medical Research Building, Philadelphia, Pennsylvania, 1957-60.

175. "Typical Floor Plan Towers A, B, & C & Erection Scheme," 21 May 1958 (LIK, 030.I.C.490.9).

The inclusion of the column in the prefabrication system coincides with the transformation of its profile. Column, core and flange remain as its fundamental parts, but they are now proportioned in a different way, for the purpose of fabrication at the workshop and "erection" at the worksite.[63] The main beam and edge beam of the floor system rest on the large rectangular sector, the true pillar. The length of the core is drastically reduced (it corresponds to about one half of the pillar). The thickness of the flange is half that of the pillar, so that once the beams are placed on the pillar their joint will be at the middle and the whole final structure will appear modular.

The arrangement of the holes for the stiff post-tensioning bars begins to approach its definitive design. The pillar has two rows of circular holes for the bars that run vertically through the trunks of the columns: the row of four serves for the passage of the bars that cross the main beam, and the row of two for the passage of the bars that cross the edge beam. The flange has two pairs of holes at the ends for the passage of the bars that connect the overlaid columns.

Kahn designs the lower extremity of the column so that the flange and core are longer than the pillar. Precisely the extended flange and core define a new basic concept of the column in the prefabricated structure: they seal and reinforce the connection between the top of the column and the beams, and permit positioning of the columns partially above and partially against the floor system, leaving them in a vertical position with the sole aid of an epoxy resin that guarantees stability prior to the attachment of the post-

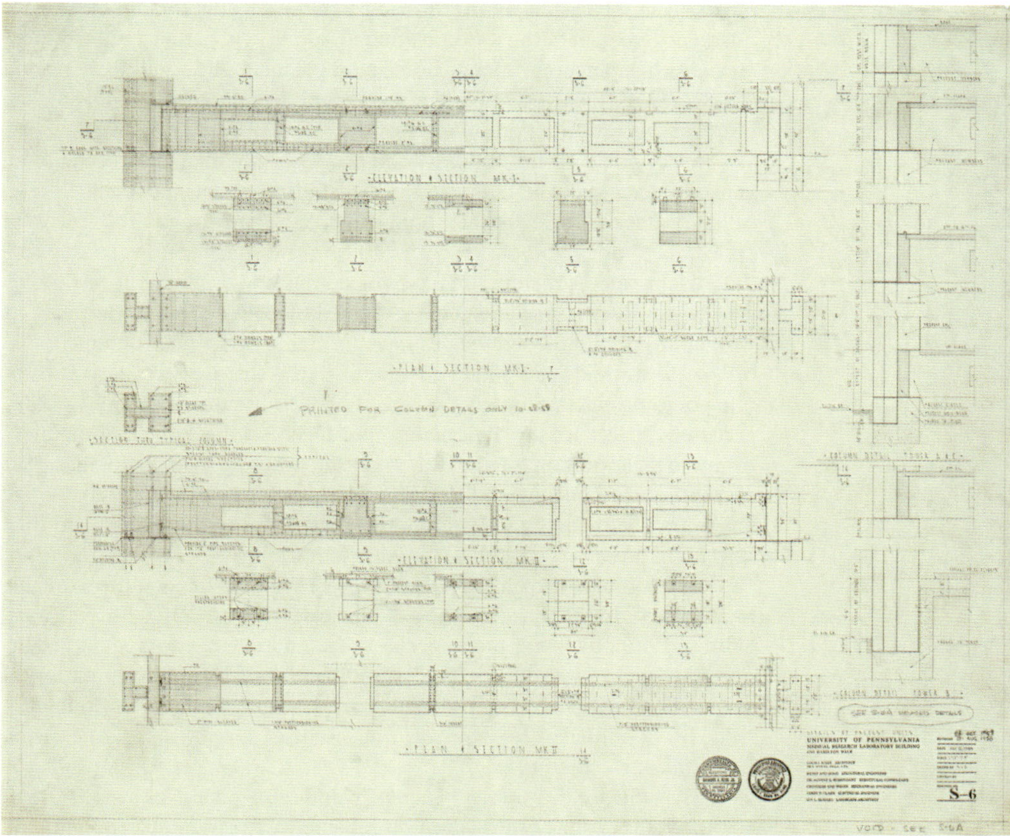

176. "Details of Precast Units," 21 May 1958 (LIK, 030.I.C.490.9).

tensioning bars, without having to make use of temporary shoring. Once again, Kahn designs every detail to integrate all the phases of the assembly as fully as possible in the configuration of the pieces of the structure, to make the entire process become visible. After the stepped profile of the column, at the top of each tower Kahn creates a special torso, also precast and at the height of the floor, to crown the sequence of overlaid columns.

The choice of the material for the making of the precast concrete parts is decisive in order to obtain a concrete with a "surface of smooth appearance, free from fins."[64] The materials suggested by Kahn have to have characteristics that ensure "leakproof rigid molds of sufficient strength to withstand deformation of the units."[65] For these molds, Kahn proposes concrete or steel, or a combination of the two, or "plastic faced plywood" to mount on a steel structure.[66] The mold must be free of "irregularities," "dents" and "sags." To facilitate extraction of the piece, the mold surface is to be treated with "mineral oil or other approved bond breaking agent."[67]

Before pouring the concrete into the molds, depending on the various

[63] See the drawing "Typical Floor Tower A, B, & C Erection Scheme," 21 May 1958, modified 8 August and 28 October, LIK, 030.I.C.490.009.
[64] Kahn office, "Specifications for the Construction of New Medical Research Laboratories," 19 May 1958, LIK, 030.II.A.24.6.
[65] *Ibid.*
[66] William Schmoll, report, 17 October 1958, LIK, 030.II.A.25.1.
[67] Kahn office, "Specifications for the Construction of New Medical Research Laboratories," 19 May 1958, LIK, 030.II.A.24.6.

pieces to be precast, steel prestressing bars or sheaths for the passage of post-tensioning bars must be positioned, along with other reinforcement elements like "anchors, anchor bolts, strands, metal ties, dowels, hangers," the sleeves of the physical plant systems and the special steel rings for the lifting and positioning of the pieces.

Every phase of the prefabrication process is planned to achieve homogeneous density of the concrete. To get "uniformly dense concrete" Kahn calls for "internal or external vibration." "True," "unwarped" and "pleasing appearance" are some of the terms used in the specifications to describe the prefabricated concrete surfaces. The compound to pour into the molds is to be made with Portland cement, treated with an additive to improve its workability, while the aggregates have to be no larger than one inch. "No broken or warped units shall be installed in the structure. Honeycombs or other defects, unless properly dressed, shall be cause for rejection." The only accepted defects are the "hair cracks" produced on the surface of the piece by rapid evaporation of the water during setting. Also to increase the strength of the concrete and to permit slow, continuous setting, each piece has to first be treated with steam, or hot water, for twelve hours, and then kept humid and protected from sudden temperature changes or impacts.

The columns, which are perforated to accept the bars on the worksite, must be kept in an "upright position" during transport. All the pieces must be raised only by using their special rings, and during storage and transport care must be taken to avoid chipping corners and the formation of cracks. The transport from the factory to the worksite is to be done with "vehicles of sufficient length providing point of support at two feet from each end of the casting."

The study of the precast concrete includes the phases of installation of columns and beams, or the "erection scheme." The assembly is done clockwise. In the first phase, immediately after the pouring of the base slab, the eight columns are erected, followed by their post-tensioning and grouting. In the second phase, first the two long prestressed main beams are installed, then the three segments of the other two main beams, followed by welding and grouting of the points of contact between the beams, and then by post-tensioning of the segments of the two beams. The third phase begins with assembly of the longer and then the shorter edge beams, followed by their connection at the corners; then comes "grouting the joints" and post-tensioning between the columns and the segments of the main beams; the Vierendeel beams – long and in segments – are then installed, after which the precast concrete slabs are placed. The operation concludes with the laying of a metal web and the pouring of the topping.

Almost all the indications made in the specifications are complied with in the production of the precast parts.

Poured in place concrete: the meanings of smooth and lined surfaces

Apart from that of the topping in the floors system of the laboratories, not visible in the ceilings, the poured in place concrete in the Richards Laboratories, described in the specifications of May 1958, can be clad with brick or left exposed. The indications in the documents offer glimpses of different solutions from the ones actually implemented.

The poured in place concrete is envisioned for the skeleton of the conduits of the air circuit and the safety staircases of Towers A, B and C of the laboratories, and for the walls, conduits and volumes of the staircases and elevators of Tower X. Tower X is the new name given to Tower D, after the project of extending the towers to make the Biology Building, with the aim of clearly distinguishing the two towers in concrete poured in place, the X and the Y, in the two contiguous complexes, from the series of other towers in precast concrete, from the A to the E. The differentiations among the various constituent parts of the organism, which Kahn had introduced in the preliminary phases of the project, continue to influence the choice of surface finishes.

In Kahn's terminology, to indicate the different quality of the concrete surfaces left exposed in relation to the degree of finishing, systematic use is made of the distinction between "exposed concrete," which is all the visible concrete to be controlled in detail in the assembly of the formwork, and "unexposed concrete," namely the concrete without cladding in secondary and often inaccessible places, such as the elevator shaft, as well as the zones to be covered in brick.

According to what is indicated in the specifications of May 1958, the concrete to pour in place is composed of Portland Cement with additives: an air entraining agent (Darex AEA or Vinsol Resin) and a densifying retardant, Plastiment or Sikacrete, to use depending on the season during which the pouring is done.

To avoid the formation of surface defects, like honeycombs or efflorescence, the proportions of the various components of the conglomerate (aggregates, cement, water) are studied to produce a compound that is easy to deploy and not prone to phenomena of segregation of aggregates or migration of water towards the surface. To guarantee proper deployment of the concrete, certain techniques are recommended for the protection of the surfaces after pouring and during setting.

For the concrete surfaces of the walls, "without painting or any other treatment," and built in such a way as to obtain "the desired finish without need for patching," Kahn looks for a "dense uniform concrete, the surface of which shall require no reworking when forms are removed, and which shall be free of voids or honeycombs."[68] The main tool for the making of this type of surface is the formwork, its materials and finishing characteristics. The materials are boards and panels of plywood or metal. Depending on the material of the formwork and the qualities of the concrete surface – "unexposed concrete" or "exposed concrete" – different types of facings are indicated for the surfaces of the formwork in contact with the pouring. "Plywood and board forms" are to be deployed, for the exposed concrete, with a "plastic form coating" like Formfilm produced by the A.C. Horn Company, "or approved equal." When plywood or wooden boards are used in the unexposed concrete, their recommended treatment is "form oil or other approved bond breaking agent." In this case the boards are assembled without smoothing – "undressed lumber." For possible metal formwork panels, a "thin coating of approved form oil" is indicated, to apply prior to positioning of the reinforcements, because the oil should not touch the steel rods.

The choice of the formwork with panels or boards with which to shape the exposed concrete of Tower X depends on the pursuit of a graphic surface texture, and on questions of hygiene in the spaces, with the goal of avoiding the accumulation of dust and organic material on surfaces. Kahn wants to set these spaces apart precisely by means of the pattern and size of the formwork: where laboratory animals are raised, the concrete is to be

[68] *Ibid.*

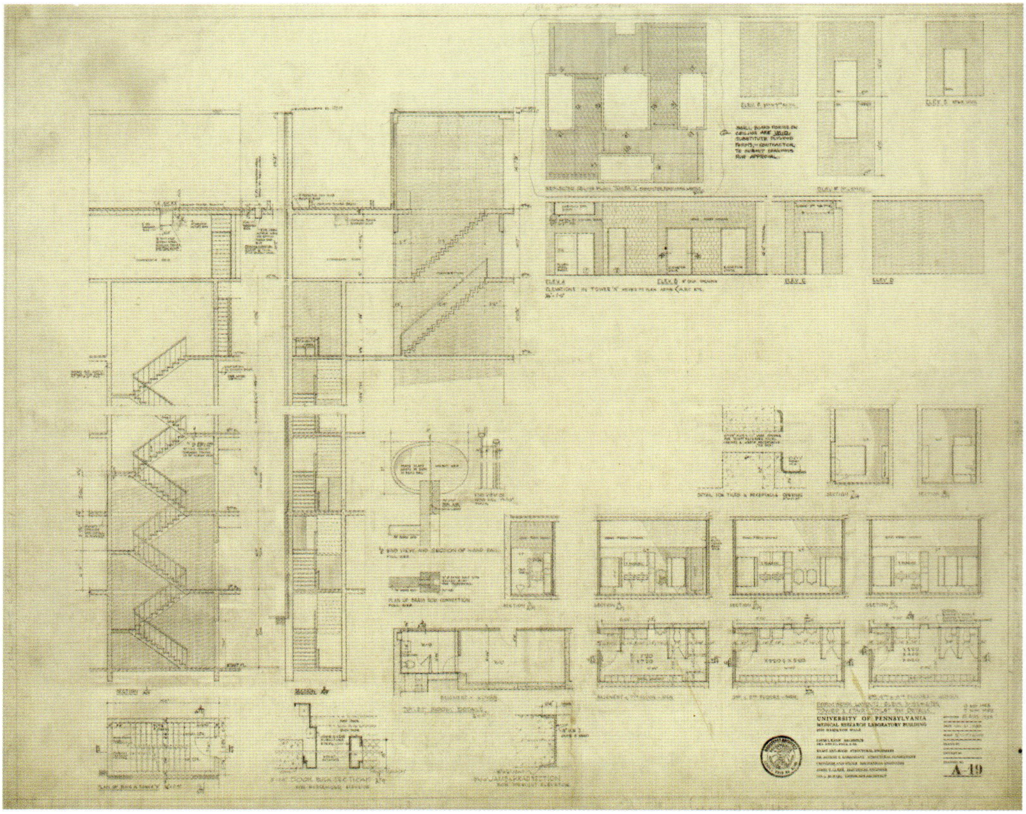

molded inside large panels of steel or plywood with a plastic facing, to obtain an absolutely smooth surface with a minimum of joint imprints, precisely for purposes of hygiene. The other spaces in exposed concrete are clearly distinguished from the previous thanks to the minute pattern of the surfaces, which also covers those of the ceilings, all formed by wooden boards of 3-4 inches. Kahn introduces true graphic motifs to imprint with panels of boards assembled vertically and horizontally. Thus in "Corridor and Stair," for the walls and ceilings the formwork are made with boards of the "tongue and groove floor boards" type, to arrange in "horizontal and vertical patterns."[69] The panel of drawings of the "Formwork layouts" for the exposed concrete contains the patterns of groups of boards specially designed to obtain, also in the ceilings, refined "concrete form marks," also with indications of the positions of the joints in the vertical assembly of the boards.[70] In these parts of the Richards Laboratories, where the use of panels of wooden boards is planned, Kahn demonstrates that he is still following certain principles of the early *béton brut* of Le Corbusier.

After Kahn's assessment of the optical effect of the holes of cones on the perception of the defects of concrete surfaces for the Mill Creek towers, another consideration arises, again connected with the perception of concrete surfaces and the effects of textures on possible defects in the pouring. Precisely in the specifications of May 1958, the aesthetic reasoning is stated for the panels of vertically and horizontally arranged boards: "The imprint of these boards is used to conceal imperfections in the surface of the con-

177. Alfred Newton Richards Medical Research Building, Philadelphia, Pennsylvania, 1957-60. "Formwork Layouts Elevs. X-152 to X. 752. Tower 'X' Stair – Toilet Rm. Details," 21 May 1958 (LIK, 030.I.C.490.5b).

crete." This statement reveals Kahn's drive to achieve optical control of all the aspects of concrete construction.

Beyond the materials of the panels, their sizes and surface treatments, Kahn specifies all the other aspects of the formwork system, from spreaders to different kinds of joints or strips, outlining the constructive principles of the vocabulary of his exposed concrete.

Unlike what was done in the Yale Art Gallery, in the making of the formwork Kahn forbids the use of beveled strips or chamfers to apply in the inner corners of the formwork, as he now wants to have sharp corners. Similar instructions will also be made for the work on his other buildings in exposed concrete. In place of the chamfers, to obtain corners without flaws in the concrete he urges particular care in the assembly of the panels: "forms for outside corners shall be reasonably tight to prevent forming of rough fins."[71] It is clear that the disappearance of the chamfers is not just a technical detail, but the clue to a desire to start to control the whole formwork system, to achieve a new American type of architectural concrete, clearly distinguished from international Brutalism and the leanings of some of its exponents, as emphasized by Banham, towards an "abstemious under-designing of the details."

For the spreaders of the formwork of the exposed concrete, Kahn and his team suggest the Tyscru type patented and produced by the Richmond Screw Anchor Corporation, Inc. – the Richmond Tyscru®. The arrangement of the spreaders in horizontal and vertical rows guarantees stability of the formwork and, above all, creates that regular pattern on the surface of the concrete that is destined, in Kahn's work, to become the basis of what he now simply indicates as "appearance." It is no coincidence that precisely in order to underline the pattern of the system, wooden Tycones are placed in the ends of the spreaders "at the exposed faces," and the holes they produce are left "unfilled." To confirm the hierarchy Kahn seeks in the design of the formwork system, for the spaces of the laboratory animals and the areas of unexposed concrete he does not specify the type of spreaders, simply requesting that the holes be sealed.

The sealing of the holes, where it is specified in the spaces with exposed concrete, is to be done with a mixture of sand and two types of cement, to apply immediately after stripping, with the surface of the concrete saturated with water. The two cements serve to obtain a suitable color: "White cement should be used with the gray cement to match the color of the wall." To avoid hollows, the seal has to be slightly protruding and leveled only after one or two hours, by means of "wood float" and "cheesecloth." The indications on the sealing of the holes of the spreaders also describe the timing of the mixing of the compound, the type of finish treatment and the time required for setting.

In the overall design of the concrete to be poured in place, Kahn includes the various phases of the pouring, some of which are controlled by means of construction joints, and the expansion joints between the sectors

[69] *Ibid.*

[70] "Formwork layouts – Elevs. X-152 to X-752 Tower 'X' stair – Toilet em. Details," 21 May 1958, modified 8 August, 7 and 13 November 1958, LIK, 030.I.C.490.005a.

[71] Kahn office, "Specifications for the Construction of New Medical Research Laboratories," 19 May 1958, LIK, 030.II.A.24.6.

of the structure.[72] For the construction joints, Kahn simply envisions the cleaning of the surface of contact between one pouring and another, both vertically and horizontally, to reduce possible drips of concrete and obtain the most continuous surface possible. "[...] where a joint is to be made, the surface of the concrete shall be thoroughly cleaned and all laitance removed. In addition, vertical joints shall be thoroughly wetted and slushed with a coat of neat cement grout immediately before the placing of new concrete," we read in the specifications. Different solutions are applied by Kahn for this type of joint on other worksites, also following observation of the defects that develop in the Richards Laboratories with the simple cleaning of the surfaces between pourings.

For the concrete poured in place, after stripping only certain operations of cleaning of the surface are permitted, in order to eliminate "oil remaining from the forms, laitance, mortar drippings, or other foreign material," done with cloths, brushes and water. If the concrete is still not clean after these operations, or if there are perceptible variations of color, the surfaces can be washed with chemical detergents, like diluted hydrochloric acid.

The concrete of the envelope of Tower X, entirely clad in brick, is part of the unexposed concrete. The walls have slender bands of concrete generated by the overhang of the floors system. Again in this case, as in that of the edge beam of the laboratory tower, the bands do not fully correspond to the whole structure (in fact just one floor out of two protrudes, and there are also exceptions to this rule). It is as if Kahn were designing, with concrete, not so much the eventual anchorings of the large sectors of brickface, as the fundamental lines of a weave that binds together the towers of the Richards Laboratories thanks to similar graphic expedients, always used by Cope and Stewardson, to create a graphic texture corresponding to their buildings along the Hamilton Walk, entrusted precisely to the slender stone cornices, like string courses running through the brick facades.

Blocks in concrete and terrazzo

As in the earlier works, the concrete of the Richards Laboratories also appears in the form of precast blocks to be left exposed, for which Kahn selects the compound, its working, its color and the color of the mortar, as well as the design of the joints.

The specifications call for certain walls of blocks of the size 8 x 8 x 4 inches.[73] The compound to make these blocks is a "lightweight aggregate" selected to obtain "uniform *texture* and *color,*" and the color Kahn wants is "gray-blue." The Primecrete-White produced by Bethayres Concrete Products Company, Inc., is indicated as a possible solution. The same qualities of "material, texture" are sought for the "concrete block lintels." The joints between the blocks are set at ½ inch. The way of obtaining the color of the mortar for the laying of both the blocks and the bricks represents another proof of Kahn's focus on the truth of materials, without altering their constitution through the addition of pigments. The color indicated by Kahn varies according to the material utilized for the masonry (block or brick) and has to be obtained not by additives, but through selection of an essen-

[72] For the "expansion joints" Kahn's office produces a series of panels that indicate the exact positioning of each line. In the "Specifications" it is indicated that the "exposed edges of finished floors" must be "neatly finished and slightly rounded" (*ibid.*).

[73] *Ibid.*

tial part of the compound: the sand. "No mortar color shall be used" is the significant and brusque warning written in the specifications.[74]

The exposed concrete poured in place is also found in some of the floorings. For the flooring and steps of the portico, the surface is to be smoothed with wood floats, to achieve the "desired texture." For the surfaces of the floorings of corridors and rooms, and for the steps of the staircase, a concrete finish of the Steel Troweled Type is indicated, done with three successive operations of flattening and polishing involving multiple tools: the float, the "motor driven float of the disk type" and a special steel trowel. After these operations, three coats of a product like Lapidolith are applied to harden the surface and protect it against dust (the concrete of the steps is excluded from this treatment and protected with slip-free products).

The other form of expression of concrete, the terrazzo so highly prized by Kahn and also used in the Richards Laboratories, is selected for certain surfaces of the floorings and stairs. Kahn and his team determine the percentages of marble chips and "gray or white Portland cement" to use in the composition of the terrazzo, and indicate deployment techniques to obtain a "uniform composition" (about 70% of the chips should be visible in the outermost layer). Only in the making of the terrazzo is the addition of a pigment permitted to obtain a given color of the cement: "Where the cement is to be colored to obtain a desired shade, the color pigments shall be limeproof and of a non-fading quality." The terrazzo surface is the result of work after the laying of the compound, including compacting with "heavy stone" or "metal roll," smoothing by hand with the trowel, polishing with "abrasive stones" and grouting. The entire terrazzo surface on the floorings is to be subdivided into compartments by means of metal bands, or "dividing strips," to reproduce a pattern resembling that of a stone pavement.[75] The composition of the terrazzo and its surface finish are tested with samples, submitted to Kahn for approval.

The definition of the guidelines for the various types of work and expression of concrete for the Richards Laboratories comes before the selection of the construction firms, prior to the imminent start of the work.

New dilemmas on concrete: precast or poured in place?

After the preparation of the project and the specifications in May 1958, at the end of the month Kahn applies to the Department of Licenses and Inspections of the City of Philadelphia for a permit for the "precast, prestressed multi-story concrete construction of three buildings forming a part of the New Medical Research Laboratories."[76] The permit application is justified by the novelty and complexity of the prefabricated structure. At the same time, bidding is accepted to determine the contractor, with the winning tender coming from Joseph E. Farrell, Inc., based in West Conshohocken, Pennsylvania.

The question of the dual nature of the concrete of the Richards Laboratories continues to generate uncertainties in Kahn and his team. A note from June, made following discussions with Gianopulos and Leidigh, renews the focus on the construction process of the concrete for the floors

[74] *Ibid.*
[75] The color of the terrazzo is requested from the supplier in May 1959 (Albert P. Flanagan, letter to Louis I. Kahn, 25 May 1959, LIK, 030.II.A.25.23). For the design of the fake plates of terrazzo see the drawing "Terrazzo strips, typical layout," 14 September 1959, LIK, 030.I.C.490.004.
[76] Louis I. Kahn, letter to Barnett Leiberman, 29 May 1958, LIK, 030.II.A.25.19. Barnett Leiberman is the Assistant Commissioner of Building of the Department of Licenses and Inspections of the City of Philadelphia.

178-179. Alfred Newton Richards Medical Research Building, Philadelphia, Pennsylvania, 1957-60.

178. "Reflectes Ceiling Plan of Concrete Structure," 14 July 1958 (LIK, 030.I.C.490.9).

179. Sketches and notes (LIK, 030.II.A.58.51).

system of the laboratories: "poured in place lab.[oratories] construction."[77] Certain particulars are also reconsidered, to the point of raising doubts about the choice of the precast concrete that had guided the definition of the main parts of the skeleton of the laboratory towers.

In July, to make the floor of the laboratories lighter and to simplify its construction, modifications are made to the system of the precast concrete, compromising Kahn's principle of integration of the various components of the concrete construction in the final structure. In the new project the measurements of the pattern of the secondary beams are definitively changed, passing from nine to four compartments, because of the elimination of the precast concrete slabs use as permanent formwork. At this point the plan is to assemble, on the secondary beams, wooden boards to make the formwork to be dismantled after the construction of the top slab. In the sequence of the assembly of the precast parts indicated on the drawing of 14 July, it is no coincidence that a fourth phase appears, "form and pour slab," which was not included in the previous "erection scheme." The fact that the layer of concrete of the top slab, previously called "topping," is now a true "slab" in Kahn's terminology, confirms the structural change that happens with the elimination of the precast concrete slabs to support the poured topping. The framework of the secondary beams in relation to the precast concrete slabs, as in the project in May, or in relation to the wooden boards, as in the project in July, demonstrates that the design of the structure of the top slab obeys the logic of the formwork system. When the construction system changes with the arrival of the new top slab, the static role of the secondary beams is challenged. Informed directly by Kahn, perhaps, regarding the particular nature of the secondary framework, William H. Jordy writes a significant passage on the floor system of the Richards Laboratories: "Structurally these secondary trusses serve only as minor stiffening, but they were important during the erection of the building since they supported forms for the floor slabs, thus eliminating the need for propping in this operation."[78] So, as in the case of the floor system of the Yale Art Gallery, that of the Richards Laboratories is also like a homogeneous spatial weave, though it is not exactly a continuous load-bearing structure in all its parts.

After having introduced, in the version of the floor system outlined in July, the reinforced concrete top slab poured in place in wooden formwork, with imprints on the ceiling to be left exposed, it should come as no surprise that Kahn and his team reconsider the possibility of eliminating all the precast parts, to make the structure of columns and beams out of one kind of concrete, poured in place.

On 8 August, following a discussion with representatives of Joseph E. Farrell, Inc., to re-examine every technical detail of the project in the light of the pursuit of maximum economy in the construction processes, Kahn and his team express their willingness to rethink certain fundamental points of the specifications of 19 May. In the document entitled "Tentative changes," which lists possible modifications of materials and physical plant systems, the first point of the "general construction" has to do with the fabrication of the concrete for the towers of the laboratories: "Change all the precast concrete to poured-in-place reinforced concrete in accordance with drawing to

[77] Handwritten notes, 25 June 1958, LIK, 030.II.A.25.15.
[78] William H. Jordy, *Medical Research Building for Pennsylvania University, Philadelphia*, in, "The Architectural Review," vol. CXXIX, February 1961, No. 768, (pp. 98-106), p. 99.

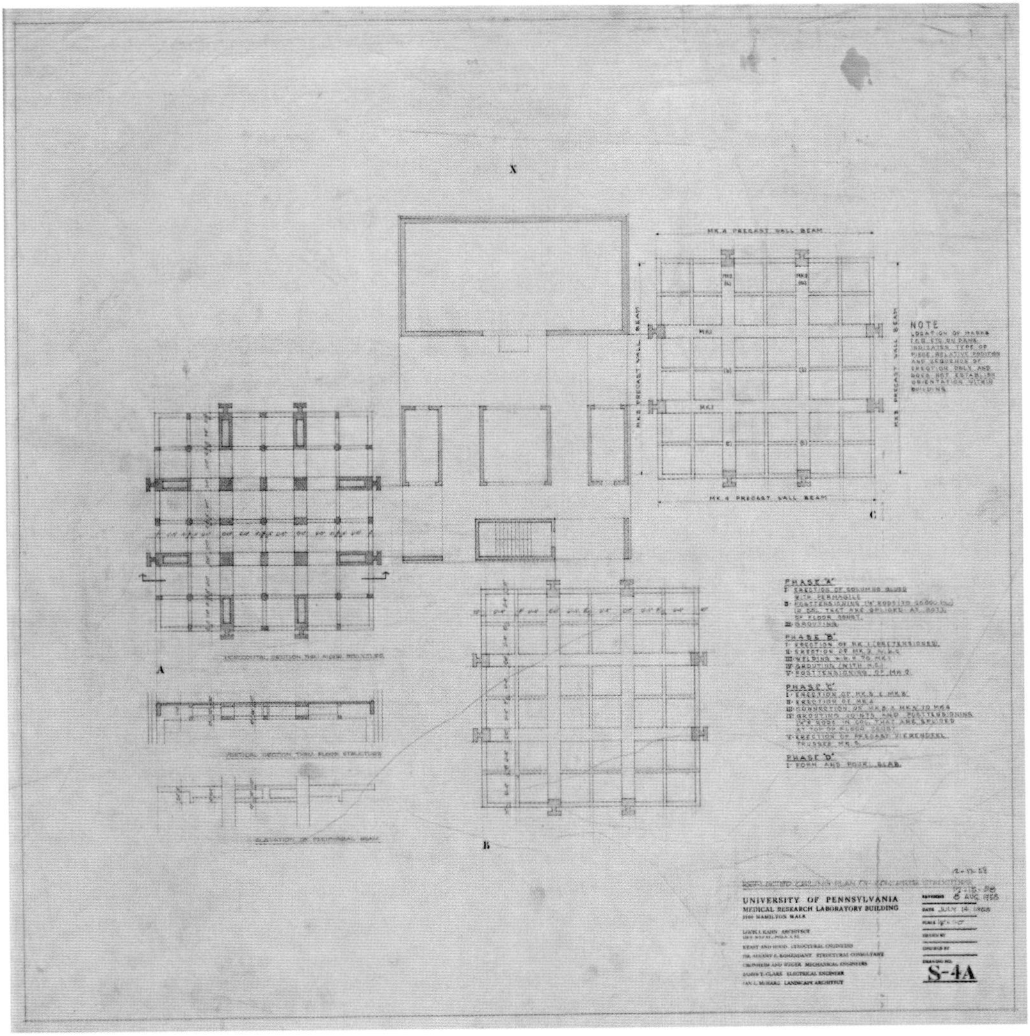
X
NOTE
PHASE "A"
PHASE "B"
PHASE "C"
PHASE "D"
A
B
C
UNIVERSITY OF PENNSYLVANIA
MEDICAL RESEARCH LABORATORY BUILDING
S-4A

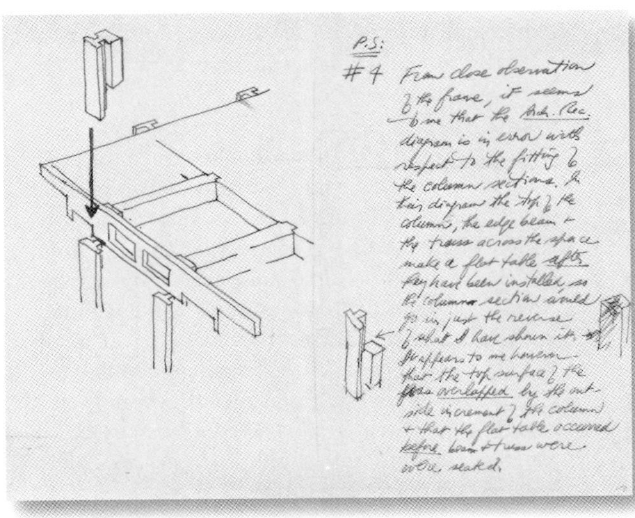
P.S:
#4 From close observation
of the frame, it seems
to me that the Arch. Rec.
diagram is in error with
respect to the fitting of
the column sections. In
this diagram the top of the
column, the edge beam +
the truss across the space
make a flat table after
they have been installed so
the column section would
go in just the reverse
of what I have shown it.
It appears to me however
that the top surface of the
truss overlapped by the out-
side increment of the column
+ that the flat table occurred
before beam + truss were
were seated.

be developed."[79] The proposal demonstrates that the question raised by Gianopulos and Leidigh is still a timely one. The fact that the "Tentative changes" are not an integral part of the "Addendum" to the specifications prepared that same day confirms that the use of poured in place concrete is still just a possibility, not a hard and fast decision. The option of using only concrete poured in place is proposed by Joseph E. Farrell, Inc., to reduce construction costs, at least according to what is reported by Komendant. The fact that Kahn, in any case, includes poured in place concrete among the "tentative changes" and that the question remains current at least throughout the month of August might lead us to believe that Kahn wants to achieve the goal of unifying the concrete of the Richards Laboratories. We should not forget that since 1957 he has himself been considering the possibility of building the floor structure with concrete poured in place.

The entries on concrete in the "Addendum" to the specifications acknowledge only a modification of a practical nature, probably suggested by the technicians of Joseph E. Farrell, Inc., to the formwork system proposed by Kahn. Instead of making formwork with wooden boards as Kahn had indicated for the ceilings of the corridors in Tower X, to obtain graphic continuity of surface texture with the walls, now the use of "plywood or other material subject to Architect's approval" is indicated.[80]

While the worksite for the foundations of the Richards Laboratories gets underway, the Department of Licenses and Inspections, on 12 August, asks Kahn's office to provide supplementary calculations and structural tests to guarantee the "method of design and construction" proposed for approval in May, that of the structure in precast concrete.[81] But following discussions held on the days after 8 August, the choice of the construction method for the concrete remains an open issue, so the response to the requests of the municipal authorities are delayed. "[...] the final plans and calculations not be submitted to Mr. Dawson, Chief of Construction section, until it has finally been determined whether the construction will be poured-in-place or precast as designed," we read in a letter dated 14 August.[82]

The crucial resolution to "change all the precast concrete to poured-in-place reinforced concrete," if taken, would lead to a radical change of value of the whole Richards Laboratories, reducing the structure invented by Kahn to a variation of the poured in place structure of the AFL Medical Services Building. But at least in the month of October, the construction system definitively selected remains that of the dual solution of precast concrete and poured in place concrete.[83]

Color trials

Kahn controls the quality not only of the precast and poured in place concrete, but also of the concrete that takes the form of blocks for the construction of the walls of the Richards Laboratories. The brick and mortar are also carefully selected to obtain a given color and perfect deployment. In November 1958 two trial masonry sectors are made to test the "appearance" and "water tightness" of walls made with a layer of brick for the external face and one of concrete blocks for the internal face.[84]

The laying of the blocks, their color and that of the mortar, are examined and approved by Kahn on 20 February 1959: "It was felt that the block backing was good. The contact between the gray-blue block and warm mortar harmonizes well. The non-struck joint was pointed out by the Architect as acceptable to him."[85] Certain criticisms have to do with the "workmanship," the "cut joints" and the color of the laid brick. The documents confirm that the brick has been chosen by Kahn to create a structure with a relationship to the buildings of Cope and Stewardson. That same day, Kahn proposes making other brickmasonry trials with "darker sand and mortar" – they are approved in May.

Even after the first trials for the walls in brick and blocks, the structure proposed by Kahn to make the turreted volumes enveloping the conduits and stairs is challenged by the technicians of Joseph E. Farrell, Inc. These discussions lead to the definitive solution for the tower volumes with the continuous walls of reinforced concrete in place of the originally planned skeleton.

Following the Farrell proposal, Kahn evaluates the possibility of "replacing concrete column and beam design of 2 Feb. for stair and flue towers with 8" concrete wall exposed inside towers and covered with brick outside, eliminating concrete backer block." Kahn is aware that the Farrell proposal leads to greater "unity in appearance" in the rooms of the Richards Laboratories, thanks to the use of only exposed concrete. It is also noted that the elimination of the "unformed brick and block walls" would guarantee more effective protection against infiltration of rainwater through the walls – an aspect tested in the first masonry trials.[86]

The blocks used in the Richards Laboratories are supplied by the Grays Ferry Brick Company, based in West Conshohocken.[87] For the bricks, Kahn chooses the type produced by one of the main American suppliers, the Sayre & Fisher Brick Company based in Sayreville, New Jersey (the supplier in Philadelphia is the O.W. Ketcham Company, Inc.).[88] The mortar for both types of wall is made with one part Nazareth Dark Portland Cement, two parts lime and eight parts yellow sand.

The laying of the blocks and bricks is governed by indications that attempt to achieve uniform joints and their precise design. The mortar of the joints has to be finished, where indicated, with a special tool, a wooden wedge – striking, as in the trial – and spreading to obtain a constant thickness: "keeping the cross-joints (vertical joints) uniform in size, and the same size or smaller than the bed joints, however, vertical bond should be maintained as closely as possible."[89] John B. Kelly, Inc., based in Philadelphia, handles the laying of the brick and blocks.

Where the block walls have to be crossed by piping, special pieces are included, the "lead-lined cinder blocks," also left exposed.[90] Nevertheless Kelly, president of John B. Kelly Inc., is against leaving this type of block exposed to construct the walls in the laboratories because their image is disfigured by the "overlap of lead around the periphery of the block" and by the

[79] "Tentative changes," 8 August 1958, LIK, 030.II.A.25.30.
[80] "Addendum 4," 8 August 1958, LIK, 030.II.A.25.30.
[81] William H. Lindsay Jr., letter to David P. Wisdom, 12 August 1958, LIK, 030.II.A.25.19.
[82] David P. Wisdom, letter to Keast & Hood, 14 August 1958, LIK, 030.II.A.25.19.
[83] See Raymond A. Hood, letter to William H. Lindsay, 24 October 1958, LIK, 030.II.A.25.19.
[84] David P. Wisdom, "Minutes of Job Meeting, Friday November 21, 1958," LIK, 030.II.A.25.6.
[85] Louis I. Kahn, "Meeting at the Job Site, February 20, 1959, 11:00 AM," LIK, 030.II.A.25.6.
[86] David P. Wisdom, "Job Meeting, Friday 27 February 1959," LIK, 030.II.A.25.6. "Mr. Farrell made comparative estimates of costs in the presence of the Architect," is written in the report of the meeting on 27 February, "of Air towers constructed of concrete columns and beams surrounded in lower three stories with brick and backer block in accord with Architect's design of 26 Feb., and of a concrete wall in first three stories in place of columns and beams on south wall with concrete exposed on inside face. The cost and appearance are acceptable to Architect. He will re-design and forward it Monday 2 March for form pricing for submission to University for approval" (*ibid.*).
[87] John B. Kelly, letter to Benjamin A. Palladino, 11 May 1959, LIK, 030.II.A.25.23.
[88] Jacob Mandel, William Kleinfelder, 19 May 1959, LIK, 030.II.A.25.23.
[89] John B. Kelly, letter to Benjamin A. Palladino, 11 May 1959, LIK, 030.II.A.25.23.
[90] John B. Kelly, letter to Joseph R. Farrell, 25 June 1959, LIK, 030.II.A.25.23.

"hammering of the lead to set the block." He also criticizes the size of the "lead-lined cinder blocks" as "contrary to that specified for this Project." For these reasons, Kelly suggests plastering over the cinder block walls.[91]

Development of construction systems

The concrete of the Richards Laboratories is done by two contractors, Joseph E. Farrell, Inc., for the poured in place concrete and the supervision of the whole worksite, and the Atlantic Prestressed Concrete Company, based in Trenton, for the precast concrete. Development of the details of the two concrete types continues in the last months of 1958 and, in the case of the precast concrete, into the next year, with ongoing discussions among Kahn and the office staff, Komendant, the engineers of Keast & Hood and the technicians of the two contractors.

An initial, important meeting on the technical questions of production of precast concrete parts is held on 16 October 1958 at Kahn's office, with (among others) Komendant, Hood, Farrell, Morrison of Joseph E. Farrell, Inc., and William Schmoll of the Atlantic Prestressed Concrete Company.[92] During the meeting there is discussion of the possible methods for installation of the columns, and it is decided to use "grout and dry-pack" for the joints; for the beams, the group approves the "Freyssinet unsheathed wires," but has not yet decided on the way of joining the pieces, simply urging a "good appearance." They also discuss the materials proposed by Kahn in May for the molds and finishing of the concrete surface not shaped and in contact with the mold: "Plastic faced plywood was considered equal to steel as a form finish on exposed concrete and the use of a hard troweled finish was considered acceptable for vertical surfaces of interior concrete." The question of the construction of the top slab returns, after Kahn's studies on precast slabs and formwork made with wooden boards. On this point, a type of anchor is examined to attach the support elements of the top slab to the edge beams of the floor system: "The ledge on the spandrel beam to support the floor slab was discussed and it was agreed that it could be attached with dowels and poured in place after initial precasting."[93]

Once the reservations regarding the way of building the structure of the floors system of the laboratories have been clarified, the Department of Licenses and Inspections of the City of Philadelphia is notified by Keast & Hood, on 24 October, that E. A. Meister of the Jersey Testing Laboratory will supervise all the phases of "manufacturing," "shop assembly" and "testing" of each "precast and prestressed member" in Kahn's structure.[94]

The excavation for the foundations continues in the fall of 1958. In November Joseph E. Farrell, Inc., begins the first work for the construction of the basement of Tower X (assembly of the plywood formwork for the walls and the pouring of the concrete). At the same time, the concrete wall trials are conducted, to assess the quality of the surface obtained with the formwork of the poured in place concrete. The report on the worksite visit

[91] *Ibid.*
[92] William Schmoll, report, 17 October 1958, LIK, 030.II.A.25.1.
[93] *Ibid.*
[94] Raymond A. Hood, letter to William H. Lindsay, 24 October 1958, LIK, 030.II.A.25.19.
[95] David P. Wisdom, "Minutes of Job Meeting, Friday November 21, 1958," LIK, 030.II.A.25.6.
[96] Thomas J. Leidigh, letter to Joseph R. Farrell, 15 December 1958, LIK, 030.II.A.25.19.
[97] See Mary Krumboltz Hurd, *Formwork for concrete*, Detroit, American Concrete Institute, 1963, p. 153.
[98] Thomas J. Leidigh, letter to Louis I. Kahn, 9 March 1959, LIK, 030.II.A.25.19. See also David P. Wisdom, letter to R. McCray, 11 March 1959, LIK, 030.II.A.25.13.

180-181. Alfred Newton Richards Medical Research Building , Philadelphia, Pennsylvania, 1957-60.

180. Construction site photograph, 2 January 1949 (LIK, 030.IV.A.490.12.32).

181. Construction site photograph (LIK, 030.V.A.3.20).

of 21 November states: "Mr. Kahn approved sample concrete wall for concrete finish Scheduled C1. At the builder's request, Mr. Kahn's exact words are quoted, 'That's the nuts.' Builder will immediately order the screw ties used in sample."[95] The phrase "That's the nuts" uttered in response to the perplexities expressed by the contractor shows that for Kahn a tolerable margin of error exists in the fabrication of the concrete on site, which does not compromise the quality of the surfaces in his view.

In December it is decided that the conglomerate of the poured in place concrete, during periods of frost, would contain Plastiment with calcium chloride and not Sikacrete (the two products are indicated in the specifications of May). In the concrete zones exposed to weather an air-entraining agent, Darex, is added to improve workability and make hydration faster, as well as to obtain concrete that stands up to large temperature shifts.[96]

For the construction of the walls of Tower X a formwork system in wood with large panels and incorporated aerial scaffolding is used, to reposition thanks to a crane after setting, the so-called Ganged Panel Forms.[97] The concrete is brought to the level of the pouring by means of a container running on a vertical truss.

For the application of the brick cladding a light metal structure on two levels is created, on which to mount the decks for the masons. The structure remains in position and is raised thanks to cables attached to a special temporary framework with posts protruding from the top of the conduits and walls to be clad. In March 1959 the walls of the first two levels of Tower X are under construction. That same month, for the floors slabs poured in place, it is suggested that the strength qualities of the concrete to be utilized in periods of frequent rain be improved, to avoid slowing the setting after pouring.[98]

Prefabricated concrete by the Atlantic Prestressed Concrete Company

At the outset of 1959, Kahn is so intensely engaged in the development of the concrete that when the review "Architectural Record" requests information on the Richards Laboratories he sends a note in which he describes only the structural elements, in terms of poured in place and precast concrete, without ever making mention of brick. "The structure," he writes, "is a Vierendeel Truss, 45' x 45' 'square span,' devised to permit the passage of pipes, conduits, ducts, and hood exhausts within the confines of the structure. Each 'square span' rests on four pairs of columns (eight columns) placed at the center of each 45' side, 15' apart. The depth of the Vierendeel is 3'. All members of the truss and the columns are prefabricated. The building is composed of three towers, each seven stories high. This group of towers is attached to a core tower constructed of concrete poured-in-place."[99]

All the details of the columns and beams in precast concrete are the responsibility of the technicians of the Atlantic Prestressed Concrete Company. They develop the working drawings, starting with those supplied by Kahn's office, and have to examine in detail all the difficulties of assembly to achieve a regular geometry of the structures of the three towers, as well as their perfect positioning with respect to Tower X. The fact that the structure is composed of both prestressed and other elements has an influence on the precast concrete – changes of length "due to creep, shrinkage and plastic flow"[100] in the case of the prestressing – and therefore generates difficulties in the calculation of the dimensioning of the individual members in order to obtain, in the final assembly, a structure with limited margins of geometric imperfection. The prestressing takes place in two ways, and these too have an influence on the dimensioning of the members, given the fact that the main beams are composed of members prestressed both in the workshop, just 48 hours from the pouring, and on the worksite, two or three months after they were made. There is no doubt that the proposal of a prestressed structure, with elements subjected to post-tensioning after installation, introduces higher degrees of difficulty in the prefabrication of the concrete.

Certain elements of Kahn's structure undergo design modifications starting in the phase of specification of the system of joining of the parts; others are altered to facilitate their fabrication in molds. Every modification made by the Atlantic Prestressed Concrete Company is discussed with Kahn, the company technicians and the engineers involved in the worksite. On 7 January the representatives of Farrell and Atlantic, the engineers of Keast & Hood, Komendant and Kahn with his staff examine the technical drawings prepared by the Atlantic Prestressed Concrete Company – showing only the dimensions of the pieces without reinforcements – and a graphic diagram of their assembly, and discuss the positioning of the formwork for the pouring in place of the top slab ("formwork must be lower"), the joints between the columns and the brick sectors, and the configuration of the edge beam in terms of thermal insulation.[101] Also in January, Kahn's office requests two supplementary types of slender rectangular bands, or "fascias," of the same height but two different lengths, prefabricated to be inserted in the floor sectors that connect the towers of the laboratories to Tower X, at the points of contact between the floor and the columns, and therefore where the two types of concrete meet.[102] Those fascias serve to disguise the finish of the poured in place concrete, making it appear precast.

The edge beams are those that undergo the most transformation, due to the technical definition of the connections between the floor system members and the installation of the floor structure. In the passage from Kahn's drawings to those of the Atlantic Prestressed Concrete Company, and following the specifications of all the details of assembly on the worksite, a series of housings for the beams and for the formwork of the poured in place top slab appear on the inner face of the edge beams. The construction of these housings has a decisive impact on the type of mold and the working of the precast concrete.

One of the critical points of the assembly system, to the point of imposing limited tolerance margins in the prefabrication, is the node in which the holes to contain the post-tensioning bars of the two overlaid columns, the main beam and the edge beam converge. In all the other cases, the junction nodes between main and secondary beams and between secondary and edge beams are done with special recesses, with holes for bolting on the worksite. Considering all the variables, the margin of tolerance is considered acceptable and reabsorbable in the joints and recesses when it is about 3/8 inch of the length of the members, while in the case of the edge beams it is drastically reduced to obtain acceptable joints in the points of contact at the extremities of the beams themselves, along the corners of the towers. "Variations in these corners would have been very noticeable in the finished structure," the technicians of the Atlantic Prestressed Concrete Company write. The margins of tolerance of defects are also reduced in the fabrication and installation of the columns, because they represent the main visible and characteristic elements of the design of the towers in the view of the Richards Laboratories. "The columns," Atlantic states, "form a principal feature of the exterior architectural treatment, and any variation, apparent to the eye, would be objectionable."[103]

The complexity of the design of Kahn's structure is summed up in the definition, proposed by the Atlantic Prestressed Concrete Company, of the "true shape," namely that with the smallest possible margin of error in the prefabrication, taking all the variables into account, from the shrinkage of the poured concrete in the molds to prestressing on the worksite. Precisely in the context of the precasting, the question of the joint, in Kahn's structural design, takes on a particular value in the phase of fabrication at the factory and installation on the site, to the point of becoming, in the final work, both the symbol of perfection and the visible sign of the process of fabrication and assembly.

The construction of the molds and every operation on the compound, from its constitution to its treatment after setting, have to take all the worksite assembly phases into account.

Steel panels are used to make the molds, after the hypotheses formulated in 1958. The technicians of the Atlantic Prestressed Concrete Company, in any case, underline the possibility of obtaining perfectly smooth concrete surfaces with wooden molds coated in plastic film as well; but this type of mold cannot be used, due to certain processes of treatment of the concrete during setting desired by Kahn.[104]

[99] Louis I. Kahn, letter to Robert E. Fischer, 10 February 1959, LIK, 030.II.A.9.2.
[100] *Ibid.*
[101] "UMPRL. Job Meeting Wednesday, January 7, 1959," LIK, 030.II.A.25.6.
[102] See the drawing "Precast concrete fascias," 16 January 1959, LIK, 030.II.A.22.17.
[103] Atlantic Prestressed Concrete Company, "Draft," n.d., [1962], Auguste E. Komendant Archive, AAUP, Philadelphia.
[104] "Contrary to the opinions of many people," assert the technicians of the Atlantic Prestressed Concrete Company, "it has been our experience that carefully made wooden forms constructed using plastic coatings and facings in some instances can produce results in every way as pleasing as can be obtained with concrete or steel forms" (*ibid*).

182-186. Alfred Newton Richards Medical Research Building, Philadelphia, Pennsylvania, 1957-60.

182-185. Photographs taken in the factory of the Atlantic Prestressed Concrete Company, Trenton (LIK, 030.IV.A.490.14.20; 21; LIK, 030.V.A.10.5; LIK, 030.IV.A.11.2).

186. Construction site photograph (M. D. Meyers Archive, AAUP, Philadelphia, 136.67.41).

The choice of the steel panels opens up the economic question of amortization of investment costs. The Atlantic Prestressed Concrete Company imposes certain condition on the design proposed by Kahn for the individual parts: on the one hand, the company requests modifications to be able to reutilize the steel of the molds in the succession of the making of the various structural members of the Richards Laboratories, so that once all the edge beams are made, the panels of the molds can be adapted to the molds for the rectangular section of the secondary beams and columns; on the other, the design of the sides of the molds of the main beams is studied to be able to reconvert the panels "into standard universal bed side-forms such as are used for building bridge girders" and to configure the molds of the single special pieces as "pockets, seats, recesses," so they can stand up to the wear of use and remain available to the company after the construction is finished. Other elements needed for the molds are purchased from those on the market, again with an eye on later reutilization: "On the other side of the ledger, we decided to use somewhat lighter liner plate for the bulk of the forms than we would buy for regular plant use."[105] The construction of the molds is assigned to both a specialized company (probably for the complex molds needed for the edge beams) and a local firm usually contacted by Atlantic Prestressed Concrete Company to make metal molds.

For the compound, the Atlantic Prestressed Concrete Company uses ordinary Portland cement, identical to that used for the concrete poured in place. The concrete is supplied by the Lone Star Cement Corporation, with headquarters at 100 Park Avenue, New York.[106] The choice of making a quick setting compound for the precast concrete takes the construction schedule into account, with the aim of producing and delivering individual parts ten days after pouring in the mold.[107] The compound is made to be dense and free of water bubbles, and to produce "uniformity of color in all pieces" and the palest color possible, to achieve with ordinary cement.[108] The fact that it is specified that the pale color of the precast concrete should not be the result of the use of white cement would lead us to suppose, unless this was a purely economic issue, that Kahn wants to keep faith with a certain ordinary nature of the cement, in both mechanical and chromatic terms, so the Richards Laboratories will be the expression of just one cement, though worked in two different ways. In the future too, Kahn makes use of white cement only in grouting work and as a component of the mortar for the laying of bricks and blocks.

The desire to achieve a smooth surface finish, without cracks caused by shrinkage, and a pale color of the cement, suggests treatment of the prefabricated parts with vaporization. This is done when the concrete is still in the molds, and it corresponds to the usual curing phase – "moist curing" in this case. This process makes the choice of metal panels obligatory for the construction of the molds. "It would also ruin wooden formwork," the technicians of the Atlantic Prestressed Concrete Company explain, regarding vaporization.[109]

Kahn supplies precise indications of an aesthetic nature on the quality of the surfaces of the individual prefabricated parts, in relation to their visibility; these indications are used by the Atlantic Prestressed Concrete Company as the guiding principles in the fabrication of the various faces. Thus, according to Kahn, "all exterior surfaces of the building have formed finishes," while in the view from below of the floor structure, also inside the laboratories, all the inner faces of the beams must be smooth, that "underside of the members" which is considered the "most important from an appearance standpoint."[110] These indications form the basis, in part, for the construction of the molds and, above all, their positioning at the factory during the pouring phase.

The main problems raised by the aesthetic principles of Kahn have to do with the fabrication of the edge beams that must have a polish exposed face – therefore, for aesthetic reasons, to be made in direct contact with the mold – and the face towards the floor fitted with special housings for the main and secondary beams, the formwork of the top slab and the connections at the corners between the two edge beams. For technical reasons, this face too has to be made in direct contact with the mold. As troweling cannot be done on the face bearing these connections, the Atlantic Prestressed Concrete Company decides to create a mold containing the two different faces, one smooth and the other fitted for the connections, and to pour the compound with the mold not laid down but crosswise, thus also obeying Kahn's indication of the smooth exposed "underside," since the only side of the beam not shaped by contact with the mold becomes the upper one. The technical risk of pouring concrete inside a mold placed in such a way – due to the slender size of the beam and the presence of reinforcements – is the possible formation of large honeycombs. The main beams too, both the monolithic ones and those in three pieces, are made by placing the molds crosswise.

In the case of the columns the position of the mold is dictated by the desire to obtain a homogenous concrete surface. To this end, the mold is placed vertically and then reinforced. The pouring of the compound is from above, without the danger of segregation of aggregate and the formation of horizontal lines as happens with horizontal molds. The corners of each side of the mold are stiffened, overlaid and nailed to create a "rigid box."

But the vertical position of the mold required to achieve high quality for the concrete demands careful study on the part of the Atlantic Prestressed Concrete Company regarding the configuration of the column, in order to make a piece that will be easy to extract after setting. The technicians discard the hypothesis of using a mold with parts that can be opened, the "collapsible boxes," to be moved with a crane, as they fear it will not be possible to reposition the parts perfectly, to produce columns with exactly the same measurements. The intervention of Kahn and Komendant makes it possible to develop a closed mold, or "rigid box," from which to extract the column thanks to the special tapering introduced on the internal faces of the narrow flange of the column.

To obtain the cavities for the steel bars to be installed on the worksite, in the mold of the columns flexible plastic tubes are inserted, stiffened inside by steel tubes and wrapped in a light reinforcement, again made of steel.

After pouring, setting and vaporization curing, a crane is used to vertically extract the column from the mold, thanks to two rings attached at its top. But for storage, since it cannot be kept upright due to the stepped configuration of the base and the rings at the top, the column is first placed on

[105] *Ibid.* The molds are made both by a specialized company and a company working with Atlantic.
[106] *Cf. Concrete for Kahn's 'ivory towers,'* in, "Architectural Forum," vol. CXIII, July 1960, No. 1, advertisement.
[107] "Report on Inspection of Medical Research Building University of Pennsylvania," n.d., [August 1959], LIK, 030.II.A.25.19. The engineers of the company working on the structure are Sandy Smythe and Harald Laupa.
[108] Atlantic Prestressed Concrete Company, "Draft" cit.
[109] *Ibid.*
[110] *Ibid.*

187-188. Alfred Newton Richards Medical Research Building, Philadelphia, Pennsylvania, 1957-60.

187. Construction site photograph (*8-Story Rigid Frame Structure Assembled from Precast Members*, in, "Architectural Record," September 1959, No. 126, p. 232).

188. Construction site photograph (LIK, 030.IV.A.490.11.1).

a pile of white sand and then tied with ropes to be raised and stacked horizontally, one column atop the next.

To facilitate setting all the structural members are treated with vaporization. The heaviest prefabricated parts reach weights of 19 and 22 tons. Certain techniques are required to avoid chipping of the corners. "When the columns are handled," Wisdom writes, "they are set on site in piles of sand [...]. In shipping, members are set on plywood a little narrower than the concrete."[111]

The program is developed to optimize the fabrication cycle of the individual parts. To keep daily production constant, the Atlantic Prestressed Concrete Company takes several steps: it readies "additional pilot liners [...], by fleeting sideforms," to use for the production of edge beams and the many secondary beams;[112] and it installs an "entire universal bed" on which to pour either six prestressed beams or eighteen post-tensioned beams (the two types are thus produced in pairs of alternating weeks).

[111] David P. Wisdom, "Job Meeting Wendesday May 20, 1959," LIK, 030.II.A.25.6.

[112] Atlantic Prestressed Concrete Company, "Draft" cit.

189. Construction site photograph (M. D. Meyers Archive, AAUP, Philadelphia, 136.67.62).

After the definition of the details and the various discussions of Kahn with Komendant, the engineers of Keast & Hood and the technicians of the contractors to specify the working modes of all the phases, from construction to transport to installation, production of the pieces of precast concrete begins at the start of March 1959. The pieces are made in Trenton, at the Atlantic Prestressed Concrete Company plant, about 35 miles away from the worksite in Philadelphia. The process calls for 14 workmen, a welder, a foreman and 8 workers specialized in the fabrication of the reinforcements of the concrete of all the elements. The production is divided into two sectors, "precast and prestressed areas," one for the prefabricated parts with conventional reinforcement – the edge beams and secondary beams – where the 14 workmen operate, and the other for the prestressed and post-tensioned parts (the "universal bed"), where precisely due to the complexity of the reinforcement, the 8 specialists of steel bars work, also overseeing the pouring of the concrete for those parts. The plant is capable of producing two edge beams, four secondary beams and two columns per day. The definitive delivery of all the prefabricated parts is slated for 1

190-194. Alfred Newton Richards Medical Research Building, Philadelphia, Pennsylvania, 1957-60.

190. Construction site photograph (*8-Story Rigid Frame Structure Assembled from Precast Members*, in, "Architectural Record," September 1959, No. 126, p. 233).

191. Construction site photograph (M. D. Meyers Archive, AAUP, 136.67.51).

192. Construction site photograph (LIK, 030.IV.A.490.11.3).

193-194. Construction site photographs (A. E. Komendant Archive, AAUP, 027.III.32.17a; 17).

July,[113] but their assembly begins when about 60% of the members have been produced. The Atlantic Prestressed Concrete Company assigns the assembly work to Cornell & Company, Inc., based in Westville; but Atlantic handles the finishing – "drypacking and pointing of joints and grouting of the ducts in which the post-tensioned bars were encased."[114] To test the prefabricated structure, Kahn has a prototype built.[115]

A visit by Leidigh to the workshop in Trenton at the beginning of March allows him to observe the first defects in the making of the precast pieces. Leidigh decides to discard the columns just molded because he realizes, from the quality of the color and surface, that the treatment indicated in the specifications of vaporization or hot water in the curing phase has not been applied. "Please advise us," he writes to the Atlantic Prestressed Concrete Company, "as to what steps you propose to take to cure the precast concrete members to meet the requirements of the specifications. We do not want to appear unreasonable in this matter, but, in our opinion, the curing specified is the only way uniform and acceptable finishes can be required."[116]

The technicians of the company and Komendant meet on Wednesday 20 May to discuss the phases of assembly of the precast pieces. Due to certain problems of production of the precast concrete, the Atlantic Prestressed Concrete Company proposes starting assembly of the parts on the site first on Wednesday 27 May, then on 1 June 1959;[117] but Farrell and the other technicians set the date for Tuesday 26 May. It is planned that on 21 and 22 May, and on Monday the 25th, the equipment required for assembly should be brought to the worksite, and that on the evening of Monday the Atlantic Prestressed Concrete Company will load the first precast members onto special trucks for transport to the worksite the following morning, along a pre-set route due to the special cargo and exceptional length of the vehicle.

There are conflicting opinions on the number of floors that can be assembled every two weeks – three, according to Komendant, or four according to Tate of Cornell & Company. As of this date the total number of weeks set aside for the completion of the assembly is 7 or 8. The number of floors that can be assembled every two weeks determines the logic of the progression of the assembly of the three towers; therefore it is decided to assemble the precast pieces of a floor system in succession, from one tower to the next, to permit the operations of pouring of the floor. Joseph R. Farrell, represented by Morrison, guarantees that in four hours from the completion of the assembly of the framework of a floor system, it will be able to put in place the formwork for the top slab, and that two days are required to install the sheaths and then to pour the concrete.

By 20 May it is clear that the Atlantic Prestressed Concrete Company, out of a total of 92 edge beams, is making twelve, of which two will be ready that same day. Seven days prior to the beginning of assembly on site, the production of two of these beams per day, as planned, has not been achieved.

[113] Wisdom, "Job Meeting Wednesday May 20, 1959" cit.
[114] Atlantic Prestressed Concrete Company, "Draft," n.d., [1962], Auguste E. Komendant Archive, AAUP, Philadelphia.
[115] "Meeting at Job Site – Wednesday, June 3, 1959," LIK, 030.II.A.25.6.
[116] Thomas J. Leidigh, letter to Atlantic Prestressed Concrete Co., 9 March 1959, LIK, 030.II.A.25.19.
[117] Farrell, instead, proposes a date, 26 June. The time estimated for the erection of the structure is 8 weeks (Wisdom, "Job Meeting Wednesday May 20,

That same day, 18 of the short main beams – the post-tensioned ones – are ready, while six are about to be removed from their molds. Given the fact that six days of each week are devoted to the production of the precast concrete, the company trusts it can gain time thanks to the fact that the assembly procedures at the worksite will be carried out for only five days a week; pauses in the work could also be caused by rainy days. According to what has been decided, only six and a half weeks remain for the delivery of all the columns; but the Atlantic Prestressed Concrete Company requests postponement of the final delivery deadline. The technicians agree that eight weeks will be needed for the assembly of all the floors systems of the three towers.

During transport by truck, the precast pieces have to be laid down on strips of plywood to prevent chipping of the corners. When they reach the worksite, the plan is to store the columns "in piles of sand to prevent edge chipping," and to lie down other members on "fire hose." It is recommended to eliminate any rust deposits on the steel plates of the post-tensioning cables prior to their use, to avoid marking the concrete.[118]

As soon as the columns reach the worksite and assembly begins on 26 May, Kahn decides, on 28 May, to eliminate the granite cladding of their bases as envisioned in the project (other granite panels are also eliminated).[119] Unexpected events prevent the work from proceeding precisely as discussed and planned by the various technicians. Problems are caused by the need to safeguard neighboring buildings and trees. "The site of the work was such as to provoke some sense of frustration," is the telling comment of the Atlantic Prestressed Concrete Company.[120] When the long trucks arrive, it becomes clear that their movements for unloading and stocking of the precast pieces will cause difficulties. The space set aside for the storage is not sufficient, which often means that the longer pieces must be assembled in place by taking them directly from the trucks.[121] All these unexpected issues lead Kahn to imagine all the phases of fabrication, including assembly, for future works in prefabricated concrete.

To set the cycle of deployment of the members, the Atlantic Prestressed Concrete Company relies on the discussions of 20 May on the timing of the pouring of the top slabs and the operations of positioning of the "post-tensioned bars." The sequence of assembly starts with Tower A and proceeds from Tower B to C[122] (in June the first two levels of Towers B and C are under construction).

"The sequence," writes the Atlantic Prestressed Concrete Company, "finally determined as best was based on working two towers at a time. One tower was erected until the post-tension bar sequence was reached. Then erection of the second tower proceeded during the interval. As soon as possible, the erection of floor of the first tower was completed and work began on the third. The third and second towers were worked on one at a time, with work transferring from one to the other to accomplish the grouting without holding up the work on either. Once this procedure was instituted the work proceeded smoothly with the general contractor being able to

1959" cit.).
[118] Wisdom, "Job Meeting Wednesday May 20, 1959" cit.
[119] David P. Wisdom, letter to Joseph R. Farrell, Inc., 28 May 1959, LIK, 030.II.A.25.23.
[120] Atlantic Prestressed Concrete Company, "Draft" cit.
[121] The maneuvers for the installation are in any case complicated ("Report on Inspection of Medical Research Building University of Pennsylvania," n.d.,[August 1959], LIK, 030.II.A.25.19).
[122] Wisdom, "Job Meeting Wednesday May 20, 1959" cit.

195. Alfred Newton Richards Medical Research Building, Philadelphia, Pennsylvania, 1957-60. Construction site photograph (M. D. Meyers Archive, AAUP, 136.99.a1).

pour his floors without interference in whatever tower was not being erected on that day."[123]

The operation of assembly of the pieces is done with a mobile crane. Kahn's comment on the functioning of the crane provides proof of the importance he attributes to the construction process and its recognizability.

"One day," he recalls, "I visited the site during the erection of the prefabricated parts of the building. The crane's 200' boom picked up 25-ton members and swung them into place like match sticks moved by the hand. I resented the garishly painted crane, this monster which humiliated my building to be out of scale. I watched the crane go through its many movements all the time calculating how many more days this 'thing' was to dominate the site and building before a flattering photograph of the building could be made. Now I am glad of this experience because it made me aware of the meaning of the crane in design, for it is merely the extension of the arm like a hammer. Now I began to think of members 100 tons in weight lifted by bigger cranes. The great members would be only the parts of a composite column with joints like sculpture in gold and porcelain and harboring rooms on various levels paved in marble."[124]

The individual elements are raised by means of special anchors and deployed in keeping with the pre-set sequence.[125] For the column, at the factory the passage from the vertical position of the pouring to the horizontal position of storage happened by means of piles of sand. Now, to move from the horizontal position of truck transport to the vertical position of its installation by the crane, a mechanical device is used that makes it possible to incline the column until it has reached the desired position. The presence of workmen during the maneuver is in any case fundamental for the exact positioning of one column over another.

"Another interesting device," the Atlantic Prestressed Concrete Company explains, "was the column plumbing jig. This consisted of a steel frame clamped to the column to which could be attached either a jack post or a come-along. With this jig, it was possible to rotate, translate or tilt the column a small amount. A separate force of two men carefully aligned each column while the main erection work was going on elsewhere, and this system contributed to the excellent control of dimensions maintained throughout the erection work."[126]

For the positioning of the columns and post-tensioning bars, slender aerial bridges are mounted with the crane, with corbels against the columns. In the assembly, priority is placed on the exact geometric arrangement of the columns and edge beams, which are the most visible parts of the structural system. For the post-tensioning of the bars, the workmen also climb up to the top of the columns, in worksite situations similar to those of the assembly of metal girders. For the positioning of the column of segments of the main beam, at least one worker, in certain cases, remains seated on the column to make sure the bars match with the holes. Given the fact that the post-tensioning bars are long and stiff, the presence of Tower X and the John Morgan Building nearly touching Tower A influences the

[123] Atlantic Prestressed Concrete Company, "Draft" cit.
[124] Louis I. Kahn, *Voice of Africa*, LIK, box 67.
[125] For the description of the assembly of the structure see, Robert M. Price, "Report on Inspection of Medical Research Building University of Pennsylvania," n.d., [August 1959], LIK, 030.II.A.25.19.
[126] Atlantic Prestressed Concrete Company, "Draft" cit.

framework of the main beams – those with three segments cannot have their ends near the existing walls, as they must have room for the long bars to enter after they have been put into position. The assembly of the long main beams is also influenced by the position of the crane, which in some cases has to avoid existing neighboring buildings. Therefore in certain towers the orientation of the framework of the main beams varies, also from floor to floor, based on the maneuvers to be done on the worksite with the bars and the crane. Wooden formwork is used for the pouring of the top slabs. The panels mounted in the single compartments of the floor structure are reinforced at the center by means of a metal rod.

During the first week of work, the whole precast structure of the first floor of Tower A is erected. On 3 June, Kahn, Komendant, Wisdom, Leidigh, Farrell, Morrison, Smythe, Sargent Laupa, Gordon and Tate visit the sites to observe the quality of the work – both that done in Trenton and the work done on worksite. On this occasion, certain defects in the making of the precast concrete are observed: the hand finishing of the exposed sides of the

secondary beams are not sufficiently clean and uniform; some surfaces made in contact with the molds have "blisters," whose cause is identified as "premature steaming"; finally, some pieces are found with corners chipped or broken during handling, transport or installation (some pieces have emerged with such defects straight from the molds). So steps are taken regarding both the work of precasting done by the Atlantic Prestressed Concrete Company, and the assembly done by Cornell & Company: all the prefabricated parts have to be "as uniform as possible in color and texture";[127] it is decided to clean and smooth the pieces at the factory and to apply a layer of silicone, probably to ensure qualities of resistance to water, instead of the "moist curing"; the installation of damaged precast parts is forbidden (therefore the flawed pieces already stored at the worksite must be discarded); in the case of the columns and slab already under construction, it is decided to repair the observed defects with grouting and to treat all the surfaces with cleaning and silicone. Where the sealing of the joints is concerned, which has not yet been done at the time of the visit, it is indicated that in the corners of the edge beams and between the columns the joints must be done according to the prototype; instead, between the main and secondary beams they are to be done with a "1/4 inch setback, square joint" except for the "bolt joints" (the indication explains, "recesses for bolts to be filled flush"[128]).

A photograph of the truck with the edge beam taken at the Trenton factory shows a worker correcting a corner that was perhaps chipped during movement of the piece. Every further operation of repair in Trenton, after the shipping of the pieces of the first floor of Tower A, happens in revised ways. Probably as a result of the defects observed during the visit to the first sector of the precast structure and the subsequent discussion on improved procedures at the workshop to correct fabrication defects, it is decided to proceed with a second vaporization operation in Trenton, rather than following the indications laid out on 3 June, involving the treatment of the surfaces with silicone.

196-197. Alfred Newton Richards Medical Research Building, Philadelphia, Pennsylvania, 1957-1960. Construction site photographs (LIK, 030.IV.A.490.12.49; 52).

Thus the steaming of the elements designed for the Richards Laboratories, like all the other works, from the molds to the deployment of the bars, becomes a special operation precisely due to the aesthetic conditions imposed by Kahn. After the first vaporization during the curing, after removal of the pieces from the molds, the members are grouted where defects are found, and these repairs are then disguised by the second phase of vaporization. Precisely this second treatment of the concrete has fundamental consequences for the definitive color of the pieces, producing that very pale concrete made without the use of white cement. "It also whitened the color somewhat further," is the observation of the Atlantic Prestressed Concrete Company regarding the effects of the second vaporization.[129] The photographs from the period, which show a clearly darker color for the first series of columns of the portico, confirm the chromatic results of the second vaporization applied after the first shipment of precast parts.[130] It is not known whether Kahn approved of the final pale color of the precast concrete. What is certain is that the color is the mechanical result of the process of prefabrication at the workshop, unavoidable to eliminate perception of the repair of defects. It is worth noting, given Kahn's focus on the neighboring constructions in the Jacobean Style, that the concrete lightened by the new procedures perfectly matches the color of the stones used by Cope and Stewardson.

On 16 June, over a span of three weeks, only three floors have been assembled – one in Tower A and two in Tower B. At this rate assembly of all the towers will take until October, at least three months behind schedule, and it is pointed out that the arrival of the cold season could lead to further delays.[131] When the assembly of the precast parts is in an advanced phase, strength testing is done on one floor of the lower levels to test the solidity of the structure, in keeping with the requests of the City of Philadelphia, possibly using bricks on the worksite as a test load.[132]

Following the inspection on 3 June of the floor of Tower A, and after the observation of the defects, during the early days of the month Kahn studies the possibility of a colored ceiling to apply in the compartments between the beams of Tower B, near the light ducts, to obtain a luminous multicolored ceiling in the portico. For the pattern of the slender squared sheets (sky blue or red) arranged in staggered bands like masonry and suspended by means of a frame, also colored (green),[133] Kahn observes the work of Saarinen – the name is written down on a sheet of sketches for the ceiling pattern[134] – and, in particular, the General Motors Technical Center in Warren, Michigan, with its luminous ceilings of plexiglas panels and the play of colors for the walls in glazed brick (the word "bricks" is also seen on the same sheet).

Kahn's study of the portico ceiling is of the greatest importance because it demonstrates the pursuit of a chromatic articulation of the compartments of the floor system, and a desire to conceal the imprints of the boards on the concrete top slab. That ceiling, had it been built, would have played a role similar to that of the tapestries applied on the concrete in the church in Rochester.

[127] "Meeting at Job Site – Wednesday, June 3, 1959," LIK, 030.II.A.25.6.
[128] *Ibid.*
[129] Atlantic Prestressed Concrete Company, "Draft" cit.
[130] See the worksite photographs dated 27 August 1959, LIK, 030.IV.A.490.12.42.
[131] Joseph E. Farrell, letter to the Atlantic Prestressed Concrete Company, 16 June 1959, LIK, 030.II.A.25.23.
[132] David P. Wisdom, letter to Joseph E. Farrell Inc., 15 July 1959, LIK, 030.II.A.25.23.
[133] See the drawing "Revised tile and lighting pattern, Tower B 1st Fl. reflected ceiling," 29 June 1959, LIK, 030.I.C.490.004.
[134] See the study drawing, n.d., LIK, 030.I.C.490.004.

198-200. Alfred Newton Richards Medical Research Building, Philadelphia, Pennsylvania, 1957-60. Construction site photographs (LIK, 030.V.A.6.15; LIK, 030.IV.A.490.12.47; LIK, 030.IV.A.490.14.10).

The installation of all the precast parts of the three towers is completed at the end of August. That same month, an inspection reveals that certain joints between the "secondary pieces" are chipped and irregular, and that the surfaces of other joints already patched seems "rough" as opposed to the "normal smooth finish of the precast surfaces"; furthermore, other parts are stained. "The ragged appearance of these joints is to be corrected," "this roughness is to be eliminated," "additional effort is to be made to remove these markings" are the orders issued by Leidigh to the contractor to correct the defects.[135] It is clear that the precast concrete has introduced surface values that cannot permit the visibility of those imperfections that, instead, are typical of poured in place concrete.

Once the precast structure has been completed, the volumes of the staircases and conduits, or the "peripheral utility towers,"[136] are made with concrete poured in place. As in the case of Tower X, here too the system of Ganged Panel Forms is used. For the application of the brick cladding

the same principle is used of suspended aerial scaffolding at the top, as in Tower X.

The installation of the network of physical plant systems completes the graphic and chromatic construction of Kahn's concrete. The attempt to introduce polychrome effects in the portico is pursued in the choice of the colors for the pipes in the cavities of the floors. Like Le Corbusier, Kahn too wants to identify the function of each tube thanks to a color coding system: red for gas, blue for air, white for vacuum, yellow for hot water, green for cold water, colorless for waste, violet for radioactive waste, orange for steam.[137]

At the end of the summer of 1959, Kahn visits the south of France, headed for Otterlo, where from 7 to 15 September the XI Congrès International d'Architecture Moderne (CIAM) will be held. The sketches of Carcassonne are made during this trip, also in an attempt to seek confirmation for the Richards Laboratories, from the arches of different radii of the bridge, drawn repeatedly by Kahn perhaps because he recognizes in them the theme of the preliminary studies of his floor system, to the dripstones inserted in the merlons, which are like those on the back of Tower X, to the crenellations themselves, powerfully depicted with a calculated off-scale effect to point to the affinities with the large towers of his work. It is the revelation that unfolds before his eyes at Carcassonne of the unexpected parameter of historicism that has taken form in his Richards Laboratories through the process of definition of the constituent organs of that complex contraption.

The Richards Laboratories are opened on 19 May 1960. The specialized press immediately reports on the qualities of the construction, the different natures of the concrete and the textures of bricks and blocks in the formal definition of a work that goes beyond any form of Brutalism.[138]

[135] Thomas J. Leidigh, "Inspection Memo," 26 August 1959, LIK, 030.II.A.25.19.

[136] Jordy, *Medical Research Building for Pennsylvania University* cit., p. 99.

[137] David P. Wisdom, letter to Robert McCray, 9 May 1960, LIK, 030.II.A.25.13 – the same letter is also dated 24 June 1969. See also David P. Wisdom, letter to Robert McCray, 9 May 1960, LIK, 030.II.A.25.20. All these pieces are produced by the W. H. Brady Company.

[138] "Burly as this skeleton is, there is no 'new brutalism' affection of muscle-flexing, nor are there the pseudo-Corbusian, wilfully crude surface and junctures. Nor is there a suggestion of the current Japanese mode of consciously archaic complication and exaggerated bulk, although its disassembled components resemble nothing so much as the gargantuan notched and seated puzzle which goes together to make the traditional Japanese house" (Jordy, *Medical Research Building for Pennsylvania University* cit., p. 104).

201-206. Alfred Newton Richards Medical Research Building, Philadelphia, Pennsylvania, 1957-60.

201-202. Details of a service tower (photographs by the author, 2010).

203. Detail of a laboratories tower (photograph by the author, 2010).

A comparison of photographs of two details of the concrete, one precast, the other poured in place, serves to underline "the clear visual distinction between the two concrete constructions used."[139] After what Banham had written about the Yale Art Gallery, the Richards Laboratories become an example of the distance between Kahn's concrete and international Brutalism. "Much of his recent work, including this building, has been linked to the movement called *the new Brutalism*. Such a suggestion Kahn brushes impatiently aside. For him the term connotes an attitude, an intellectual posture external to the act of designing and consciously applied to it, like a coat of paint."[140]

But only Jordy conveys the image not only of the just opened work, but also of the worksite and the fundamental phases of the making and assembly of the reinforced concrete, precast and poured in place, of the visible graphic contrasts of the two types of concrete, contrasts that are reminders of the lesson of Le Corbusier. It is significant that at the outset of the scenes showing the worksite of the Richards Laboratories, Jordy calls

into play the person who more than others had contributed to inscribe, for centuries to come, the apologia of construction and its details, an artist beloved to Kahn and shown here in the unexpected guise of a spectator delighted with the construction process staged in Philadelphia.

"Piranesi would have relished such a building spectacle," Jordy write, "the cavernous poured concrete core, against which the crane lifted the giant sculptural components of the frame; then the climb of the wooden forms for the peripheral pours outside the frame. And this sense of the building process remains in the completed complex. It is especially evident in the contrast of smooth and rough concrete surfaces on the interior of the building. Here the exposed sections of the poured work reveal the impress of wooden forms and even the holes where the forms were tied into the pouring. On the other hand, the skeletal components reveal the smooth surfaces, sharp edges and precise tolerances possible only with metal forms, steam curing and factory control. The two types of concrete, in turn, contrast with the unfinished concrete block of the interior partitions (except where a few adamant scientists have demanded their painting), while the purplish-red brick of the exterior shows through the windows. The brick exterior is partly a cavity wall, especially around the animal tower, but mostly a harmonizing veneer which leaves to the interior the fullest revelation of its construction."[141]

[139] Joseph W. Molitor, *Art serves science. Alfred Newton Richards Medical Research Building [...]*, in, "Architectural Record," August 1960, (pp. 149-156), p. 156.
[140] James Marston, *A building of rugged fundamentals*, in, "Architectural Forum," vol. CXIII, July 1960, No. 1, (pp. 83-86, 185), p. 86. The definition of archaic is echoed in *'Arcaismo' tecnologico*, in, "L'architettura," vol. VI, October 1960, (pp. 410-411), p. 411.
[141] Jordy, *Medical Research Building for Pennsylvania University* cit., p. 104.

204. Detail of the porch (photograph by the author, 2010).

205. Detail of the porch. On the bottom of the column are visible the pracast concrete fascias (photograph by the author, 2010).

206. Study drawing for precast concrete fascias, 16 January 1959 (LIK, 030.II.A.22.17).

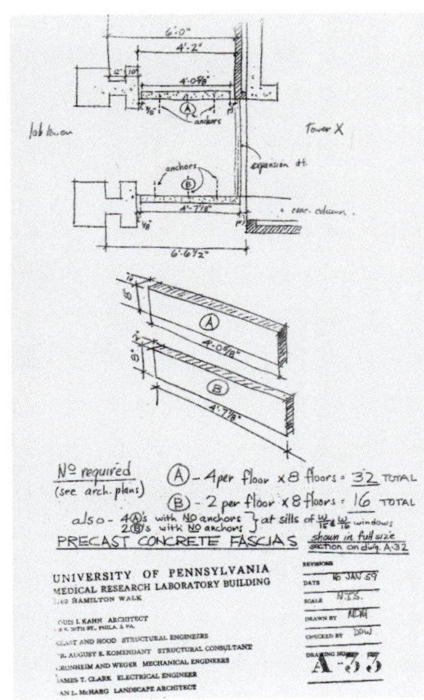

207-211. Alfred Newton Richards Medical Research Building, Philadelphia, Pennsylvania, 1957-60.

207. Porch (photograph by the author, 2010).

208. Detail of the floor structure of the porch (photograph by the author, 2010).

209. Interior view (LIK, 030.IV.A.490.3.1).

210-211. Details of the staircase (photographs by the author, 2010).

Crevices, cracks and repairs of the surfaces of the Richards Medical Research Laboratories

The opening of the facility does not put an end to the work on the Richards Laboratories. At the same time, Kahn is continuing with construction along the Hamilton Walk, with the Greenhouse Service Building and the Biology Building. Normal cleaning of the Richards Laboratories is done in 1961, when the order is given to remove the nails of the formwork from the concrete surfaces poured in place.[142] But the most urgent and difficult problems come from a series of cracks already visible at the start of 1961 in the brick claddings, caused by different degrees of thermal expansion between the brick and the underlying support.

Leidigh proposes the creation of "caulked joints" to facilitate movement of the different materials,[143] though he is aware that such an intervention would alter Kahn's calculated design of the joints, commissures and textures of all kinds. "Lou will certainly have something to say about caulk joints because they affect the appearance of the tower," it is warned.[144] A few days later, Kahn's office makes the project of the restoration of the brick claddings with the "control joints" suggested by Keast & Hood.[145] Another type of defect appears in the openings made in the walls for the passage of the conduits, where the blocks are chipped and the openings are too large with respect to the piping – all consequences of Kahn's radical choice of leaving the materials nude.

At the end of August, United Engineers & Constructors, Inc., called in by the technicians of the University of Pennsylvania, prepares a detailed report on the defects found in the blocks and bricks of the walls, on the concrete of the top slabs poured in place and the prefabricated concrete of the columns,[146] and proposes techniques and materials for the restoration of the surfaces: fill-

[142] "University of Pennsylvania. Service Laboratory Building," 9 June 1961, LIK, 030.II.A.25.8.
[143] Thomas J. Leidigh, letter to Louis I. Kahn, 25 January 1961, LIK, 030.II.A.25.19.
[144] David P. Wisdom, letter to Thomas J. Leidigh, 23 February 1961, LIK, 030.II.A.25.19.
[145] "Control joints [...] as suggested on Louis I. Kahn's drawing A-49, revised March 8, 1961" (John Dallas Jr., "Report of Structural Observations," received by the Kahn office on 31 August 1961, LIK, 030.II.A.22.12).
[146] John Dallas Jr., "Report of Structural Observations," received by Kahn office on 31 August 1961, LIK, 030.II.A.22.12.

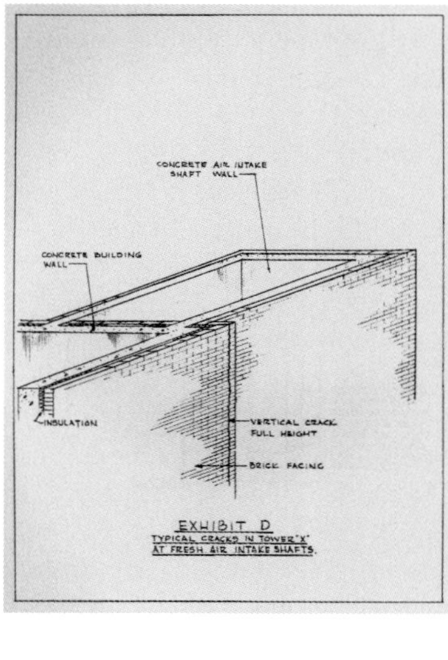

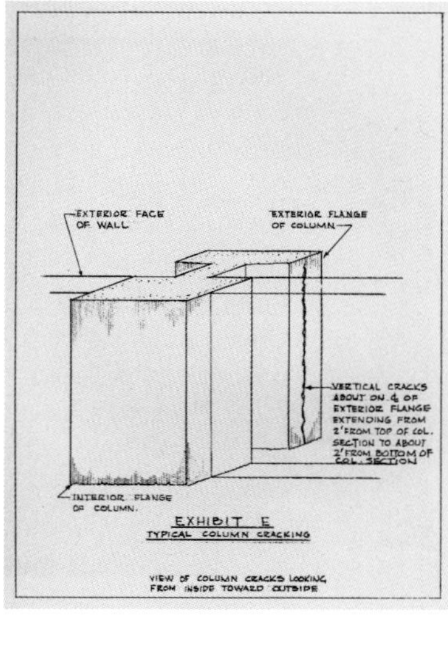

212-215. Alfred Newton Richards Medical Research Building, Philadelphia, Pennsylvania, 1957-60.

212. Photograph of the construction site (LIK, 030.V.B.29.12).

213. Elevator tower of the Tower X (photograph by the author, 2010).

214-215. "Typical Cracks in Tower X" and "View of Column Cracks" (J. Dallas, Jr., "Report of Structural Observations," 31 august 1961, LIK, 030.II.A.22.12).

ing the crevices in the walls, floors and columns with "transparent synthetic rubber coating such as DuPont 'Hypalon' or other similar product";[147] preventive covering of all the concrete surfaces of the floors structures to avoid water seepage and to make the cracks less visible; and elimination of the cracked masonry, reconstructing the courses in keeping with the "control joints" approved by Kahn, then sealing the new joints with Thiokol Sealant.[148]

The analysis and proposed remedies of United Engineers & Constructors, Inc., are discussed during the month of September.[149] While the control joints at the position of the cracks in the brick walls are accepted by Kahn,[150] doubts are raised regarding the modes of intervention on the concrete surfaces, due to fears of definitive damage. For this reason, Keast & Hood advises sealing the cracks in the pillars not with Hypalon but with "clear silicone."[151] Kahn expresses his view on the proposals of the technicians called in by the University of Pennsylvania: "I do not approve of the coating of the edges of columns on the Richards Building as specified by United Engineers & Constructors. The cracks are too small to clean. They should be carefully filled with a crystallizing silicone or equal. Dr. Komendant has great experience in this isolated repair, rather than trying to disguise it."[152]

While the repair of the cracked masonry sectors is already under way in the fall with the addition of the control joints, it is decided not to proceed with that of the columns prior to testing both with "flexible coating," with a Hypalon or Urethane base, and with "silicone waterproofing" to ascertain "what effect the seal coating would have on the appearance of the concrete."[153] The technicians suggest the use of "flexible coating" with which to obtain the best protection of the precast concrete, but with that compound the intervention turns out to be visible, also altering the "sheen" of the surfaces. Instead, the tests with silicone-base products turn out to be "virtually invisible." "We hesitate to recommend the use of any material which would tend to alter the appearance of the concrete by the addition of a color variation or a sheen which would result in a somewhat reflective surface," is the significant comment of Keast & Hood.[154]

Given the characteristics of the "clear silicone" it is advised to protect the surfaces of the precast concrete exposed to weather of the Richards Laboratories by means of application of a layer of "waterproofing compound containing not less then five per cent solids, similar to 'Cecosil' as manufactured by the Cheeseman [sic: Cheesman]-Elliot Company, Inc."[155] This type of material also meets with the approval of the client.[156] The application of the Cecosil comes after grouting of the holes and cracks of the concrete with cement-base mortar.[157] Treatments with silicone are also recommended for the brick walls.[158]

The repairs are done by United Engineers & Constructors, Inc., under the supervision of Komendant. But the process of alteration of certain surfaces of the precast concrete is still in progress in 1962. A commission formed by Komendant, engineers of Keast & Hood and technicians of United Engineers & Constructors, Inc., observes, during an inspection on 6 April, the presence of marks and "fine cracks" in the columns. "Most of these," they report regarding the fine cracks, "were characterized as 'dead,' they had already sealed themselves with free lime. Others need filling. Two dark stains have appeared. Some rust spots have appeared in column faces and some on spandrel beams."[159]

To eliminate the defects without modifying the chromatic quality of the surface of the precast concrete, cleaning of the marks with a wet solution of hydrochloric acid is suggested, followed by rinsing with "clear water" and then treatment of the surface with water containing lime capable of neutralizing the acid (the lime also ensures repair of small cracks). To eliminate rust marks, a "cold chisel" is suggested (specifying that the marks should be removed down to a depth of about ¼ inch). After all these operations, two waterproofing layers of "'Cecosil' silicone material" are recommended for the columns, followed by hand filling of the larger cracks with a compound formed of 75% white Portland cement and 25% gray Portland cement (the compound is first spread on the palm of the hand, and then applied to the crack with the thumb). Finally, once the repairs have dried, one more coat of Cecosil is urged.[160]

Other important interventions are those needed to replace the coverings of certain exposed reinforcements. Again in this case, the advice insists on the need to avoid altering the chromatic continuity of the concrete surface. To this end, it is recommended that the repairs be entrusted to a "concrete finishing specialist workman," like the specialists that work for the Joseph E. Farrell, Inc.

[147] *Ibid.*
[148] *Ibid.*
[149] Already in a letter of 11 September, Keast & Hood expresses perplexities regarding certain contents of the report (Keast & Hood, letter to Louis I. Kahn, 11 September 1961, LIK, 030.II.A.22.12).
[150] "Meeting held in Mr. Pemberton's office," 28 September 1961, LIK, 030.II.A.22.12.
[151] Keast & Hood, letter to Louis I. Kahn, 11 September 1961, LIK, 030.II.A.22.12.
[152] Louis I. Kahn, letter to Office of the Construction Engineer, 13 October 1961, LIK, 030.II.A.22.12.
[153] John Dallas Jr., letter to George H. Turner, 8 December 1961, LIK, 030.II.A.22.12. Tests are also conducted on the absorption capacity of the precast concrete parts and the bricks. In both cases the surfaces appear extremely porous, therefore proving that no type of insulating material has been applied to protect the materials during construction.
[154] *Ibid.*
[155] *Ibid.*
[156] David P. Wisdom, letter to August E. Komendant, 13 March 1962, LIK, 030.II.A.25.11.
[157] John Dallas Jr., letter to George H. Turner, 8 December 1961, LIK, 030.II.A.22.12. Information on the method and equipment for the sealing of the cracks is requested from Kleinfelder and John B. Kelly of Brick Work, Inc. The work is supervised by United Engineers & Constructors, Inc., under the guidance of Komendant (David P. Wisdom, letter to August E. Komendant, 13 March 1962, LIK, 030.II.A.25.11).
[158] John Dallas Jr., letter to George H. Turner, 8 December 1961, LIK, 030.II.A.22.12.
[159] David P. Wisdom, letter to August E. Komendant, 10 April 1962, LIK, 030.II.A.25.11.
[160] *Ibid.*

chapter four archaic concrete pattern

Precast stones for the Jewish Community Building

Certain works by Kahn designed in 1958-59 outline a true system of expression of concrete in the form of prefabricated blocks laid in masonry sectors, as opposed to structures in reinforced concrete, often designed to be prefabricated and assembled: the later projects for the Jewish Community Building, the Tribune Review Building in Greensburg, and the United States Consulate Buildings in Luanda, Angola. Emerging since the early 1900s as an alternative to the skeleton, also capable of imitating stone constructions, concrete blocks had not yet met with recognition in the catalogue of avant-garde construction materials, as constructive parts endowed with their own particular, crude expressive force. In this perspective, the Yale Art Gallery and the Bath House are simply preludes to the breakthrough Kahn makes precisely in the works designed at the end of the 1950s. In this period, also thanks to the extension of the expressive range of concrete made possible precisely by those blocks, he invents a personal vision of *béton brut*, by now

216. Community Building, Jewish Community Center, Trenton, New Jersey, 1954-59. Model (LIK, 030.V.A.2.6).

totally independent with respect to the more or less recognizable international variants that can be traced back to Le Corbusier. The choice of including blocks in the range of concrete for the construction of walls, no longer just for service areas, also has an impact on the developments of *béton brut* in Europe, starting with the construction of the Municipal Orphanage in Amsterdam by Aldo van Eyck, from 1958 to 1960.

The project for the Jewish Community Building in Trenton continues to develop during the construction of the Richards Laboratories. In 1958 Kahn further modifies the design, precisely in the light of his experiences with the Richards Laboratories. Now a large room, the Auditorium, becomes the counterpart of the gymnasium, in place of the Social Hall. The vertical structures of the Auditorium and the gymnasium become colossal paired protruding pillars. In some drawings Kahn depicts the faces of the pillars with vertical stripes, like the imprints of the boards of formwork left on poured in place concrete (the first columns of the Richards Laboratories are to be made on site, with formwork with vertical boards). The beams interlock at the top of the pillars. The section of the beams is designed to have two flanges that form the housing for the end of the pillars at the base – a carpentry detail. A lengthened, slender core and a flange, also thin and large, generate the unusual T-shape of the beam seen against the sky. There is little doubt that this section comes directly from the singular one created in 1957 for the columns of the Richards Laboratories, which precisely in 1958 is modified by Kahn for that building and instead used for the roof of the Community Building. Like those of the Richards Laboratories, the beams of the Community Center are in prefabricated and prestressed reinforced concrete, to be left exposed. The beams are topped by tall sections with a truncated pyramid form. The fact that an awareness exists of the clear ambiguity produced by a section typically used in steel constructions is also conveyed by MacAllister, when he specifies the material of the beams to Arthur Teich, chairman of the committee behind the work: "The apparent steel girder-shapes you have noticed are actually smoothly finished concrete beams which support the gymnasium and auditorium roofs and are brought to the outside in the classic manner where a structure is both functional and handsome."[1]

Where the roofs are flat, the design continues to rely on prefabricated parts – in place of the Flexicore slab, Kahn gets information about the Airfloor produced by the Airfloor Company of California, Inc.[2] The prefabri-

217-218. Community Building, Jewish Community Center, Trenton, New Jersey, 1954-59.

217. Perspective drawing (LIK, 030.I.C.430.1.3).

218. "Auditorium & Gymnasium Sections 9," 3 June 1958 (LIK, 030.I.C.430.1).

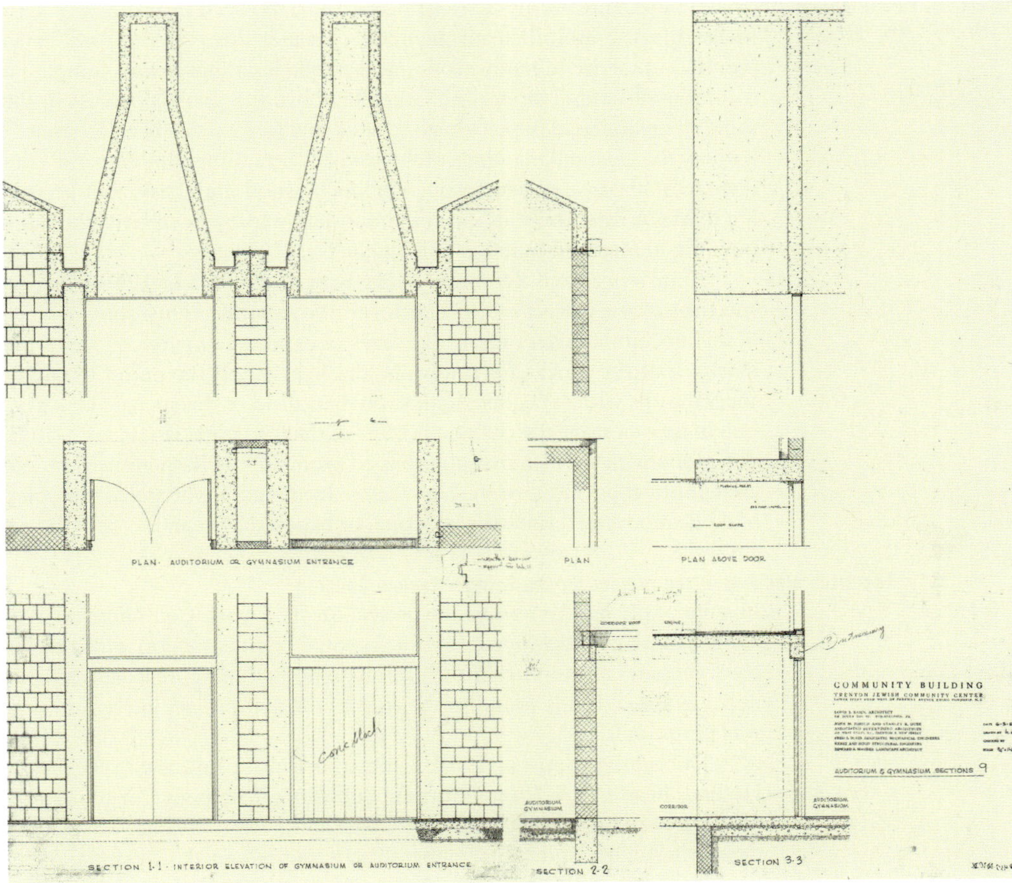

cated floor type guarantees savings, due to the construction process, and a smooth surface. All the walls are to be erected with 8 x 8 inches concrete blocks, the size with which Kahn also wanted to build the Bath House. The building becomes the result of an assembly of prefabricated concrete parts, with the exception of the columns that are to be poured in place. During construction, the workers would use the typical operations of buildings made with stone or brick.

At the end of 1958, Kahn tries to simplify the materials as much as possible, to reduce construction costs and comply with the wishes of the client committee. The main modification is in the roofs of the gymnasium and the auditorium, due to the elimination of the precast and prestressed beams, which are replaced by wood carpentry. Then some of the pillars are eliminated, leading to major savings on the concrete of the foundation. The "precast dome" system is still included, to be installed on the wooden beams. Instead of the sheets of marble and wood panels for the walls, concrete blocks appear, while the painting is also drastically reduced.[3] The

[1] John E. MacAllister, letter to Arthur Teich, 12 December 1958, LIK, 030.II.A.9.2.

[2] Louis I. Kahn, letter to W. M. Anderson & Company, 22 September 1958, LIK, 030.II.A.9.2.

[3] Marian Company, "Contract No. 1, General Construction," LIK, 030.II.A.9.2.

219-224. Tribune Review Publishing Company Building, Greensburg, Pennsylvania, 1958-62.

219. North elevation (LIK, 030.I.A.505.15).

square blocks measuring 8 x 8 inches are also replaced by ordinary 8 x 16 inches blocks.[4] Kahn's team members consider this exposed masonry a "common practice in many economical public buildings,"[5] and compare it to blocks of natural stone: "All concrete above the floor is precast in the same technique used in precast stone work and looks much like limestone."[6] By now the essential materials of the Jewish Community Building are "concrete (cast-in-place, precast), concrete block, wood and glass," but according to MacAllister this modification imposed by costs, "represents the same building in area and function and character."[7]

Kahn's project is definitively filed away at the start of 1960. The perspectives of the project of 1958, prior to the simplifications, show a work of architecture marked only by the lines of exposed concrete, those of the masonry of the blocks for the walls, the beams and the grand pyramid-shaped roofs with skylights. Against the backdrop of this archaic construction in precast stone the changing colors of the landscape would have stood out, along with the also pyramid-shaped profile of the Bath House and the metaphorical pavilions of the Day Camp, forming a monumental complex of edification based on artificial stone, or exposed concrete.

The brick cladding of the project for the Tribune Review Building

In spite of the lack of results, the project for the Jewish Community Building is a work that demonstrates the potential of construction with precast concrete parts as outlined in the prototype of the Bath House, and refined in the graphic details found in the studies for that project in Trenton that was never built.

During the course of 1958, Kahn comes into contact with Robert B. Herbert, head of the Tribune Review Publishing Company, for the design of the newspaper's headquarters in Greensburg, Pennsylvania. Kahn visits the site in August, and in September the contract is signed. Construction begins in the fall of 1959, handled by the Westmoreland Construction Company of Greensburg.

The building represents the apotheosis of exposed precast concrete, in the form of brick masonry, blocks, beams and ties. The concrete parts poured on site, apart from the foundations, are so exceptional that they become sculptural. Every part of the structure, from the pillar to the wall panel, reveals the different logics of the constructive processes of concrete in its textures and measurements, and every part of every piece is selected in keeping with its particular role in the overall structure. The fact that the concrete pieces have to be laid leads to careful study of all the graphic patterns of the joints. It is no coincidence that the mortar becomes an essential part of Kahn's pursuit of an absolute, archaic truth of concrete.

After a series of studies, guided by the desire to identify the best form of openings for diffused lighting of the rooms, screened from direct sunlight, Kahn reaches the solution of the long pre-tensioned beams in reinforced

220. Plan, with sketch of the Chapel of Notre-Dame-du-Haut, Ronchamp (LIK, 030.I.C.505.1).

Following pages:

221. Section sketch, 20 January 1959 (LIK, 030.I.C.505.1).

222. Plan and section, 20 January 1959 (LIK, 030.I.C.505.1).

223. "South Alcove & Loading Dock Details," 11 August 1959 (LIK, 030.I.C.505.2).

224. "Exterior Stair – Northeast Corner," 22 January 1960 (LIK, 030.I.C.505.2).

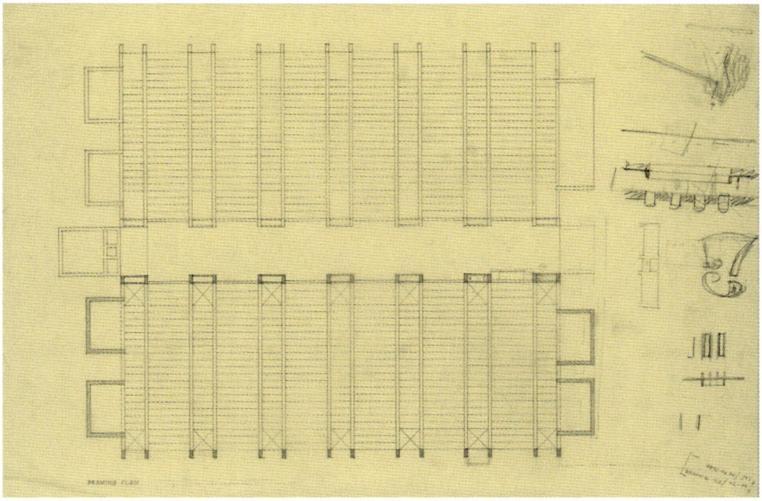

concrete, organized in two rows of large spans and separated by the middle strip of the service spaces, in keeping with the scheme of the Yale Art Gallery.

The question of the nature of materials is crucial in the evolution of the character of the Tribune Review Building. In the initial phases of the project, the building is composed of a skeleton formed by columns in ordinary clay brick covered with bricks also in clay, but of a superior quality, and by the exposed reinforced concrete beams of the roof, with a triangular section (their length determines one span). The infill of the skeleton is in brick on the outside, like the bricks of the pillars, and concrete block on the inside, again based on the model of the wall of the Yale Art Gallery[8] – "facing and back-up," separated by a cavity and connected together.

The openings are arched along the short sides, and triangular along the others; the first form comes from the assembly of the bricks, while the second is based on the configuration of the beams. The bricks are of standard size, with the exception of those "for structural arches and jambs of triangular windows" which Kahn specifies in clay, but with special measurements. Brick would also be used to build the "floor of the porch" and some surfaces of internal partitions.

The laying of the hollow concrete blocks of the internal facade of the perimeter walls includes special pieces in the corners of the vertical openings, to make a groove in which to fit the flush casements of the wall. In the specifications these blocks are indicated as "jamb blocks"; there are also other special pieces, the "corner blocks." The full blocks are to be used, again according to the specifications, "around mechanical room, transformer vault and in load bearing piers where shown."

The type of concrete block envisioned by Kahn in this phase of the project is still the Waylite variety already used for the Bath House, in a size of 8 x 8 x 16 inches. There are no particular indications in the specifications for the mortar compound with a base of Portland Cement. The design of the

[4] *Ibid.*
[5] John E. MacAllister, letter to Arthur Teich, 12 December 1958, cit.
[6] *Ibid.*
[7] *Ibid.*
[8] See the sketch published in, McCarter, *op. cit.*, p. 138.

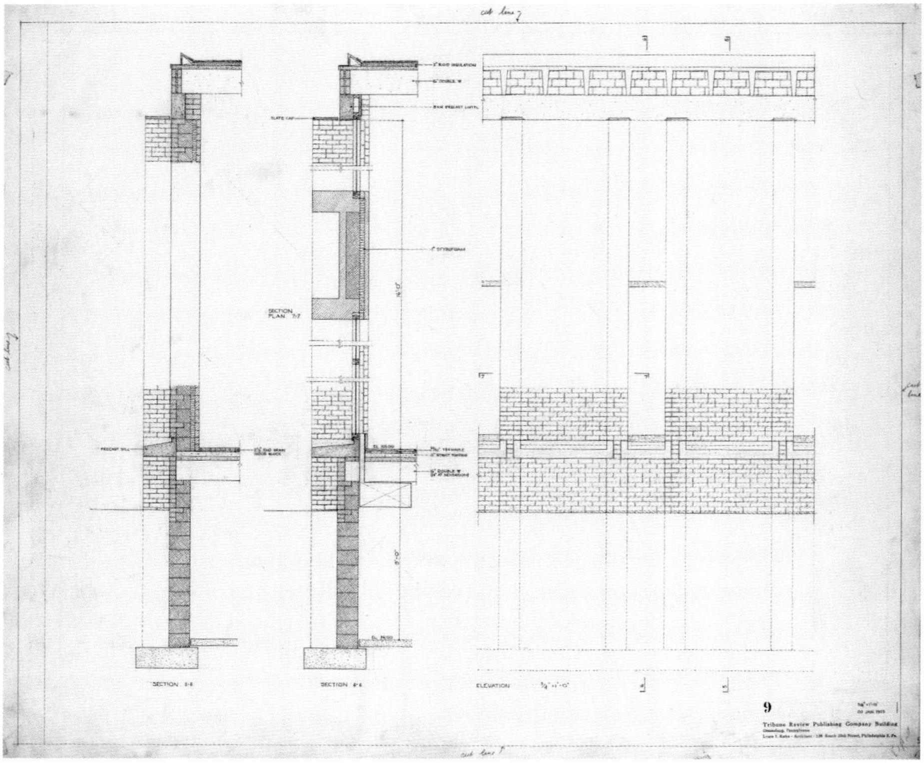
SECTION
PLAN 7-7
ELEVATION
9
Tribune Review Publishing Company Building

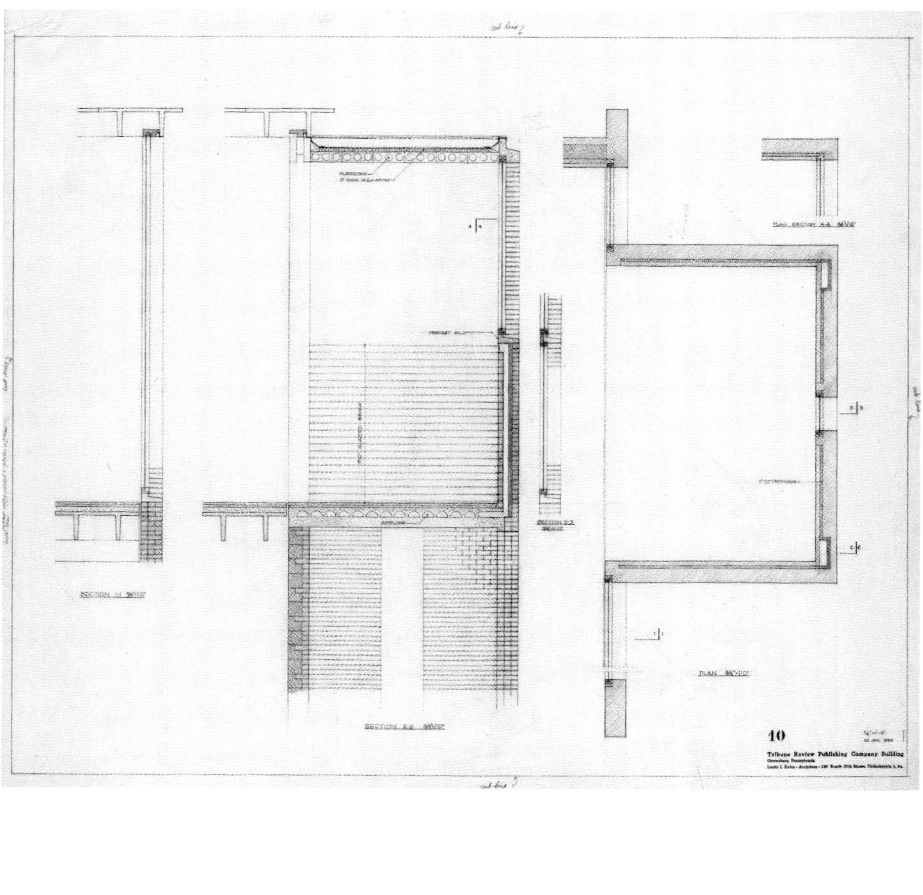
PLAN
10
Tribune Review Publishing Company Building

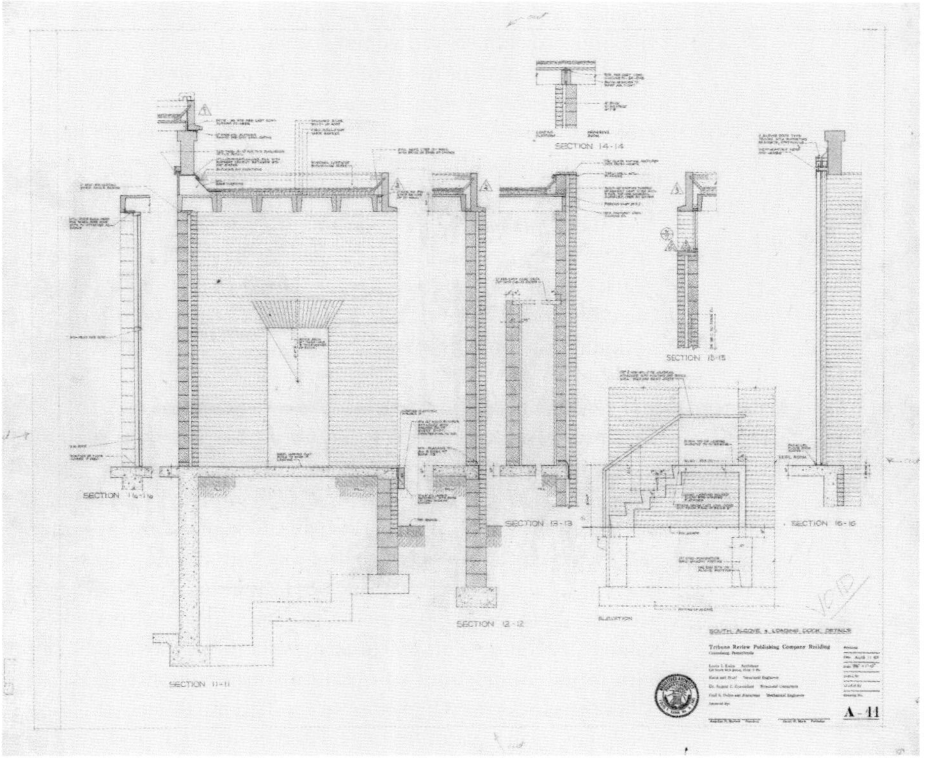
SECTION 14-14
SECTION 15-15
SECTION 11a-11a
SECTION 13-13
SECTION 16-16
SECTION 12-12
SECTION 11-11
ELEVATION
SOUTH ALCOVE & LOADING DOCK DETAILS
Tribune Review Publishing Company Building
A-41

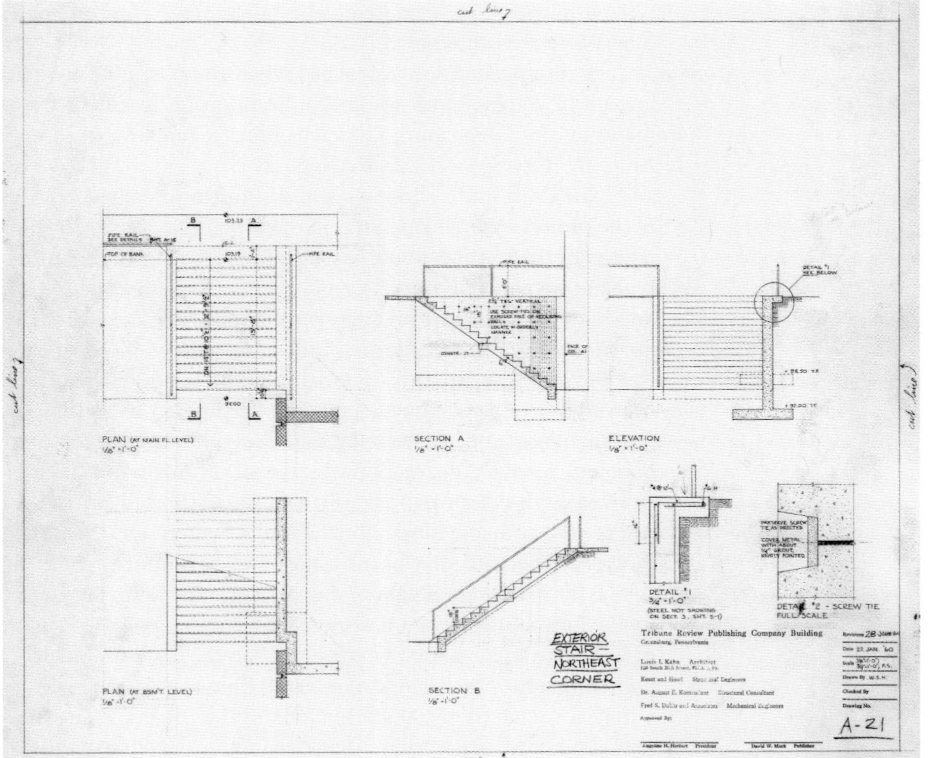
cut line
PLAN (AT MAIN FL. LEVEL)
1/8" = 1'-0"
SECTION A
1/8" = 1'-0"
ELEVATION
1/8" = 1'-0"
PLAN (AT BSM'T. LEVEL)
1/8" = 1'-0"
SECTION B
1/8" = 1'-0"
DETAIL #1
3/4" = 1'-0"
(STEEL NOT SHOWING ON SECT. 3, SHT. S-1)
DETAIL #2 - SCREW TIE
FULL SCALE
EXTERIOR STAIR — NORTHEAST CORNER
Tribune Review Publishing Company Building
Greensburg, Pennsylvania
Louis I. Kahn Architect
Keast and Hood Structural Engineers
Dr. August E. Komendant Structural Consultant
Fred S. Dubin and Associates Mechanical Engineers
Approved By:
Date 22 JAN. '60
Drawn By W.S.H.
Checked By
Drawing No.
A-21
Title : EXTERIOR STAIR

joints is not underscored by grooves in the installation: "Exposed joints shall be struck flush and wiped with a coarse cloth."

All the parts of the roof are to be made in exposed precast concrete, with prestressed structural members. The question of the mold for their prefabrication becomes essential to obtain a clean surface design. Kahn proposes various materials for the molds ("concrete, steel, or a combination of both. Plywood on steel framing accurately made is acceptable").

The finishes give rise to a building with walls in clay brick and concrete beams, and black lines formed by sheets of black slate (Pennsylvania Black Slate), such as those for the window sills, and touches of white marble with grain (Vermont Eureka Danby Marble), used for the steps and the coping of the entrance platform.

Metamorphosis of clay into concrete

When the Tribune Review Building project enters its definitive phase, the clay bricks are eliminated to make a building entirely out of concrete. This is the phase of project development when the section of the specifications regarding the brick is cut, and at the entry "brick" the adjective "concrete" is added. When Kahn decides to do away with the clay brick, he can no longer use Waylite block, due to its lack of formal quality. Precisely in the Tribune Review Building, Kahn experiments with the construction of blocks of ordinary size, made with machinery available on the market, but designed by him in its composition to obtain a specific quality of color and texture. With the blocks used by Kahn for the Tribune Review Building a new chapter opens in the evolution of exposed concrete. It is no coincidence that in the specifications the possibility left for the contractor of indicating an "alternate proposal for using concrete block made with an aggregate other than Waylite" is eliminated.

In the specifications corrected when Kahn eliminates the clay bricks, instead of Waylite block we find indication of a "standard concrete block" made by the Greensburg Concrete Block Company of Joseph Repasky, Anthony Repasky and Edward Repasky.

The decision to eliminate the clay bricks already comes in the phase of the project with the openings in the form of arches, as a result of the correction of the specifications, in which in place of "ground block" indication is made of "concrete brick" for the arches. The subsequent removal of the arches depends in part precisely on the transformation of the wall materials. Having introduced bricks of the same size as those of clay, but fabricated in concrete, prompts rethinking of the original logic of the wall infill in two layers separated by a gap – the "facing and back-up." Given the fact that the material of the two layers is the same, Kahn decides to use a single block for the construction of the wall. At this point the building is organized with concrete bricks for the pillars, based on the original clay bricks, and concrete blocks for the infills. As a result, the form of the window is changed to that of a T, since arches cannot be built with blocks. Again, for a coherent design of the various parts, Kahn decides to also modify the form of the triangular beams, to determine on the long sides a form of window similar to that which has taken the place of the arch. It is worth underlining how the change of material from clay to concrete has generated a logical sequence of transformations of the project, to reach a form consistent with the new system of bricks and blocks in concrete.

The prefabricated beams of the Formigli Corporation

The long roof beams of the Tribune Review Building are made in precast concrete by the Formigli Corporation, based in Berlin, New Jersey. This is the company hired to make the prefabricated parts in reinforced concrete for the Biology Building, taking the place of the Atlantic Prestressed Concrete Company that had made all the pieces for the Richards Laboratories.

From the start of June 1960, Kahn envisions closing the cavities made in the ends of the beams for the post-tensioning using steel cables with special precast concrete blocks,[9] "cast true and natural, without surface corrections."[10] Nevertheless, certain alterations and flaws caused by operations on the worksite convince him to modify the solution. On the one hand, Kahn has eliminated the slate of the grooves along the pillars, replacing it with white marble, "Georgia White Cherokee." On the other, Paul Formigli has intervened at the ends of certain beams to correct the finish with cement-base mortar that is carefully smoothed but visible, especially on rainy days, when the surfaces get wet and take on different colors. "[...] experience," Kahn explains, "proves that corrected concrete does not withstand weather and even if your methods are guaranteed, all the corrections show up in the rain."[11]

The alternatives proposed by Kahn to Formigli's corrections consist in either eliminating the grouting, sculpting the ends and then reshaping them with concrete poured into special molds, or in modifying the material of the plugs to introduce a chromatic note that will distract the eye from the repairs. This is the same use of color applied by Le Corbusier to correct the flaws of the *béton brut* and other *malfaçons* of the worksite. The force of attraction to the gaze of the chromatic inlays is far superior to that of the shadows of the holes left behind by the formwork cones that also served to distract the observer from seeing defects, as Kahn stated regarding the Mill Creek towers.

"If you can guarantee," Kahn explains, "that corrected concrete will not sprawl from change in temperature, then maybe we may try the introduction of another, more important material to possibly take the eye off the corrected imperfections. [...]. This sharp, precise marble would contrast with the soft, flour-like appearance of the rubbed structural girder and then detract the eye from its visual weakness."[12]

The four entirely visible beams in the two facades are made with careful workmanship and a different design, since they are the elements of the portal exposed for the whole length, a sort of triumphal porch leading to the structural system of the whole building. Kahn later makes a more evident and monumental porch for the Kimbell Art Museum.

Kahn plans to have the intrados of the beams of the Tribune Review Building perfectly horizontal and not curved, and asks Komendant and Keast & Hood, on 17 June 1960, to make special calculations for the pretensioning. Furthermore, he wants to conceal the imprint of the joint of the mold halfway down the beam, by means of a cavity placed to correspond to the joint itself and to then to be filled with a marble plug, or "joint cover."[13]

[9] Louis I. Kahn, letter to Westmoreland Construction Co., 6 June 1960, LIK, 030.II.A.9.7.
[10] Louis I. Kahn, letter to Formigli Corporation, 20 June 1960, LIK, 030.II.A.9.7.
[11] *Ibid.*
[12] *Ibid.*
[13] Marshall D. Meyers, letter to Westmoreland Construction Co., 1 August 1960, LIK, 030.II.A.9.8.

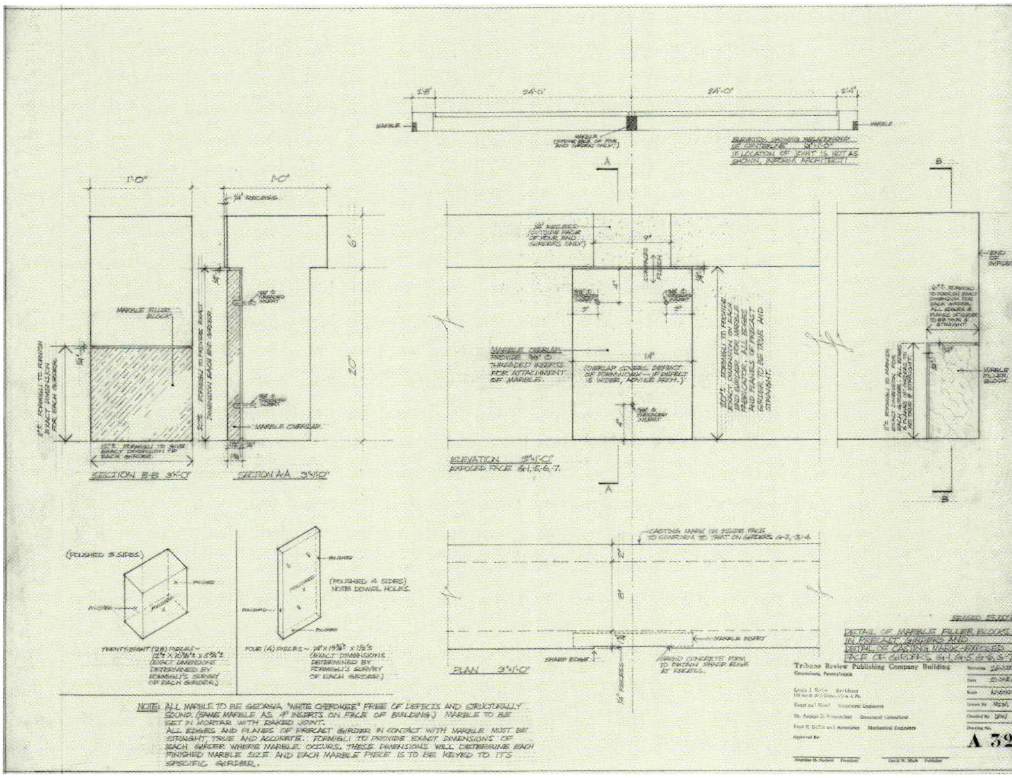

225-226. Tribune Review Publishing Company Building, Greensburg, Pennsylvania, 1958-62.

225. "Details of Marble Filler Blocks in Precast Girders and Detail of Casting Mark – Exposed Face of Girders G-1, G-5, G-6, G-7," 20 June 1960 (LIK, 030.I.C.505.2).

226. Construction site photograph (LIK, 030.IV.A.505.4.50).

"Because the entire length of these girders is exposed to the weather, no patching or rubbing will be permitted," Kahn writes to the Formigli Corporation. "The casting procedure should conform to that discussed with Paul Formigli in our office on Friday, 17 June 1960, in order to obtain a good beam directly from the mold. Furthermore, these peripheral girders must be cured according to the specifications. To obtain the most uniform color and the best finish for weathering, steam curing gives the finest result. Enclosed is a sketch of the proposed method of handling the unsightly mark caused by the formwork joint at the center of these peripheral girders. The width of the marble and the recess in the flange is dependent upon the width of your spacer piece."[14]

The four special beams are made in August, and before they are installed Kahn and his staff check on the quality of their surfaces.[15]

The concrete bricks and blocks of the Greensburg Concrete Block Company

The concrete bricks and blocks of the Tribune Review Building are produced by the Greensburg Concrete Block Company with a special machine installed at the worksite.[16] The geometry of the laying is calculated by Kahn to also leave cavities along the vertical edges of windows and doors, to insert the flush-mounted casements. As for the version in clay brick, these cavities are supplied with special "jamb blocks" that are not made, in the realisation.[17] Other special blocks are planned for the corners, the "corner blocks,"[18] again as in the previous brick version.

The walls of the basement against the terrain are made in reinforced concrete with plywood formwork (the walls have outcroppings that form the bases of the brick pillars). For the visible exterior faces the walls are clad with concrete blocks with a square section to create a pattern that is the same as the one on the upper level, obtained by laying blocks with a square section and two cavities. This is a detail with a value similar to the one implemented by Kahn in the Richards Laboratories, with the precast facing slabs to conceal the poured in place concrete. The rectangular blocks with three holes are laid in the basement sector, remaining visible with their rectangular sections only on the inside.

In December 1959 Kahn selects three concrete block samples on the worksite. In December and January trial masonry sectors are made with bricks and blocks. The focus is on their color and that of the mortar. Kahn wants to obtain chromatic uniformity of surfaces and lines. To this end, the sand to be used in the compound is also specified, supplied by the Pennsylvania Sand Glass Corporation, based in Maryland.[19] On 19 January 1960 William S. Huff, a member of Kahn's staff, informs the company of Kahn's indications:

"In order to achieve the desired coloring in the mortar, which is to approximate the color of the block itself, he has suggested that you might use a product of Universal Atlas known as 'Masonry Cement.' This is the same

[14] Louis I. Kahn, letter to Formigli Corporation, cit.
[15] Marshall D. Meyers, letter to Formigli Corporation, 1 August 1960, LIK, 030.II.A.9.8. "Regarding your statement about 'no finish or patching will be done on these girders after casting,' this is Mr. Kahn's intent. However, as he stated during our visit, if honey-combing or a non-weatherable surface occurs because of some unavoidable condition, then he shall have to accept a rubbed and patched end girder. Please inform us of the earliest possible date that we can see these girders to inspect their surface."
[16] David W. Mack, letter to Louis I Kahn, 24 February 1960, LIK, 030.II.A.35.24.
[17] See also William S. Huff, letter to Westmoreland Construction, Co., 2 July 1961, LIK, 030.II.A.9.25.
[18] *Ibid.*
[19] William S. Huff, letter to Westmoreland Construction, Co., 10 February 1960, LIK, 030.II.A.9.5.

cement used in the sample I gave you on January 18. However, since this sample seems to be a little too yellow in color, he further suggests that you might achieve the desired effect with white sand rather than the bar sand you used in the sample. Of course, we will welcome any suggestions you might have in order to achieve the effect he wishes."[20]

The mortar is to be used in all the masonry sectors between the pillars and in the "interior of the south west wall of the basement Reel Room."[21]

Particular care goes into the making of the concrete bricks with the machine supplied by Repasky, observing their grain, edges and sizing, to guarantee homogeneity between blocks and bricks. After having seen the new trials, Kahn remarks that "there is not a very great improvement."[22] The fact that Repasky's machine does not produce pieces of the same identical quality is also confirmed by what David W. Mack, of the Tribune Review Publishing Company, writes: "Repasky, the block bird is going to make some 16" blocks with as fine a mixture as he can – 16 inches so he can sell them if you don't like them. […]. I don't think we will ever get absolutely perfect blocks consistently. The full 8" bricks also present a problem […]."[23]

Kahn suggests laying the blocks in such a way as to place towards the bottom the notch produced by the mold of the machine during fabrication.[24] At the end of February a batch of 16 inches blocks of the type with holes is produced, to see "if they could get a fine enough texture to satisfy."[25]

During the trials done in April to determine the color of the mortar, Kahn advises the use of "more lime" to make the compound "more workable, give better water impermeability and help give better color to the mortar."[26] The thickness of the joints in the laying of the bricks and blocks is calculated to obtain a matching texture of lines.[27] Observation of the shrinking of the mortar in the joints prompts Kahn's office to modify the entire pattern of the masonry of the two facades, thus increasing the width of the

[20] William S. Huff, letter to Westmoreland Construction, Co., 19 January 1960, LIK, 030.II.A.9.5.
[21] William S. Huff, letter to Westmoreland Construction, Co., 10 February 1960, cit.
[22] William S. Huff, letter to Westmoreland Construction, Co., January 1960, LIK, 030.II.A.9.5.
[23] David W. Mack, letter to Louis I Kahn, 24 February 1960, cit.
[24] William S. Huff, letter to Westmoreland Construction, Co., 30 April 1960, LIK, 030.II.A.9.22.
[25] Tribune Review Publishing Company, letter to Louis I. Kahn, 27 February 1960, LIK, 030.II.A.35.24 (the photograph of the blocks is attached to the letter).
[26] William S. Huff, letter to Westmoreland Construction, Co., 30 April 1960, cit.
[27] The series of calculations made by Kahn's office is significant, to establish the progression of the operations in one space and the thicknesses of the joints (*ibid.*).
[28] William S. Huff, letter to Westmoreland Construction, Co., 2 May 1960, LIK, 030.II.A.9.22.
[29] Marshall D. Meyers, letter to J. Cavalier, 28 February 1961, LIK, 030.II.A.9.11.
[30] *Ibid.*

227-228. Tribune Review Publishing Company Building, Greensburg, Pennsylvania, 1958-62. Construction site photographs (LIK, 030.IV.A.505.4.64; 46).

joints between the walls and the pillars of the structure (the width is doubled, from 4 to 8 inches). It is on this occasion that Kahn decides to change the color of the cladding material of the inside of these joints, to achieve an effect that is no longer one of shadow, as in the case of the envisioned slate sheets, but of clear display of the gap: "on considering both the color of the mortar and block, Mr. Kahn has decided that a polished white marble (Georgia or similar) will be a better choice of material."[28]

"Late last winter Mr. Kahn and William Huff of this office went to Greensburg to approve a wall sample. They rejected both the joints and the quality of the block. The erecting of a new sample wall lagged, and as the critical date to begin the masonry approached, they returned to the job site again in April 1960. On this visit, a mason laid up, in their presence, a sample panel with wiped joints and improved block. That sample panel was approved as it conformed to Mr. Kahn's intent (this included the diminishing of vertical block joints and the adding of block in the horizontal coursing [...]). After this agreement, as work was progressing back at piers C-11 and C-12, the Contractor complained that he did not like the joint that was agreed upon (including the shrinking of the vertical joint) and began to use this portion of the building as a sample panel instead of putting up new test panels as called for in the specs. He took it on his own to try out his own joints on this wall without consultation with the architect."[29]

The new joint is approved by Kahn and Huff during a worksite visit on 6 May 1960.[30]

The concrete compound of the blocks and bricks produces an overall color bordering on the yellowish, the same color as the mortar used for the

229-233. Tribune Review Publishing Company Building, Greensburg, Pennsylvania, 1958-62.

229-230. Construction site photographs (LIK, 030.IV.A.505.4.77; 81).

231. View (LIK, 030.IV.A.505.2.49).

232. Detail of a facade (photograph by the author, 2010).

233. Detail of a facade in which is visible the texture of concrete blocks and bricks (photograph by the author, 2010).

234-239. Tribune Review Publishing Company Building, Greensburg, Pennsylvania, 1958-62.

234. Detail of a pillar in front of the facade (photograph by the author, 2010).

235. Detail of a concrete sill (photograph by the author, 2010).

236. Detail of the surface of the concrete bricks (photograph by the author, 2010).

237. Detail of the surface of the concrete blocks (photograph by the author, 2010).

238. Detail of the roof structure (photograph by the author, 2010).

239. Detail of staircase (photograph by the author, 2010).

240. Tribune Review Publishing Company Building, Greensburg, Pennsylvania, 1958-62. View (LIK, 030.V.A.4.13).

laying, but with perceptible shadings – more yellowish in the bricks and the mortar. The small stone aggregate used for the compound has ochre, brown, beige, green, gray and reddish pieces; these aggregates are also found in the mortar. The grain in the two types varies (it is finer in the brick). In spite of the pursuit of uniformity, the batches of blocks differ perceptibly from one another, and the materials are deployed without attempting to make a design. An interruption of construction caused by a strike leaves a visible trace in the color of the concrete blocks, because when work is resumed "all of the new concrete blocks were a slightly different shade and you could see a definite 'line' where the new blocks met the older ones."[31]

The concrete of the prefabricated parts of the roof stands out from that of the bricks and blocks for its grayer color and perfect smooth surfaces. The beams are shaped so that the two ends have measurements corresponding to those of the pillars, and to create along the sides a vertical line that continues that of the pillars and then runs horizontally, parallel to the upper edge of the beam, creating an abstract moulding. In the end of the beams Kahn inserts a plug of white marble that seals off the prestressing attachments. Reversed U-shaped pieces, also precast in reinforced concrete, are mounted on the beams; placed side by side, they form the roof structure on which the final surface is placed. A series of precast reinforced concrete slabs completes the assembly of blocks, bricks and beams – these pieces are used as sills, to seal the upper course of the infilling blocks, or to shelter the horizontal T-shaped windows from the penetration of sunlight. The whole structural system is left exposed, without claddings or ceilings, offering a view of the framework and its material from the offices.

The canopy, a tribute to Le Corbusier

All the parts of the exterior organization of the Tribune Review Building are poured in place and left exposed, from the steps to the canopies to the terracing walls. One indication of the ambition to make a work that is the result of prefabricated elements is that Kahn prefers, for the main staircase, "concrete steps laid up individually like big stones instead of poured-in-place."[32] The image of the "big stones" is emblematic of a worksite using pieces of concrete, laid in keeping with the systems of stone construction.

In the concrete parts poured in place, Kahn establishes a dialogue with the more fashionable forms of international *béton brut* and shapes the surfaces with the play of vertically and horizontally placed boards. The most significant element of this decorative interpretation of *béton brut* is the canopy with drip placed against the room for the post office. Thanks to the working of the concrete through the arrangement of the vertical and horizontal boards of the formworks (width 2 inches), the canopy becomes a true work of sculpture in the manner of those of Le Corbusier. The form forecasts the structure Kahn designs for the platform of the organ in the First Unitarian Church.

No other concrete feature in Kahn's architecture offers a better example of a formalist game of composition of the boards like the canopy of the Tribune Review Building. The decorative pattern of boards of the top slab of the AFL Medical Services Building and certain studies of panels with vertical and horizontal boards for the Richards Laboratories are the antecedents. The fact that the definitive drawing for the canopy is dated 25 November 1959 suggests that it may be a personal interpretation of certain forms of *béton brut* that Kahn sees in September at the chapel of Ronchamp, like the fontain and the pulpit.[33]

Regarding the deployment of the concrete, Kahn is critical when the canopy turns out to be defective and starts to distort: "The architect again wishes to restate his great displeasure with the resultant distortion of the designed shape of the canopy, the excessive honeycombing and the dark color of the concrete."[34] It is precisely regarding the exposed concrete of the canopy that Kahn uses terms similar to those used by Le Corbusier at the sight of all the *malfaçons* in the *béton brut* of the Unité d'Habitation of Marseille. Kahn too, like Le Corbusier, does not want corrections made to the concrete surfaces after stripping: "Regardless of the imperfections, the canopy is safe. Corrections to these imperfections cannot be made gracefully. It is therefore recommended that, short of tearing the whole thing down and doing it over again, the imperfections should remain. The doctoring up of any of these imperfections would appear artificial and look worse to the discerning eye."[35]

The design of the retaining wall, dating back to January 1960,[36] is another demonstration of Kahn's desire to create an ordered formal system of *béton brut* with the assembly of vertical boards and the holes made

[31] Marshall D. Meyers, *Interview with Alessandra Latour, [...], March 19, 1983*, in, Latour, *Louis I. Kahn, l'uomo, il maestro* cit., p. 81.
[32] Marshall D. Meyers, letter to Verne Norman, 14 June 1960, LIK, 030.II.A.9.7.
[33] Kahn takes part in the conference of Otterlo from 10 to 14 September 1959 (the whole conference lasts from 7 to 15 September). The flight from Paris to Geneva on 15 September seems to indicate that Kahn reaches Ronchamp by way of Switzerland, after having been in Otterlo. For the dates of the flights recorded on Kahn's passport, see Eugene J. Johnson, Michael J. Lewis, *The Travel Sketches of Louis I. Kahn*, Cambridge, Massachusetts, London, MIT Press, 1996, p. 132, note 236.
[34] Louis I. Kahn, letter to Westmoreland Construction, Co., 15 July 1961, LIK, 030.II.A.9.25.
[35] *Ibid.*
[36] See the drawing "Exterior Stair – Northeast Corner," 28 June 1960, LIK, 030.I.C.505.002.

241-245. Tribune Review Publishing Company Building, Greensburg, Pennsylvania, 1958-62.

241. Photograph of the canopy construction (LIK, 030.IV.A.505.4.1).

242. View (LIK, 030.V.A.4.12).

by the spreaders. "Use screw ties on exposed face of retaining wall located in orderly manner," is written on the drawing. The platform with the entrance steps is also decorated by vertical and horizontal board imprints, like the canopy.

In the summer of 1961 the surfaces of one masonry sector in the Cooling Tower are whitewashed. This is the type of intervention not allowed by Kahn: "This is unfortunately a thing done – without Mr. Kahn's approval or even knowledge. Please do not cover any other exterior, exposed concrete work without the approval of Mr. Kahn."[37]

The whole Tribune Review Building presents itself in the green landscape with its range of warm grays and pattern of elements; in the international panorama of *béton brut*, it becomes an expression of the research on exposed concrete in the unusual form of masonry that began with the Yale Art Gallery and the Bath House. In no other building constructed until this date, had Kahn explored all the possible constructive forms of concrete, so much so that the Tribune Review Building can be considered the work that best exemplifies a form of expression with concrete that is original even in

243. Detail of exterior stairs (photograph by the author, 2010).

244. Detail of the concrete wall of the exterior stairs (photograph by the author, 2010).

245. Detail of the wall of the front square (photograph by the author, 2010).

the vast, complex experimentation with exposed concrete conducted by the leading figures of the 20th century, from Perret to Le Corbusier. This building is therefore capable of ushering in another chapter in the various international formulations of *béton brut*: an archaic concrete, shaped in pieces to be laid like stones.

In the debate held in Otterlo in 1959 after the conference, Kahn returns to the question of the origin of ornament in the joint between construction parts, already sketchily outlined at the start of the 1950s. In this context his reasoning seems to be limited to the connection of pieces to assembly on the worksite, such as bricks, stones and sheets. But it is clear that those connections also have to do with the imprints of formwork. Even in the idea of enlarging the gaps of the joints, we can see the impact of some of the constructive questions raised by the Tribune Review Building.

[37] William S. Huff, letter to Westmoreland Construction, Co., 2 July 1961, cit.

"The way things are made, the way they are put together, the way one thing comes to the other, is the place where ornament begins," Kahn states. "It is the glory of the joint which is the beginning of ornament. The more a man knows the joint, the more he wants to show it. The more he wants to show the joint, the more he wants to show the distance. And if he wants to have it show the distance, he wants to exaggerate and caricature things which ordinarily are small. The beginning of ornament lies also in the challenge against the elements."[38]

"Concrete in lieu of limestone": United State Consulate Buildings, Luanda

The concrete used in the Bath House and the Tribune Review Building becomes a model of construction for other buildings as well, at least until Kahn does not turn to the construction of true above-ground walls in concrete poured in place.

Exposed concrete is indicated for the walls, pillars, slabs, staircases and perforated screens for natural ventilation and shade in the project, developed from 1959 to 1963, for the United States Consulate Buildings in Luanda, to be built under the scorching sunlight and in the extreme heat of the tropical African climate of Angola. This work too no longer belongs to the genre of *béton brut* with concrete shaped on the worksite in formwork. The building is another assembly of precast concrete pieces to be left exposed, with the pattern of mortar joints, as in the Tribune Review Building. Precisely thanks to the Luanda project, the building in Greensburg takes on the value of a prototype of a particular form of American architectural concrete.

246-249. Project for the United States Consulate Buildings, Luanda, Angola, 1959-62.

246. Perspective drawing, 1960 (LIK, 030.I.A.555.28).

247. Isometric drawing (LIK, 030.IV.A.555.1.2).

248-249. Study sketches (LIK, 030.II.A.34.30; 31).

As in the Tribune Review Building, Kahn uses full bricks and hollow blocks, both in concrete, to set apart the different structural role of pillars and walls thanks to their different textures. For the supply of the concrete blocks, contact is made with the Angola facility of the Portuguese company Artefactos de Cimento, Lda., whose materials are made with the method of simultaneous vibration and compression developed by the company Rosacometa.[39]

The interlock of the slabs in the walls generates a special course, lower than the others. The last course of blocks is sealed with sheets, again as in the Tribune Review Building. The reinforced concrete parts of the slabs are also prefabricated. In the study sketches an assembly of pieces in a T form also appears, to lay like certain wedges in stone structures and to fasten on site using concrete poured in place.[40]

The two rows of four pillars each are introduced independent of the walls to form, with the four beams, the main structure of the framework of

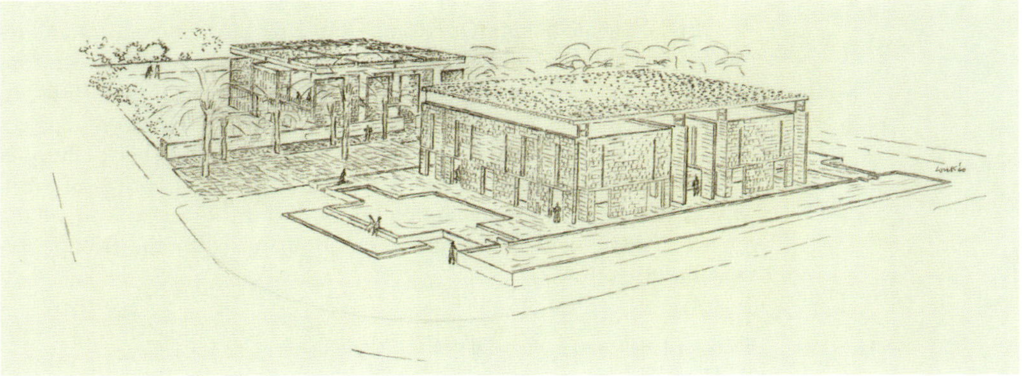

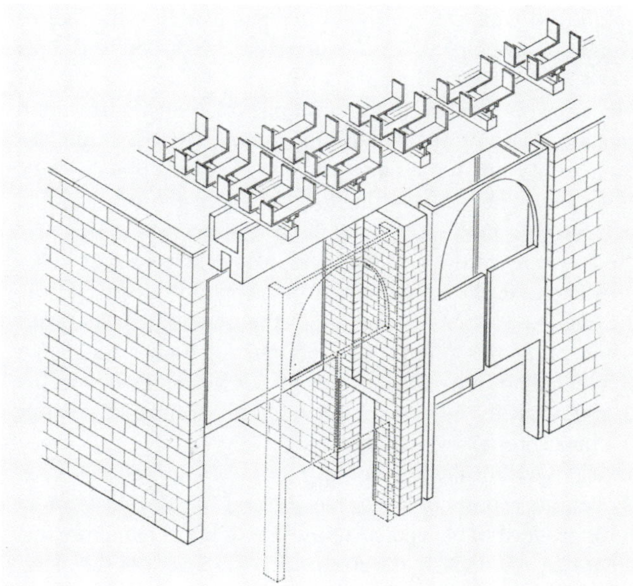

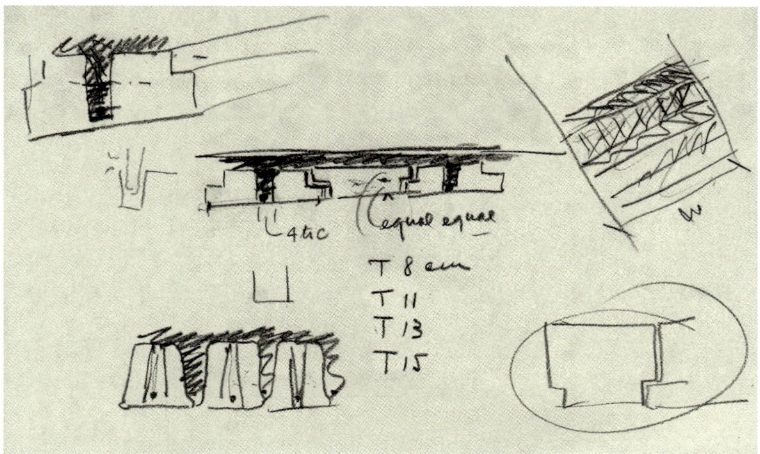

small prestressed joists with U-shaped sunscreens. This system creates the horizontal screen to filter light, "a loose sun roof" as Kahn calls it,[41] detached from the roof of the various volumes, or a "rain roof" to permit ventilation. The sunscreens were to be made "clay-tiled,"[42] to interlock at the top of the specially shaped joints, in keeping with an overturned U connection Kahn utilizes for the crowning elements of the roof of the Tribune Review Building.

To shield the sides with their large windows, Kahn invents another thin screen, with the same role of a "sun roof," to place at a distance from the windows to control the penetration of sunlight. This screen too is a pre-

[38] Latour, *Louis I. Kahn: Writings, Lectures, Interviews* cit., p. 98.
[39] Artefactos de Cimento, Lda., price list, n.d., LIK, 030.II.A.34.31.
[40] See the study sketches, LIK, 030.II.A.34.31.
[41] Jan C. Rowan, *Wanting to be the Philadelphia School*, in, "Progressive Architecture," vol. XLII, April 1961, No. 4, (pp. 130-163), p. 140.
[42] *Louis I. Kahn. A discussion recorded in Louis I. Kahn's Philadelphia office in February*, 1961, in, Latour, *Louis I. Kahn: Writings, Lectures, Interviews* cit., (pp. 122-140), p. 126.

250-251. Margaret Esherick House, Chestnut Hill, Philadelphia, Pennsylvania, 1959-61.

250. Study sketches (LIK, 030.I.A.520.18.1).

251. View (photograph by the author, 2010).

cast piece in reinforced concrete. "[…] glare screening walls which allow more modified light to envelope the enclosing walls," according to Kahn's description.[43]

Though the screen is derived from the various brise soleil forms built by Le Corbusier for works of architecture in tropical settings, Kahn describes the element as a ruin, in the same way he describes a prefabricated screen of the same period he is developing for the Salk Institute in La Jolla. The Roman arch opening in the screen for the consulate justifies this definition. Even the assembly of the elements, the various screens and brises soleil, seems intended to evoke a Mediterranean vision of ruined walls, observed from the shade of a tree's foliage, with the openings in the walls lacking in casements, at this point – "the absence of frames," as Kahn writes.[44] The colossal pillars are opened to permit passage and viewing, but perhaps also to contribute to the sense of the ruin that permeates the whole work: "I want to pass through everything […]."[45]

To convince the authorities regarding the use of exposed concrete, Kahn considers various aspects of the construction, from the existence of an important local production of concrete to the possibility of reinterpreting, precisely with that type of concrete, the constructive traditions of the place: "Luanda has a fully equipped and efficiently managed cement manufacturing plant of Portuguese and Danish ownership."[46] His statements on the possibility of imitating stone construction with concrete reveal his particular position in the evolution of *béton brut*. The imitation does not have to do so much with the working of the concrete surfaces as with the very form of the construction, based on the laying of masonry and the assembly of parts. Precisely with the works in blocks, bricks and beams of prefabricated concrete, Kahn is taking concrete back to its roots as artificial stone.

"Construction as shown is the kind of construction sympathetic to the building crafts of Luanda," it is noted in the project description. "Limestone is plentiful and worked with common labor and should prove to be a relatively economical material. The same forms could be obtained with exposed concrete in lieu of limestone. There is no craft attitude in concrete work which would warrant exposure without undue supervision."[47]

Concrete blocks and plaster finish: the Esherick House

While he is engaged in the choice of materials for the Tribune Review Building and the design of the United States Consulate Buildings in Luanda, Kahn designs, in 1959, a house for Margaret Esherick, in Chestnut Hill, near Philadelphia (construction is completed in 1961). Concrete blocks also appear in this work, but the design process reveals an evolution towards plaster facing.

Some working drawings of the Esherick House reveal Kahn's intention of using a structural system to erect with the same elements used in the Tribune Review Building. It is no coincidence that the two works have similar forms and sizes of the openings, made possible by that structural system.

[43] Louis I. Kahn, letter to D. Merle Walker, 30 August 1961, LIK, 030.II.A.34.28.
[44] *Louis I. Kahn. A discussion recorded* cit., p.123.
[45] *Ibid.*, pp. 126-127.
[46] "Note on the Program of a Proposed Office Building and Consul's Residence, Luanda, Angola," 29 October 1959, LIK, 030.II.A.34.28.
[47] "Project concept and description," n.d., LIK, 030.II.A.34.28.

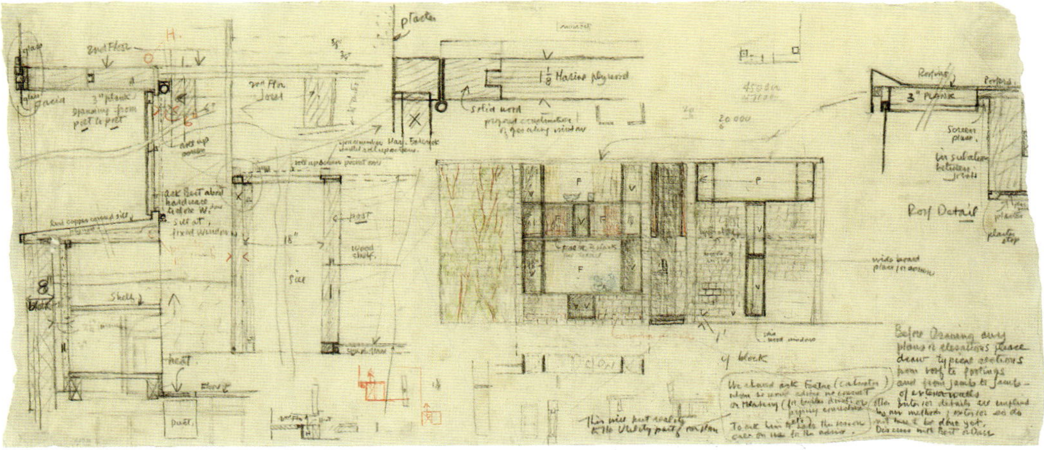
Roof Detail
3" plank
1 1/8" Harbor plywood
Before Drawing any plans & elevations please draw typical sections from roof to footings and from jamb to jamb of exterior walls

252-256. Margaret Esherick House, Chestnut Hill, Philadelphia, Pennsylvania, 1959-61.

252. Detail of the exteriorl plaster, (photograph by the author, 2010).

253. Balcony structure (photograph by the author, 2010).

The original conception of the Esherick House, like that of the Tribune Review Building, is the result of a repetition of equal spans formed by beams resting on pillars and arranged in two detached rows, to generate a middle band set aside for circulation – the service spaces, like the kitchen and bathrooms, are placed up against one span. Kahn considers the possibility of making the materials of the structural system visible, as in the Tribune Review Building: the concrete blocks of the pillars and the reinforced concrete of the roof beams. In an intermediate phase of the project the envelope of the Esherick House is entirely textured by the pattern of the concrete blocks, while a continuous roof slab joins the two series of spans and the addition with the service spaces. Precisely the homogenous pattern of the blocks creates the premises for the unification of the whole volume of the house, entrusted to the plaster facing used in the final project.

The disappearance of any concrete structural parts does not mean that the Esherick House belongs to a genre extraneous to Kahn's research on all the possible variations of concrete. Precisely the nature of the plaster, its "special color" and texture,[48] reveal the extent of the presence of concrete in a different way from that of the blocks or the skeleton. Kahn wants a plaster with a color that is obtained through the compound, just as when he selects the aggregate of blocks or concrete.

The plaster used for the facades of the Esherick House is the Penn Crete Exterior Finish with "natural off-white finish," produced and marketed by the Penn Crete Products Company of Philadelphia,[49] to be applied in three coats.[50] In July 1961 Kahn's office selects the "color chips,"[51] starting with which the company has to produce "Penn Crete samples of these colors." Among the shades, Kahn considers the hypothesis of using a "Mustard color."[52]

Concrete in the form of structural members is visible only in the prefabricated slabs of the balconies, equipped with flanges at the extremities, and similar to the architraves Kahn is making with fiberglass formwork for the Mill Creek Apartments II complex. The wooden structure of the staircase is a

254-255. Details of the elements of the wooden staircase (photographs by the author, 2010).

256. T-element in concrete placed at the entrance (photograph by the author, 2010).

veritable catalogue of pieces and interlocks, some based on precast concrete parts of the Richards Laboratories, others destined to form the basis of the assembly of what would be the most complex formwork ever created by Kahn. The recurring iconic figure in the design of the villa materializes in a sculpture Kahn places by the entrance: a concrete torso in the form of a T.

[48] "Addendum No. 1 to Specifications," 14 October 1960, LIK, 030.II.A.34.4.
[49] Louis I. Kahn, letter to William J. Salman, 7 November 1962, LIK, 030.II.A.34.2 (transcriptions of the specifications are inserted in the letter).
[50] In March 1961 Kahn's office is requested to supply samples of "sand plaster for all walls in off white color" (Galen Schlosser, letter to Thomas J. Regan, 6 March 1961, LIK, 030.II.A.34.1).
[51] A. J. Webb, letter to Penn Crete Products, Co., Inc., 5 July 1961, LIK, 030.II.A.34.1.
[52] *Ibid.*

chapter five
forms of prefabricated concrete

Formwork for the poured in place concrete of the Biology Building

The experience with prefabricated concrete that begins with the construction of the Richards Laboratories and continues on the worksite of the Tribune Review Building enters another chapter, though one still linked to the Richards Laboratories. While restoration of the surfaces of that complex is under way, Kahn is involved in the construction of the Biology Building, also along Hamilton Walk.[1] Construction begins in May 1961, when Kahn's office is still preparing the drawings and specifications[2] (a new version of the specifications is prepared in June).[3] United Engineers & Constructors, Inc., called in to repair the defects in the Richards Laboratories, takes charge of the construction. The engineers of Keast & Hood and Komendant remain by Kahn's side, offering consulting on matters having to do with concrete. Joseph E. Farrell, Inc., continues to handle the making of the concrete elements poured in place, while the precast parts are assigned to the Formigli Corporation[4] – it is not

[1] A version of the design for the Department of Biology is examined on 5 February 1958 ("Minutes of Meeting in Dr. Topping's Office, 9:00 A.M. February 5, 1958," LIK, 030.II.A.25.8). In July 1959 certain "exterior walls" planned to be made with an external layer of bricks and an inner layer of blocks are replaced by walls in blocks only, using the size 8 x 8 x 8 inches (Louis I. Kahn, letter to Joseph R. Farrell Inc., 30 July 1959, LIK, 030.II.A.25.23). The specifications in September on the concrete parts to pour in place are based on those of the Richards Laboratories ("Specifications," [21] September 1959, LIK, 030.II.A.22.2).

[2] United Engineers & Constructors Inc., "Sequence of structure construction activities Biology Building University of Pennsylvania," 31 January 1962, LIK, 030.II.A.22.21.

[3] "Animal-Insect Rooms-Service Laboratory Building. Phase B," 28 June 1961, LIK, 030.II.A.22.3.

[4] "Biology Building. University of Pennsylvania. Meeting – Tuesday 10 July 1962," LIK, 030.II.A.24.10. It is initially planned to produce the pieces in both the Berlin and the Williamstown plants.

257. Biology Building, Philadelphia, Pennsylvania, 1960-65. Detail of the porch (photograph by the author, 2010).

known why the collaboration with the Atlantic Prestressed Concrete Company was interrupted.

In August some of the foundations of Tower D have already been built. The construction of the tower foundations continues until the winter. In November the concrete is poured for the basement walls. Kahn intervenes in this phase of the work, on 4 December, to halt the pouring in metal formwork of "exposed concrete surfaces," since he has noticed unsatisfactory quality of the surfaces of the parts already made.[5] On 6 December the pouring of concrete on the formwork already in place for the slab of the first floor of Tower Y is also postponed, awaiting approval of the "flooring board forms,"[6] while possible changes to the formwork with steel panels criticized by Kahn are discussed. The pouring of the concrete for the slab of Tower Y happens halfway through January 1962. In February it is decided that the ceiling should be cleaned due to "remnants of form work clinging," while certain concrete walls have to be corrected because they are "too small and without proper anchor bolts."[7]

The definition of the technical details of the formwork for the concrete poured in place, and in particular of the "screw ties," is done by Joseph E. Farrell, Inc. The technicians of this company prepare the drawing of the "screw tie layouts" sent to Kahn on 25 May (it is the third variation submitted for Kahn's approval).[8] Kahn, in turn, sends the company his own proposal for "screw tie cones."[9] At the start of June there is discussion of the use of "red plastic tie cones," and Kahn sends Joseph E. Farrell, Inc., the drawing of the joints of the formwork of the slab of Tower Y.[10] The drawing, entitled "Cone Tie and Joint Patterns In Concrete," made by the technicians of Joseph E. Farrell, Inc., is approved by Kahn. Following the criticism of Dubler of United Engineers & Constructors, Inc., who states that the drawing does not show the "joints clearly or completely," Kahn indicates in red, on two copies of the "Cone Tie and Joint Patterns in Concrete," all the joints between panels, and sends the copies to Joseph E. Farrell, Inc., and United Engineers & Constructors, Inc.[11] In the meantime, test panels are made with Waylite blocks to be shown to Kahn (they are approved in May), while for the mortar Kahn asks Kelly to provide the formula of that of the Richards Laboratories.[12]

The formwork of the walls of the first floor of Tower Y of the Biology Building are made at the end of June. Kahn asks the technicians of Joseph E. Farrell, Inc., about the correct positioning of the spreaders, also to avoid flaws in the pouring phase; he is not convinced of the efficacy of the rows of ties inserted between the vertical joints of the panels. "Mr. Kahn telephoned the meeting that the cone ties centered on plywood joints are approved only if they are technically correctly placed to hold the forms without leakage at the ties and joints."[13] Steinbruch of Joseph E. Farrell, Inc., reassures Kahn that the position of the spreaders has been chosen precisely to prevent defects in the pouring.[14] Instead it is not possible to wait for Kahn's approval for the positioning of the formwork of the second floor of Tower Y, which Steinbruch warns must take place that same day, on 26 June, due to an unexpected strike of the workmen in charge of the installation of the reinforcements (the strike was cancelled). So Farrell proposes assembly of the formwork in keeping with what is indicated in the corrections to the drawing "Cone Tie and Joint Patterns" made by Kahn to show the positioning of the joints between the panels. It is therefore decided to apply at "joints in form panels" an "extra coat of form film."[15]

At the start of July certain particulars of the formwork are under discussion: "a. Ceiling forms at second floor and above where Farrell would like all the eight-foot dimensions to run east and west so that joists supporting forms run same direction. He will submit this for approval. b. Cone ties and joints in West wall of [tower] Y in view of new 5' 8" wide opening for future

[5] United Engineers & Constructors Inc., "Sequence of structure construction activities Biology Building University of Pennsylvania," 31 January 1962, LIK, 030.II.A.22.21.
[6] *Ibid.*
[7] "Reinspection," 15 February 1962, LIK, 030.II.A.23.8.
[8] "Meeting Tuesday, 29 May 1962," LIK, 030.II.A.24.10.
[9] *Ibid.*
[10] "Biology Building. University of Pennsylvania. Meeting Tuesday, 5 June 1962," LIK, 030.II.A.24.10.
[11] "Biology Building. University of Pennsylvania. Meeting Tuesday, 12 June 1962," LIK, 030.II.A.24.10.
[12] "Biology Building. University of Pennsylvania. Meeting Tuesday, 29 May 1962," LIK, 030.II.A.24.10.
[13] "University of Pennsylvania. Biology Building. Meeting – Tuesday, 26 June 1962," LIK, 030.II.A.24.10.
[14] *Ibid.*
[15] *Ibid.*
[16] "University of Pennsylvania. Biology Building. Meeting – Tuesday, 3 July 1962," LIK, 030.II.A.24.10.
[17] "University of Pennsylvania. Biology Building. Meeting – Tuesday, 10 July 1962," LIK, 030.II.A.24.10.

connection."[16] The concrete of the slab of the second floor is poured on 10 July.[17] Joseph E. Farrell, Inc., also shows Kahn the design of the formwork of the ceilings starting from the third floor.[18] The definitive drawings of Joseph E. Farrell, Inc., for the formwork continue to be sent to Kahn for his approval.[19] During an inspection on 7 August, Leidigh finds a significant amount of "honey-combed concrete," and Kahn issues a reminder of the fact that according to the specifications parts with these defects must be remade.[20]

The whole project process shows that the design of the formwork is done by the contractors and technicians, and developed level by level, sector by sector, during the construction phases. Kahn's office intervenes only to approve or modify the proposed solutions, not always managing to supervise all the phases of assembly of the panels. Further confirmation of the fact that the design of the formwork is not totally controlled by Kahn in the construction of the Biology Building can be found in Farrell's proposal of a rough estimate of costs of assembly of the panels in staircase A-46, since in the drawings there are no precise indications of the number and position of the ties.[21] It is no coincidence that in the concrete masterpieces built towards the middle of the 1960s, Kahn gets involved in the study of all the technical details of the formwork, from the number of boards or panels to the form of the imprint of the ties, achieving almost total control of the fabrication of the concrete, though it is still done during the worksite phases.

The experience gained working with Kahn on the construction of the Richards Laboratories allows Keast & Hood, also during the making of the Biology Building, to take suitable measures to protect the concrete surfaces. Keast & Hood intervenes, through its employee Leidigh, to prevent the concrete surfaces of the walls from being defaced by the many drawings made by the technicians to proceed with installation of the physical plant systems.[22] The expertise acquired by Keast & Hood, and also by Komendant, in the making of concrete to match Kahn's expectations, can also be seen when the technicians of United Engineers & Constructors, Inc., decide to introduce "plastic wood" in the joints between the panels of the formwork to prevent the phenomena of "bleeding of concrete"; Kahn personally intervenes to prohibit any corrections of this type without prior consent of either the engineers of Keast & Hood or of Komendant, "who know the results desired and how to attain them."[23]

After having opted for red plastic ties in the formwork, at the start of August, Joseph E. Farrell, Inc., submits a request to Kahn for supplementary payment for their removal.[24] Therefore the plastic ties that characterize the concrete walls of the Biology Building are left in place only as a result of decisions regarding the cost of their removal. Precisely this question of the ties, for Kahn, also following the episodes of the formwork of the Biology Building, becomes a crucial question for the constructive form of the exposed concrete walls he builds at the end of 1962 in La Jolla.

Precast concrete of the Formigli Corporation

The structure of prefabricated members of the towers of the Biology Building is modified with respect to that of the Richards Laboratories. The complex interlock of secondary Vierendeel beams, mainly required for the installation of the formwork for the pouring of the top slabs, is eliminated, and in its place Joseph E. Farrell, Inc., develops a different anchoring system for the formwork. The profile of the edge beams is adapted as a result of this elimination of the secondary beams. The edge beam resembles the

258. Biology Building, Philadelphia, Pennsylvania, 1960-65. Detail of the concrete surface where it is visible a spacer used in the formwork of the interior walls (photograph by the author, 2010).

one Kahn proposed for the Washington University Library. In the portico, the square compartments formed by the beams are closed with plaster ceilings to protect the lighting fixtures (to be made in "cemesto").

For the fabrication of the columns and beams in precast concrete of the floors systems, at first it is decided to use both the plants owned by the Formigli Corporation, the main one in Berlin and the other in Williamstown, also in New Jersey (about 10 miles from Berlin), though production begins, at first, only in Berlin.

In May and June 1962 the Formigli Corporation makes definitive drawings of beams and columns, in collaboration with Joseph E. Farrell, Inc.[25] The quality of the concrete of the first columns produced in Berlin, examined by Kahn and Komendant on 8 June,[26] is considered unacceptable; they order treatment of the surfaces to make it possible to still use the columns, at the least visible in points in the Biology Building. The inspection of 20 June to check on the quality of the concrete of the new columns also disappoints Kahn, Komendant, the engineers of the United Engineers & Constructors, Inc., the technicians of Joseph E. Farrell, Inc., and the representatives of the University of Pennsylvania. The main defects are an excessive percentage of large-grain aggregate, leading to rough surfaces – a composition considered "harsh" – and the loss of compound caused by improper fabrication of the joints of the molds, producing "discolors" on the surfaces of the columns.[27] Another series of columns inspected on 25 June is produced based on technical remedies recommended to the Formigli Corporation. Following the satisfactory results of this new trial, an official document is drawn up, the "Memorandum" dated 6 July and sent to the Formigli Corporation on 16 July. The Memorandum specifies the production modes to be applied, to obtain a precast concrete "as sound and as strong as it is possible to make it."[28]

The paragraphs of the Memorandum cover the making of the molds, the positioning of the reinforcements and the sheaths inside the molds, the

[18] *Ibid.* Kahn approves the design on 18 July ("Biology Building. University of Pennsylvania. Meeting – Tuesday 17 July, 1962," LIK, 030.II.A.24.10).
[19] "Biology Building. University of Pennsylvania. Meeting – Tuesday, 24 July 1962," LIK, 030.II.A.24.10; "Biology Building. University of Pennsylvania. Meeting – Tuesday, 14 August 1962," LIK, 030.II.A.24.10.
[20] "University of Pennsylvania. Biology Building. Meeting – Tuesday, 3 August 1962," LIK, 030.II.A.24.10.
[21] "University of Pennsylvania. Biology Building. Meeting – Tuesday, 17 July 1962," LIK, 030.II.A.24.10.
[22] "All on jobs are warned that concrete walls are the final finish, that necessary marks must be small pencil marks for control only," Leidigh advises ("University of Pennsylvania. Biology Building. Meeting – Tuesday, 10 July 1962," LIK, 030.II.A.24.10).
[23] "University of Pennsylvania. Biology Building. Meeting – Tuesday, 17 July 1962," LIK, 030.II.A.24.10.
[24] "Architect was asked if red cone ties are to be left in place in face of Farrell's demand for extra payments" ("University of Pennsylvania. Biology Building. Meeting – Tuesday, 3 August 1962," LIK, 030.II.A.24.10).
[25] "University of Pennsylvania. Biology Building. Meeting – Tuesday, 29 May 1962," LIK, 030.II.A.24.10, and "Biology Building. University of Pennsylvania. Meeting Tuesday, 5 June 1962, LIK, 030.II.A.24.10.
[26] "Biology Building. University of Pennsylvania. Meeting Tuesday 5 June 1962," LIK, 030.II.A.24.10.
[27] "Memorandum," 6 July 1962, LIK, 030.II.A.22.22.
[28] *Ibid.*

259-260. Biology Building, Philadelphia, Pennsylvania, 1960-65.

259. Column at the ground floor (photograph by the author, 2010).

260. Detail of a column at the ground floor (photograph by the author, 2010).

composition of the concrete, the pouring methods, removal of the pieces from the molds, treatment of possible surface defects, or dressing, steam treatment, surface cleaning and smoothing, storage at the plant and criteria of control of individual pieces both at the plant and at the worksite.[29]

After the preparation of the Memorandum, the technicians in charge of the worksite for the Biology Building intervene to speed up the making of the pieces. At the Formigli plant in Berlin, the only one active at the time, seven columns have been made in the month of June, and two more in July. Given the fact that starting on 18 July the plant will work exclusively on the production of pieces for the Biology Building, Formigli claims to be able to make two or three columns per day.[30] It is also planned to produce the parts in both of the company's plants, distinguishing between them in terms of finishing and complexity: the columns will continue to be produced in Berlin, while the main and edge beams will be made at Williamstown.[31] Kahn unsuccessfully proposes that the edge beams also be made in Berlin, as he considers them essential parts of the exposed concrete, on a par with the columns.[32] To start production at Williamstown, Formigli obtains the molds for the girders on 19 July.[33] In the meantime, the Formigli Corporation continues to make the definitive drawings that are still lacking for the pieces in precast concrete (Komendant is always in charge of checking the reinforcement details). One category of the precast concrete parts is that of the "fascias" to be placed above the windows. During July the prefabricated fascias stored on the worksite are damaged by vandals, or appear "discolored" due to lack of adequate protection.[34] Following Kahn's request, the damaged fascias are remade and sent to the worksite at the beginning of September.

The critical observations on the quality of the production of the precast pieces, after the issuing of the Memorandum, though there are no official replies on Formigli's part, do not fall on deaf ears among the company's technicians, who go to Philadelphia on 24 July to visit the Richards Laboratories and to compare their precast concrete with that of the Atlantic Prestressed Concrete Company. As they see it, the recommendations indicated in the Memorandum to obtain perfect precast concrete

were not applied in the making of the pieces of the Richards Laboratories, for which they make a true list of flaws:

"Messrs. Stagliano and Gross of Formigli had been looking over the precast work of the Richards Building and came to the meeting room as the meeting was breaking up. They are disturbed that the architect and his engineers are telling Formigli how to make the precast work and are complaining of many defects in Formigli's work which, in their opinion, are many times worse in the accepted Richards Building. Mr. Stagliano took the architect's representative and the U.E. men to the Richards Building and pointed out 'oil canning' marks left by steel forms, weld spots left by steel forms on a column, cracks and grazing of surfaces, patches on corners, slump markings, shrinkage cracks on column edges and faces, rust spots caused by stirrups being too close to the surface, form oil stains near bottoms of columns, crooked column edge liners. Formigli wants the architect to stop telling them how to achieve results, but merely tosay [sic] what is wanted and to rely on them to achieve it and meet the specifications. Formigli resents the requirement of the memorandum to wash columns down with green soap because it costs them about ten dollars per column. They agree to do it and will continue to do it but do not feel happy about it after seeing the oil stains on the Richards columns. Formigli wants Architect to judge the columns already cast by the terms of the contract and to give a specific acceptance or rejection of each piece with specific reasons for rejections, if any."[35]

On 31 July the columns of the first floor have been made, and the "stubs columns" are in progress.[36] Kahn and the engineers working on the project make a visit on 13 August to the plant in Williamstown to examine the main beams and edge beams that have just been completed. Indications are made to improve the surfaces, including a change to the molds.[37] Towards the last week in August, United Engineers & Constructors, Inc., has still not been able to ascertain from Formigli Corporation the date of the start of erection of the prefabricated parts at the worksite, though appreciable improvement is noted in the Williamstown production with the new beam molds.[38] After the discussions of the previous months regarding the various possibilities of the assembly phases,[39] United Engineers & Constructors, Inc., asks the

[29] *Ibid.*
[30] "University of Pennsylvania. Biology Building. Meeting – Tuesday, 17 July 1962," LIK, 030.II.A.24.10.
[31] *Ibid.*
[32] See "Biology Building. University of Pennsylvania. Meeting – Tuesday, 17 July 1962," LIK, 030.II.A.24.10.
[33] "Biology Building. University of Pennsylvania. Meeting – Tuesday, 17 July 1962," LIK, 030.II.A.24.10. Formigli claims it is possible to have the molds of the girders ready by 27 July in Williamstown ("University of Pennsylvania. Biology Building. Meeting – Tuesday, 24 July 1962," LIK, 030.II.A.24.10). The fact that these molds arrive from Philadelphia and can "be adapted for any length girder needed on this job" would suggest that they are the molds used for the girders of the Richards Laboratories.
[34] "Biology Building. University of Pennsylvania. Meeting – Tuesday, 24 July 1962," LIK, 030.II.A.24.10.
[35] *Ibid.*
[36] *Ibid.*
[37] "The tops of surfaces in girder openings have some depressions left by air pockets which will be filled level to make housekeeping easier. It will not be necessary to remove form marks from interior surfaces of openings. Some repairs will be permitted on bottom edge, one side, where form infill was high. This form will be corrected in subsequent pours" ("Biology Building. University of Pennsylvania. Meeting – Tuesday, 14 August 1962," LIK, 030.II.A.24.10). "Formigli poured the first two spandrels without the required camber of ¼" in top surface. Forms must be corrected and spandrels re-made" (*ibid*).
[38] "Biology Building. University of Pennsylvania. Meeting – Tuesday, 21 August 1962," LIK, 030.II.A.24.10.
[39] Although the erection of the columns and beams of the floors is assigned once again to Cornell & Company, with decisive technical consulting supplied by Keast & Hood, at first the contact of United Engineers & Construction, Inc., for questions of assembly on worksite of precast concrete parts is Stoelker, of the Formigli Corporation. Unlike the sequence employed for the Richards Laboratories, Stoelker proposes completing the assembly of all the parts of each single tower, E or D, before starting work on the other tower. But United Engineers & Construction, Inc., wants to follow the example of the Richards Laboratories in order to allow other contractors to continue their own work without interruptions ("Biology Building. University of Pennsylvania. Meeting – Tuesday, 17 July, 1962," LIK, 030.II.A.24.10).

Formigli Corporation for instructions on the sequence of erection of the pieces, taking the indications supplied by Keast & Hood into account.[40] It is the document prepared by Keast & Hood on the erection procedure that circulates among the people in charge of the worksite of the Biology Building.[41] In the meantime, Kahn has to approve the poor results in the making of a column of Tower E before it is shipped to the worksite – it is the column corresponding to the code E-1, which has a "bulge in one side of the opening through the web."[42] Also in September, steps are taken to produce the prefabricated parts needed for the construction of the sixth floor that has been added to the towers. The last member of precast concrete, one of the main beams, is produced in Williamstown on 16 October.[43] Certain fascias continue to be made, even after this date; others are replaced with concrete poured in place, in clean formwork envisioned at first in steel or Masonite, to convey the impression of precast concrete.[44] "Farrell is inserting Masonite in concrete forms," we read in a worksite report, "to give smooth appearance at exposed ends of slabs in lieu of precast fascias."[45]

The first shipment from Berlin and Williamstown of the pieces in precast concrete happens during the first days of September – on 11 September the shipment is at the facility of Cornell & Company in Woodbury, New Jersey.[46] A few days later, Leidigh provides the employees of Cornell & Company with the list of the materials needed for the grouting of the precast concrete, based on the experience gained with the Richards Laboratories.[47] The first prefabricated floor system of Tower D is installed towards 25 September.[48] At the start of October, Steinbruch explains to the worksite supervisors how Joseph E. Farrell, Inc., intends to assemble the formwork to pour the concrete top slabs of towers D and E.[49] In the meetings of the first days of October, the sequence of the pouring of the top slabs of Towers D and E is often discussed, leading to delays in the making of the top slab of the second level of Tower E, resulting in growing costs for the scaffolding at the worksite[50] – the pouring for this level is done on 26 October. "Farrell is tightening floor forms by wedging. They will also be tightened by wetting overnight before pour."[51] After the discussions on the phases of pouring, work resumes in keeping with a sequence that takes the timing of installation of the formwork and the setting of the concrete into account, and the possibility of bringing the concrete from Tower Y using concrete buggies. The timing is also impacted by the fact that Joseph E. Farrell, Inc., has readied only two of the three series of formwork envisioned, with nine compartments each. Starting from the third floor, the sequence of the pouring of the top slabs starts from Tower E and proceeds towards Tower D, so that the floor completed in Tower E can be crossed by the concrete buggies transporting the compound from its point of arrival at this level, in Tower Y.

[40] "Biology Building. University of Pennsylvania. Meeting Tuesday, 18 September 1962," LIK, 030.II.A.24.10.
[41] "Biology Building. University of Pennsylvania. Meeting – Tuesday, 25 September 1962," LIK, 030.II.A.24.10.
[42] "University of Pennsylvania. Biology Building. Meeting – Tuesday, 18 September 1962," cit.
[43] "University of Pennsylvania. Biology Building. Meeting – Tuesday, 16 October 1962," LIK, 030.II.A.24.10.
[44] "Biology Building. University of Pennsylvania. Meeting – Tuesday, 30 October 1962," LIK, 030.II.A.24.10.
[45] "Biology Building. University of Pennsylvania. Meeting – Tuesday, 6 November 1962," LIK, 030.II.A.24.10.
[46] "University of Pennsylvania. Biology Building. Meeting – Tuesday, 11 September 1962," LIK, 030.II.A.24.10.
[47] "Biology Building. University of Pennsylvania. Meeting – Tuesday, 18 September 1962," LIK, 030.II.A.24.10.

At the meeting on 30 October discussion focuses on the date when representatives of the University of Pennsylvania and United Engineers & Constructors, Inc., together with Kahn and his engineers (those of Keast & Hood and Komendant) can inspect the eventual "flaws" caused by the installation of the pieces, using drawings for each floor system procured from Kahn on which to indicate the positions of the flaws[52] (copies of the drawings are made with indications from the inspection, to distribute to the various people in charge). Given the fact that some of the edge beams are not positioned exactly at the level of the floors (they are slightly higher or lower), surveying is ordered to find out whether the defect is a result of fabrication of their curvature, or of settling (two series of surveys are recommended at the same points, when the formwork of the top slabs are still in place, one before and one after the pouring).

United Engineers & Constructors, Inc., urges the Formigli Corporation to spread silicone on the precast concrete pieces stored at the worksite, after having corrected possible defects, to make the surfaces waterproof.[53] But halfway through December the Formigli Corporation has not yet carried out these operations.[54]

When Kahn's engineers inspect the pouring of the top slabs, at the start of January 1963, the fabrication turns out to be defective, as Leidigh emphasizes, due to the fact that the prefabricated beams already installed have not been heated prior to the pouring of the concrete: "The temperature of precast girders must be brought from freezing level up to near the temperature of the concrete while being placed. This entails keeping heat overnight in the story below the floor to be poured."[55] At this date the formwork of the top slab of the sixth level of Tower E are already in place, so according to Joseph E. Farrell, Inc., it is possible to pour the flat roof, heating the girders prior to the pouring by means of the method proposed by Leidigh, starting from the floor below – the sixth, taking advantage of the presence of the formwork. The formwork of the roof of Tower E are reassembled in the roof of Tower D, which is poured at the end of January, before the sixth floor, in keeping with the same procedure used for Tower E. In January and February the pouring of the concrete for the walls, the roof and the staircases of Tower Y is completed, and in the roofs and sixth floors of Towers E and D (the formwork of the sixth floor of Tower D are those stripped from the fifth floor of the same tower).

United Engineers & Constructors, Inc., continues to complain that the Formigli Corporation has yet to take a position on the request sent in November regarding the necessary treatments to be performed on the precast concrete surfaces, and urges Kahn and Keast & Hood to meet with the contractor's technicians for an inspection. Halfway through February the Formigli Corporation commits to the treatment of the precast concrete, including the application of silicone.[56]

48 "Biology Building. University of Pennsylvania. Meeting – Tuesday, 25 September 1962," LIK, 030.II.A.24.10.

49 "University of Pennsylvania. Biology Building. Meeting – Tuesday, 2 October 1962," LIK, 030.II.A.24.10.

50 "University of Pennsylvania. Biology Building. Meeting – Tuesday, 9 October 1962," LIK, 030.II.A.24.10.

51 "University of Pennsylvania. Biology Building. Meeting – Tuesday, 23 October 1962", LIK, 030.II.A.24.10.

52 "Biology Building. University of Pennsylvania. Meeting – Tuesday, 30 October 1962," LIK, 030.II.A.24.10.

53 "Biology Building. University of Pennsylvania. Meeting – Tuesday, 20 November 1962," LIK, 030.II.A.24.10.

54 "Biology Building. University of Pennsylvania. Meeting – Tuesday, 18 December 1962," LIK, 030.II.A.24.10.

55 "Biology Building. University of Pennsylvania. Meeting – Tuesday, 2 January 1963," LIK, 030.II.A.24.10

56 "University of Pennsylvania. Biology Building. Meeting – Tuesday, 19 February 1963," LIK, 030.II.A.24.10.

261. Booklet of the G. Kreier, Jr., Inc. (G. Kreier, Jr., Archive, AAUP, box 6).

262. Mill Creek district, Philadelphia, 1951-56.
"Plans Elevations Section I BR D.U. Building 7.20.32, and of 12. Lintel Schedule," 27 May 1959 (LIK, 030.I.C.375.8.12).

The last important act of the concrete work of the Biology Building has to do with the pouring of the top slab of the sixth floor of Tower E and that of the roof of Tower D, in February 1963, under problematic weather conditions. The large temperature drop during the work prompts the engineers of United Engineers & Constructors, Inc., and Keast & Hood to request structural load testing.[57]

During the course of 1963 the Biology Building is completed with the assembly of the brick claddings, after the trials conducted since the summer of the previous year to permit Kahn to choose a combination of bricks with different shades of color, to produce a delicate effect of chromatic variation. The model remains that of the Richards Laboratories.

The records of the inspections of the trials of bricks and concrete blocks on the worksite reveal something of the personality of an artist guided by his sensitivity to the effects of the most imperceptible color variations, as demonstrated in his way of controlling the surfaces of concrete. Precisely in this same period, Kahn – on the worksite of one of his masterpieces – is radically re-examining the technical and poetic question of the color of concrete.

The fiberglass molds of George Kreier Jr. for the Mill Creek Apartments II

After the construction of the Richards Laboratories and the Tribune Review Building, Kahn no longer limits his action to the use of concrete poured in place, deploying different types of formwork with wooden boards or sheets of plywood.

The construction of the second lot of housing units, designed back in 1956, in the Mill Creek development, Mill Creek Apartments II, begins at the end of 1960, and offers another opportunity to come to grips with the technique of prefabrication of reinforced concrete. Kahn designs special architraves for the openings of the ground and first floors, to be made in precast concrete and assembled on worksite with the help of a crane, on concrete block walls clad in brick. The architraves of the ground floor also form the parapet of the first floor windows. Most of the architraves have protruding H-shaped flanges to protect the windows from sun and rain – the so-called "projecting hoods."[58] The particular height of the architrave-parapets demonstrates Kahn's desire to also obtain large pale and smooth concrete inserts in the brick cladding.

The specifications prepared in June 1960 contain the section "Special precast concrete items" which includes, among other things, "lintels, beams, skylights." The technical indications describe the possibility of making the molds in concrete or steel, with surfaces "free from irregularities, dents, and sags," and careful positioning of the joints: "No form joints in plane exposed surfaces will be accepted."[59] In spite of the indication that "edges and fins are to be lightly stoned," the concrete produced with the molds is not to be retouched: "There will be no other opportunity for providing a final acceptable finish other than that obtained by the initial pouring." To prepare the compound, "high-early strength Portland Cement" is used.[60]

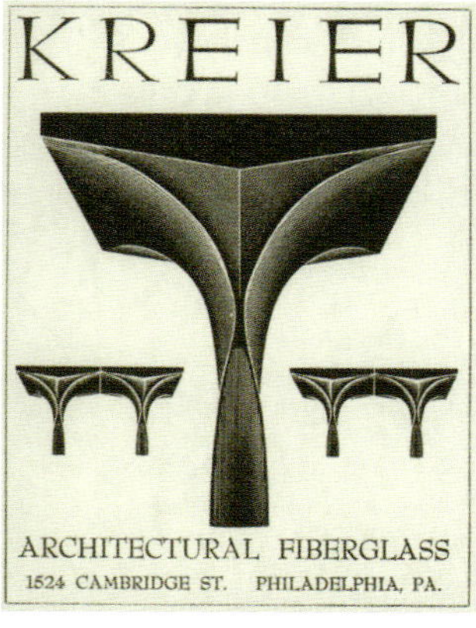

The making of the precast elements is assigned to the Herculite Corporation based in Chester, Pennsylvania. On 7 February 1961 two members of Kahn's staff, Kwo and Wisdom, together with Gianopulos and the General Contractor, John McDonnell, visit the production facility of Herculite Corporation at Raubsville, to inspect the equipment with which the pieces of the Mill Creek Apartments II are to be prefabricated.[61] On this

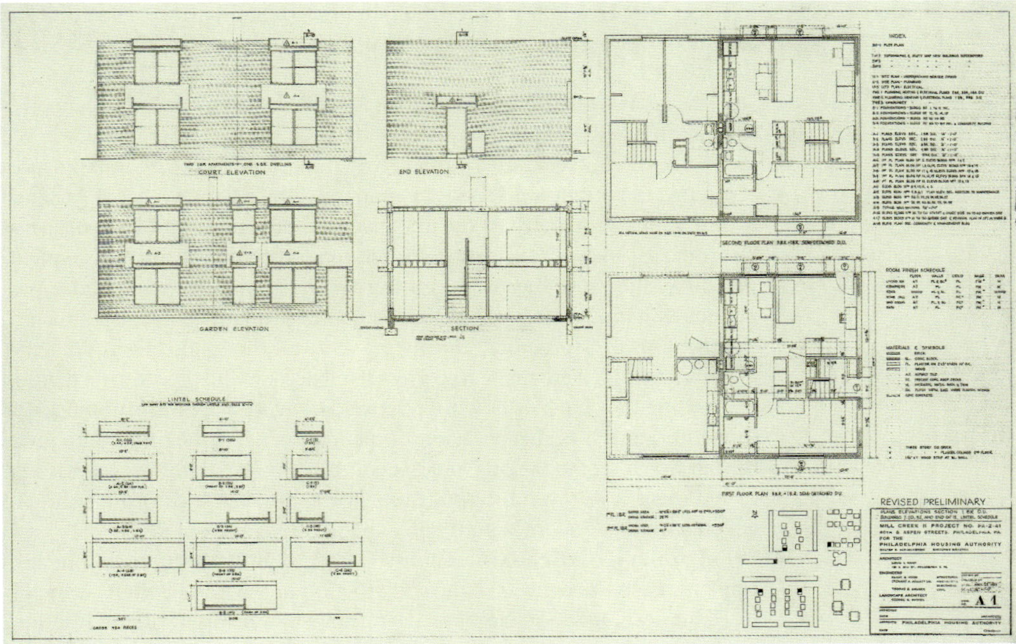

occasion they examine a sample architrave made according to the dictates of the specifications (the material used for the mold – concrete or steel – is not indicated). George Kreier Jr. is also there at the visit, and McDonnell introduces him to Kahn's staff.

In 1951 Kreier Jr. had founded an entreprise based in Philadelphia, specializing in the fabrication of molds for precast concrete pieces, and known for its use of fiberglass. Already in 1960 Kreier Jr. had contacted Kahn to show him his "fiberglass reinforced polyester" for the construction of architectural members, after having heard Kahn's lecture at the Academy of Fine Arts, on 17 November. "Like your web of glass and glass beams with stainless steel rods," he wrote to Kahn, "these materials are immediately realizable if we but have the imagination to translate our best vision into reality."[62]

"Mr. Kreier recommended the use of reinforced plastic molds for the lintels with projecting hoods," it is reported following the meeting of 7 February 1961 at the workshop of the Herculite Corporation. "We had specified steel or concrete forms, but we and our engineers find no reason

[57] "University of Pennsylvania. Biology Building. Meeting – Tuesday, 12 February 1963," LIK, 030.II.A.24.10.
[58] David P. Wisdom, letter to the Philadelphia Housing Authority, 14 February 1961, LIK, 030.II.A.9.11.
[59] Louis I. Kahn, "Mill Creek Apartments II […]. Specifications for the construction of two and three story row and semi-detached dwellings," 21 June 1960, LIK, 030.II.A.31.8.
[60] *Ibid*. The exposed blocks are in Waylite of "smooth textured and of uniform color." Also for the mortar, Kahn indicates that the color should be that of its components – "the natural color of the aggregates," "*no mortar coloring* will be required."
[61] David P. Wisdom, letter to the Philadelphia Housing Authority, 14 February 1961, LIK, 030.II.A.9.11.
[62] George Kreier Jr., letter to Louis I. Kahn, 18 November 1960, LIK, 030.II.A.9.11.

263-267. Mill Creek Apartments II, Philadelphia, 1951-56.

263. Sketches by G. Kreier, Jr. (G. Kreier, Jr., Archive, AAUP, box 2).

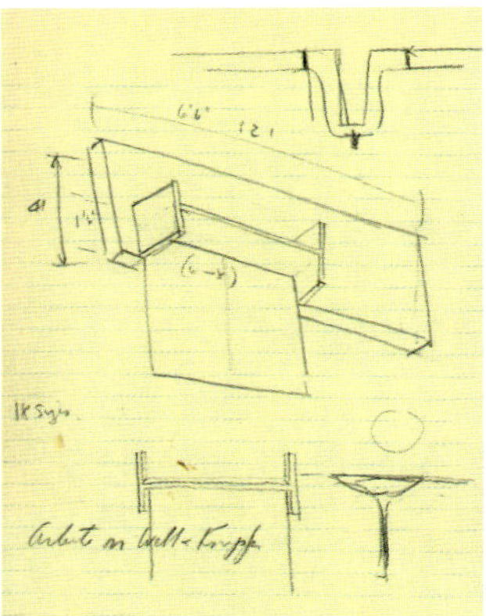

to reject reinforced plastics if Kreier and the Herculite Corp. can solve the details of casting and stripping and if their products can meet our specifications."[63]

The construction of the architraves for the Mill Creek Apartments II is done with molds supplied by George Kreier Jr. On 29 August the first sixteen "precast concrete lintels" are delivered to the worksite by the transport company, the Structural Product Company – their deployment is assigned to B.A. Lindstrom, Inc.[64] At the first inspection the pieces prove to have certain defects, like porous and dark corners, visible and not "unsteamed" patching, and chipping produced during transport. Therefore Kahn proposes involving Komendant as a consultant for quality control of prefabrication. On 5 September, during a visit to the plant in Raubsville, recommendations are made to correct the defects in the pieces – the method of grouting the chips, or how to "remove dark work edges" by rubbing with carborundum.[65] On this occasion Komendant advises the firm not to grout the "air holes" of the various "precast concrete lintels." "Dr. Komendant stipulated there is no need to grout these holes, because such is the nature of concrete that at any freezing it will expand and come out."[66]

For the roof of the Community Center, built with the same prefabricated parts, in the context of the construction of the Mill Creek Apartments II, Kahn develops a system of precast beams to assemble with an interlock of short rods whose origin dates back to the scheme outlined in the notebook of Villard de Honnecourt.[67] At the center of the framework he places a skylight, also known as a "cupola," again in prefabricated concrete. As in the Tribune Review Building, each beam end rests on a brick pillar emerging from the wall. To make the system visible, Kahn calls for openings in the wall at the position of each pillar, which were never built.

The smooth and perfectly shaped concrete from the molds of George Kreier Jr. forms a contrast with the exposed concrete of the four play areas next to the Community Center, conceived as an artistic composition of enclosures and steps, in concrete shaped by basic formwork made of wooden boards. This work is unique in Kahn's output for the elementary, primeval way the concrete is deployed.[68] In the study for the garden on the lot where the Community Center will stand, Kahn had considered constructing concrete walls that would border the outdoor spaces, partially recessed in the ground, envisioned as true ruins of a city.[69]

[63] David P. Wisdom, letter to Philadelphia Housing Authority, 14 February 1961, LIK, 030.II.A.9.11.9.11. See also George Kreier Jr., letter to Marie Kwo, 23 February 1961, and George Kreier, Jr., letter to Louis I. Kahn, 23 February 1961 (George Kreier, Jr., Archives, AAUP, Philadelphia). Kreier writes a series of notes for the molds of the Mill Creek towers, with drawings and estimates (George Kreier, Jr., letter to Marie Kwo, 17 April 1961, George Kreier, Jr., Archives, AAUP, Philadelphia). See also the sheet of sketches, "Louis Kahn. Meeting 10 Feb." (George Kreier, Jr., Archives, AAUP, Philadelphia).

[64] Marie Kwo, letter to Harry H. Brunt, 8 September 1961, LIK, 030.II.A.31.7.

[65] *Ibid.*

[66] *Ibid.* Kahn criticizes the stains in the concrete of the "cupolas"; these stains are the result of having used "'high early' light weight concrete."

[67] See period photograph, LIK, 030.IV.A.375.3. During a visit to the workshop, Kahn complains about the "blotches" in the concrete of the "cupolas" – the cause is identified as the use of "high early light weight concrete" (Marie Kwo, letter to Harry H. Brunt, 8 September 1961, LIK, 030.II.A.31.7).

[68] See the photographs (LIK, 030.IV.A.375.5, and LIK, 030.IV.A.375.3), the plan (LIK, 030.IV.A.375.1) and the axonometric dated 31 August 1962 (LIK, 030.IV.C.375.006).

[69] See the drawing LIK, 030.IV.C.375.006.

264-266. Construction site photographs (LIK, 030.IV.A.375.1.32; 45; LIK, 030.IV.B.375.21).

267. View (LIK, 030.IV.A.375.3.8).

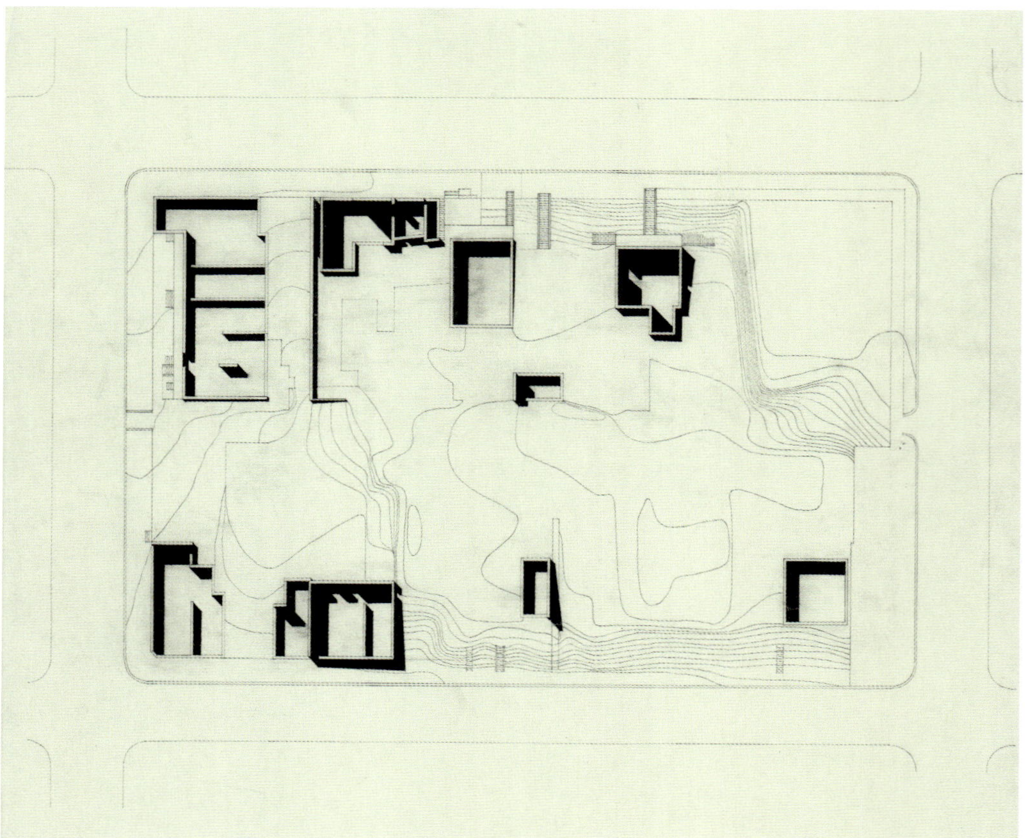

Scholastic exercises on tetrahedra prefabricated in post-tensioned concrete

Precisely when the various worksites, from Philadelphia to Greensburg, demonstrate Kahn's commitment to the development of the precasting of reinforced concrete members, research begins to spread that can be directly traced back to the experiences of these architectural projects, also with an eye on being able to transform other previous works according to the system of working of the concrete.

Kahn never manages to build a Space Frame with individual precast tetrahedra, in spite of his statements regarding the project for the City Hall Building. But an important path of research is concealed behind his statements, one that begins after the opening of the Yale Art Gallery and continues until the early 1960s, in different American universities, on the various ways of achieving Kahn's visions. When American students experiment with floors system made of precast reinforced concrete members, also through the making of experimental models, as an example they look to the floor system of precast parts of the Richards Laboratories, giving rise to a series of variations on the prefabrication of the constituent parts of a framework of Vierendeel beams.

Kahn's lectures at different American institutions of higher education and his contacts with teachers like Horacio Caminos and Eduardo Catalano lay the groundwork for the experimental research conducted by students on the construction of tetrahedra, on webs of Vierendeel beams and other similar forms for reinforced concrete floors system, of the type used in the Yale Art Gallery and the Richards Laboratories. All the parts of the various floors systems belong to Kahn's "hollow stones" typology, for the passage of physical plant conduits. The columns also obey the same principle of hollowness, or they have a cruciform section, like the one proposed for the AFL Medical Services Building, again to facilitate the passage of vertical piping. It is no coincidence that the students often use the term "hollow column."

268-269. Mill Creek Apartment II, Philadelphia, 1951-56.

268. View of the garden (LIK, 030.IV.A.375.5.8).

269. Planimetry of the garden (LIK, 030.I.C.375.6.8).

The main centers of experimentation on the prefabrication of tetrahedra in keeping with Kahn's vision are the School of Design of North Carolina State College,[70] where Kahn gives important lectures, and the Massachusetts Institute of Technology in Cambridge, Massachusetts, where Kahn acts as an advisor on doctoral theses, under the direction of Catalano. Instead, it does not appear that the educational activities of the Yale School of Architecture covered the question of the making of concrete floors system like those described by Kahn.

With the students of the Massachusetts Institute of Technology, Kenzo Tange develops, in 1959, the project of alternative *Unités d'Habitation* to those of Le Corbusier, marked by the use of the Tetrahedron Floor System, also with the inclined main load-bearing structures, as in Kahn's studies after the Yale Art Gallery. Confirmation of the spread of Kahn's geometrics at Massachusetts Institute of Technology can also be found in the master's thesis by Bruce E. Erickson, in 1960, on a "prototype floor structure" to be made by means of "truncated precast pyramids with solid sides."[71]

[70] At least until 1952, Catalano, at the School of Design of North Carolina State College, conducts studies on the space frame as part of his teaching (*Space frame discussed*, in, "Architectural Forum," vol. XCVIII, February 1953, No. 2, pp. 154-160, p. 158).

[71] Robert P. Burns Jr., Philip A. Kupritz, John E. Rudquist, "Precast Floor System. Orthogonal, Open-web Concrete Structures," "submitted in partial fulfillment of the requirements for the degree of Master of Architecture in the Department of Architecture on July 25, 1962," p. 39. See also Bruce E. Erickson, "A Prototype Floor Structure," Masters of Architecture Thesis, MIT School of Architecture and Planning, September, 1960.

Kahn's lectures at the School of Design of North Carolina State College have an impact on the educational exercises suggested by Caminos, in 1961, on the principle of how "to incorporate in the true structure, space for ducts and pipes using the total height of the space between floor and ceiling."[72] The students make models in concrete of two types of floors system, one with tetrahedra (triangular pyramids) and one with pentahedra (square-base pyramids), with the individual prefabricated parts – "maximum use of precast elements, minimum of concrete poured in place." The tetrahedron is "post-tensioned in two directions."[73] The form of the supports of the two floors system is the cruciform column. "The precast pyramids are set in place, reinforcing added, bottom grid poured, floor slabs set, and top grid poured. The edge units are poured in place. When cured, the system is post-tensioned."[74]

The construction of the floor system models with characteristics similar to those developed at North Carolina State College happens during the course of 1962 at the Massachusetts Institute of Technology, in the context of the thesis projects monitored by Catalano, and prepared by Robert P. Burns Jr., Phillip A. Kupritz and John E. Rudquist. The three projects are grouped under the heading "Precast Floor System. Orthogonal, Open-web Concrete Structures."[75] The construction systems imagined by Burns Jr., Kupritz and Rudquist are conceived on the basis of the problematic issues that have emerged starting with the construction of the Yale Art Gallery. In certain cases use is envisioned of panels made with a "mixture of wood fibers and cement," of the type made by the Tectum Corporation used in the Yale Art Gallery.[76] Among the works analyzed by Burns, Jr., Kupritz and Rudquist, there are the Yale Art Gallery and the Richards Laboratories.

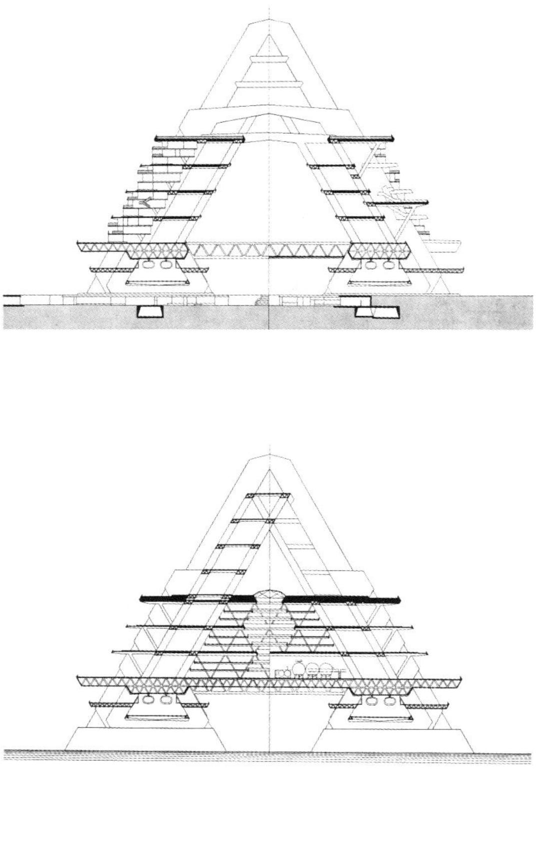

270. G. Pillorge, E. Haladay, T. Niedermann, G. Solomons, school project for an Residentian Unit, director K. Tange, School of Architecture and Planning, M.I.T., 1959 (U. Kultermann, *Kenzo Tange 1946-69 Architecture et Urbanisme*, Zürich, Verlag für Architektur, Artemis, 1970, p. 109).

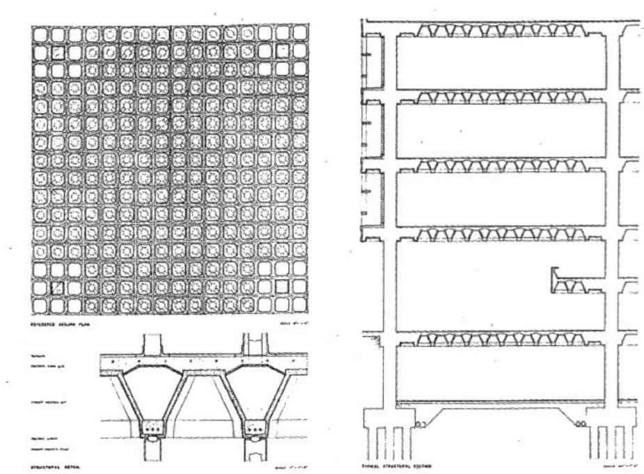

271. B. E. Erickson, "A Prototype Floor Structure," Masters of Architecture Thesis, School of Architecture and Planning, M.I.T., September, 1960.

272. H. Caminos students, two-way concrete space frame as a floor system with the column, School of Design, North Caroline State College, 1961 (J. Borrego, *Space Grid Structures, Skeletal Frameworks and Stressed-Skin Systems*, Cambridge-Massachusetts, London, The MIT Press, 1968, p. 58).

273. H. Caminos students, precast concrete tetrahedron unit, School of Design, North Caroline State College, 1961 (J. Borrego, *Space Grid Structures, Skeletal Frameworks and Stressed-Skin Systems*, Cambridge-Massachusetts, London, The MIT Press, 1968, p. 60).

Burns, Jr., as variations for the supports, considers "the solid cruciform and the hollow column."[77]

"The code of New Haven," we find written about the Yale Art Gallery, "required the architect to measure the geometry by making some of the sides of the tetrahedron continuous to achieve beam action. [...]. This triangulated grid is not applicable to the usual orthogonal partitioning methods. The triangulated voids restrict the mechanical services as does the relatively shallow overall depth. The continuous 'beams' allow only linear mechanical service, although they provide a good 'cover' for mechanical equipment."[78]

During the same year of 1962, also at the Massachusetts Institute of Technology, the master's thesis of Roberto Einaudi, who worked in Kahn's office from 1957 to 1959,[79] consists in the application to a vaulted roof type in the manner of Nervi of the same principles of prefabrication of concrete outlined by Burns Jr., Kupritz and Rudquist – Kahn intervenes, offering advice on the project.

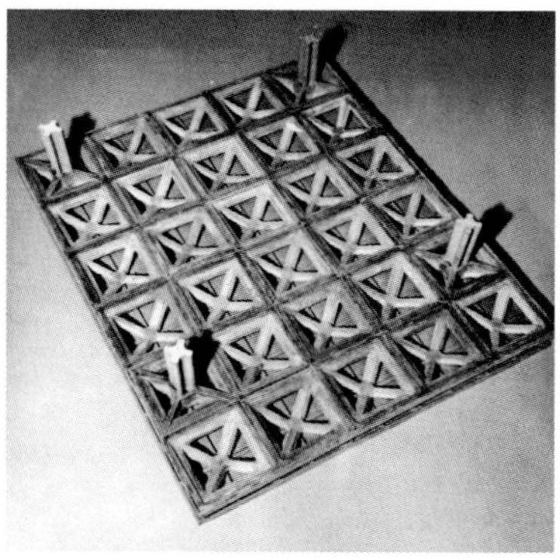

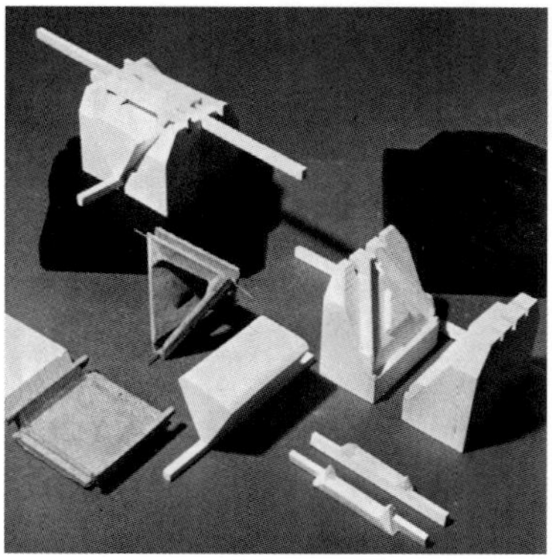

274. H. Caminos students, formwork for the precast tetrahedron unit and precast floor slab, School of Design, North Caroline State College, 1961 (J. Borrego, *Space Grid Structures, Skeletal Frameworks and Stressed-Skin Systems*, Cambridge-Massachusetts, London, The MIT Press, 1968, p. 61).

275. H. Caminos students, two-way concrete space frame as a floor system with the columns positioned, School of Design, North Caroline State College, 1961 (J. Borrego, *Space Grid Structures, Skeletal Frameworks and Stressed-Skin Systems*, Cambridge-Massachusetts, London, The MIT Press, 1968, p. 62).

276. H. Caminos students, formwork and underside of precast pentahedron unit of a concrete space frame, School of Design, North Caroline State College, 1961 (J. Borrego, *Space Grid Structures, Skeletal Frameworks and Stressed-Skin Systems*, Cambridge-Massachusetts, London, The MIT Press, 1968, p. 63).

[72] John Borrego, *Space Grid Structures, Skeletal Frameworks and Stressed-Skin Systems*, Cambridge-Massachusetts, London, The MIT Press, 1968, pp. 57-58.
[73] *Ibid.*
[74] *Ibid.*, p. 64.
[75] Burns Jr., Kupritz, Rudquist, *op. cit.* See John E. Rudquist, letter to Louis I. Kahn, 2 April 1962, LIK, 030.II.A.25.1; Louise M. Badgley (Kahn's secretary), letter to John E. Rudquist, 5 June 1962, LIK, 030.II.A.25.1; John E. Rudquist, letter to Louis I. Kahn, 12 June 1962, LIK, 030.II.A.25.1.
[76] Burns Jr., Kupritz, Rudquist, *op. cit.*, p. 55.
[77] *Ibid.*, p. 82.
[78] *Ibid.*, p. 38.
[79] Roberto Einaudi, *L'insegnamento di Pier Luigi Nervi: una testimonianza*, in, Annalisa Trentin, Tomaso Trombetti, ed., *La lezione di Pier Luigi Nervi*, Milano, B. Mondadori, 2010, pp. 45-61.

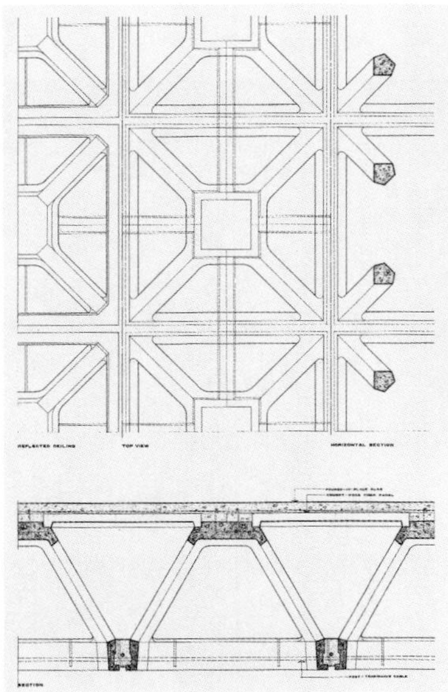

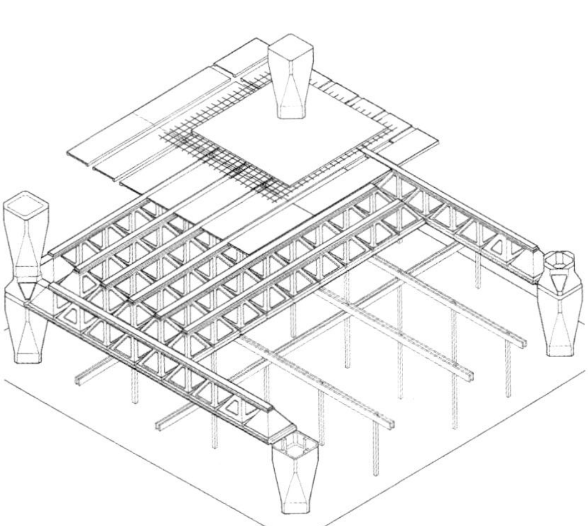

277-278. H. Caminos students, concrete space frame, School of Design, North Caroline State College, 1961 (J. Borrego, *Space Grid Structures, Skeletal Frameworks and Stressed-Skin Systems*, Cambridge-Massachusetts, London, The MIT Press, 1968, p. 64).

279-280. R. P. Burns, typical precast penthedron unit from a two-way concrete space frame for a precast floor system, school project, director E. Catalano, School of Architecture and Planning, MIT, 1962 (J. Borrego, *Space Grid Structures, Skeletal Frameworks and Stressed-Skin Systems*, Cambridge-Massachusetts, London, The MIT Press, 1968, pp. 70-71).

281. P. A. Kupritz, isometric view of the construction sequence for the two-way concrete space frame, school project, director E. Catalano, School of Architecture and Planning, MIT, 1962 (J. Borrego, *Space Grid Structures, Skeletal Frameworks and Stressed-Skin Systems*, Cambridge-Massachusetts, London, The MIT Press, 1968, p. 74).

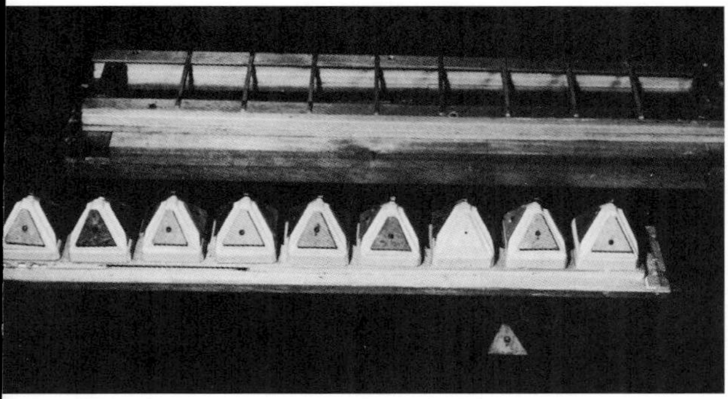

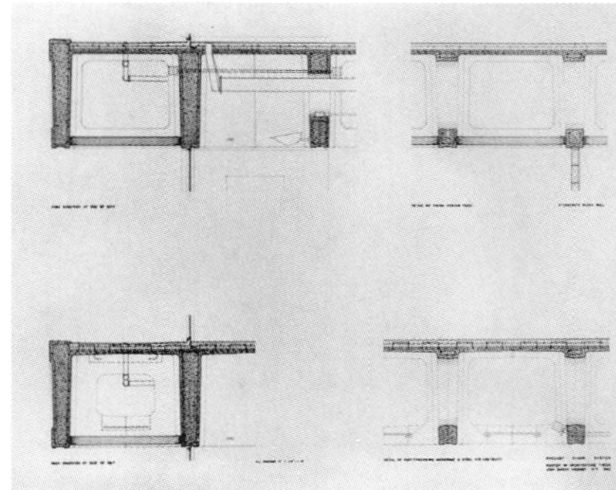

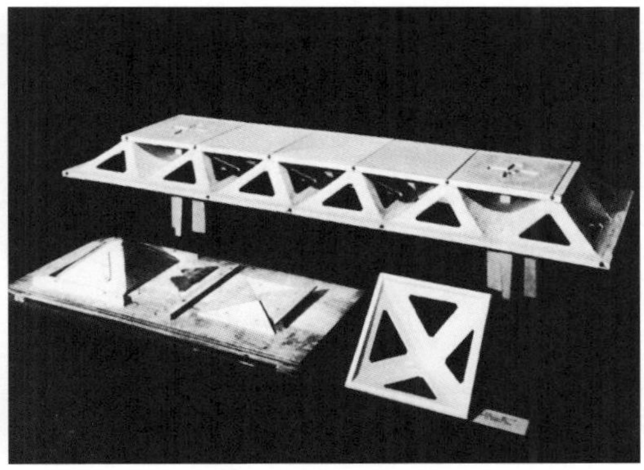

282. P. A. Kupritz, formwork and typical precast linear beam, school project, director E. Catalano, School of Architecture and Planning, MIT, 1962 (J. Borrego, *Space Grid Structures, Skeletal Frameworks and Stressed-Skin Systems*, Cambridge-Massachusetts, London, The MIT Press, 1968, p. 75).

283. J. E. Rudquist, school project, precast concrete floor structure, School of Architecture and Planning, M.I.T., 1962 (R. P. Burns Jr., P. A. Kupritz, J. E. Rudquist, "Precast Floor System. Orthogonal, Open-web Concrete Structures," "submitted in partial fulfillment of the requirements for the degree of Master of Architecture in the Department of Architecture on July 25, 1962").

284-285. H. Caminos students, two-way concrete space grid and formwork for the precast pentahedron unit, School of Architecture and Planning, MIT, 1964 (J. Borrego, *Space Grid Structures, Skeletal Frameworks and Stressed-Skin Systems*, Cambridge-Massachusetts, London, The MIT Press, 1968, p. 69).

286. R. Einaudi, model of a dome in prefabricated reinforced concrete, degree thesis, School of Architecture and Planning, M.I.T., 1962 (A. Trentin, T. Trombetti, eds., *La lezione di Pier Luigi Nervi*, Milano, B. Mondadori, 2010).

chapter six
first unitarian church and school, or the end of *béton brut*

The choice of materials

When Kahn's research on concrete is branching out into different forms of expression and different techniques, the last example of his concrete in the manner of *béton brut* is built in the symbolic heart of the First Unitarian Church and School, in Rochester, New York. It is in this work that the assembly of the boards of the formwork to mark the surfaces, which made its first appearance in the Yale Art Gallery and continued in the slabs of the AFL Medical Services Building, to then be deployed with sculptural intentions in the canopy of the Tribune Review Building, is taken to its highest level of perfection and metaphorical impact, like a harbinger of other results achieved by Kahn, later, with concrete. It is precisely during the construction of the First Unitarian Church and School that certain documents reveal other reasons for Kahn's preference for formwork made of wooden boards.

After the decision made at the start of 1959 to build a new church with a connected school in Rochester, on 19 March the organizing committee chaired by David Williams sums up the program, as well as the list of materials of the future construction, based on the results of a questionnaire distributed to the congregation: "Materials considered undesirable, in the following order from most to least were: plain cinder block, plain concrete, painted cinder block and painted concrete."[1] Kahn would have to come to terms with the aversion of the Unitarian community of Rochester to cinder blocks and exposed concrete.

[1] "Report of the Committee on Church Architecture," 1959, LIK, 030.II.A.15.3.

From the very first sketches, Kahn individuates the configuration of the roof of the main hall, or auditorium, with large beams arrayed along the diagonals of an octagon, as the structure capable of granting symbolic value to the whole building, or centrality with a powerful humanistic and even Renaissance character, representing the aspirations of the Unitarian congregation of Rochester.

The design process is marked by Kahn's visit in September to the chapel of Ronchamp, programmed precisely in relation to the construction of the church in Rochester, during his travels in Europe to reach Otterlo. His observation of the chapel by Le Corbusier in the landscape, with its roof in the form of a curved velarium in *béton brut* detached from the walls, the three higher rounded caps cut to capture indirect light, and the opening and view shaped by the thick wall, has a profound impact on the design of the First Unitarian Church.

In the first project, presented to the committee on 13 December, Kahn proposes a sequence of "spaces around" contained in a square plan, conceived not to express of a formation of Renaissance origin, but to make an abstract ideogram for a creative process capable of encompassing the widest range of communitarian wishes – the square as a "non-choice," in keeping with Kahn's definition.[2] The reinforced concrete that constitutes the main structure of the whole organism is still in the form of a skeleton with infills, as in the project of the same period for the Tribune Review Building.

In the project prepared at the beginning of 1960, to respond to the criticisms expressed by the committee regarding the lack of flexibility of the plan and its "rigid geometry,"[3] Kahn eliminates the square perimeter and the ambulatory around the auditorium, with a freer arrangement of halls, rooms and offices around the symbolic core. This crucial symbolic and functional choice generates the fragmentation of the structure and the diversification of its forms and materials seen in the final work. The reinforced concrete is now the material of the structure of the roof, broken down into units shaped like crosses, and always detached from the base to allow light to pass.

The rethinking of the roofing system, in May and June, also to respond to other doubts expressed by the committee, leads to the invention, in the summer of 1960 and at the start of 1961, of a symbolic folded structure in reinforced concrete with inclined planes arranged in an overturned cross "vault," with four towers at the corners, each with two sides open for

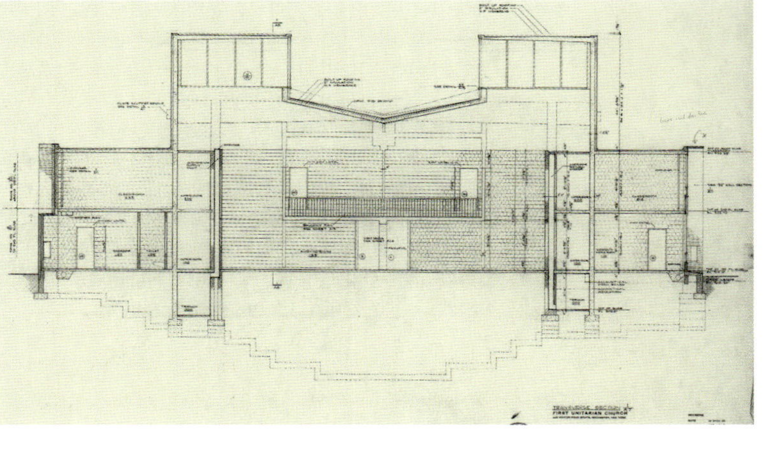

287-288. First Unitarian Church and School, Rochester, New York, 1959-69.

287. Model (LIK, 030.IV.A.525.4.14).

288. "Transverse Section," 10 May 1961 (LIK, 030.I.C.525.1).

zenithal lighting screened by blind walls. The inclined planes rest on the wall, again in reinforced concrete, that separates the corridor from the classrooms and other spaces, and on the reinforced concrete system of four pillars connected by powerful cruciform beams, visible from the auditorium. The wall of the auditorium becomes thick, like those of the fortresses admired by Kahn. The assembly of the concrete blocks generates functional cavities for ventilation conduits and openings for ducts. The conduits have air intake openings ("supply ducts" with "diffusers" at the top), and vertical openings functioning as "return slots" are opened towards the auditorium and along the corridor. The pillars are also set into that wall.

The concept of the "space around" winds up finding its symbolic representation in the materials themselves, as demonstrated by the fact that Kahn wraps the structural nucleus in reinforced concrete of the auditorium with rooms whose walls are made with concrete blocks, concealed in turn by a layer of bricks. Precisely these perimeter walls have a detailed profile that creates openings for seating and shapes the light and the view, like the thick wall of the chapel of Ronchamp.

Bricks, concrete blocks and rough concrete are arranged in a succession intended to translate that sequence of differentiated spatial layers Kahn has identified as the founding concept of a chapel. We cannot exclude the possibility that the choice of brick for the building in Rochester is based on the indications on materials that emerged from the questionnaire submitted to the congregation.

The last important modification made to the structure of the auditorium, after the publication of the project in April 1961 in "Progressive Architecture," simply confirms the influence of the chapel of Ronchamp, of its ruled surface of the roof that guides Kahn in the definition of the final configuration: the inclined planes are without ribbing, to let the light flow, and are sustained by a system of four tripartite supports, all incorporated in the concrete block walls – though the openings offer a glimpse of their trunks that rise above the wall.

Metaphorical patterns for formwork and blocks

In the spring and summer of 1961, Kahn determines the nature of the materials of the First Unitarian Church and School, develops the structures with the consulting of Keast & Hood and Komendant, prepares the specifications and contacts the construction companies in Rochester. The general works are assigned to Robert F. Hyland & Sons; Domine Builders Supply Corp., specializing in the fabrication of concrete parts, provides the concrete blocks; Weckesser Brick Company supplies Kahn with technical know-how to make the desired type of cladding brick, in keeping with the "color and texture of brick you have in mind to use for this new Church."[4]

In May Kahn indicates the characteristics of the system of formwork for the reinforced concrete of the First Unitarian Church. The definitions of "exposed" and "unexposed concrete" have to do, as usual, with the hierarchy of the spaces featuring concrete elements. Wooden boards are still used to design the surfaces of the main spaces, those in exposed concrete; instead, plywood panels are used for the ceilings of the slabs and the walls of the secondary spaces below ground level (there is also indication of the possi-

[2] Cit. in, Goldhagen, *op. cit.*, p. 152.
[3] *Ibid.*, p. 150.
[4] Weckesser Brick Company, letter to Louis I. Kahn, 11 April 1961, LIK, 030.II.A.15.20.

bility of using steel panels or "rough boards"[5] or "undressed board forms").

For the wooden boards or "board lumber" to make the formwork for the exposed concrete of the main walls, or "vertical wall surface (of exposed concrete)," Kahn indicates the usual type of tongue and groove floorboards. When he describes the measurements of the boards, he makes sure to indicate their maximum width: "Contact face of board shall be not more that 4 inches (wide)."[6] The indication of this measurement demonstrates the desire to obtain a given texture of lines. The reasons behind it depend on questions related to the perception of the *béton brut* surface, as shown by a phrase contained in the specifications: "The imprint of these boards is used to conceal imperfections in the surface of the concrete." So after the statements regarding the holes of the cones of the towers of Mill Creek, and those about the boards of the Richards Laboratories, we find confirmation of the fact that in Kahn's conception the assembly of the formwork is not a simple technical detail, but also a question of creating an illusion of lines woven on a surface, as discussed in the field of art by theorists of visual perception. The imprints on the exposed concrete surfaces serve to make flaws in the construction less perceptible, as if to reflect, in the process of fabrication of the concrete, through the imprints of the boards of the formwork, the optical questions of the studies of lines addressed by Albers in the Basic Drawing Course at Yale University.

While for Le Corbusier the assembly of the boards generates a pattern with decorative potential, with panels in alternating sequences of vertical and horizontal boards to add character to a single surface, for Kahn that same assembly is purged of its ornamental play to emerge as an illusionist texture that does not break the surface down into compartments. The symbolic as well as illusionist value Kahn assigns to the lines is revealed by the

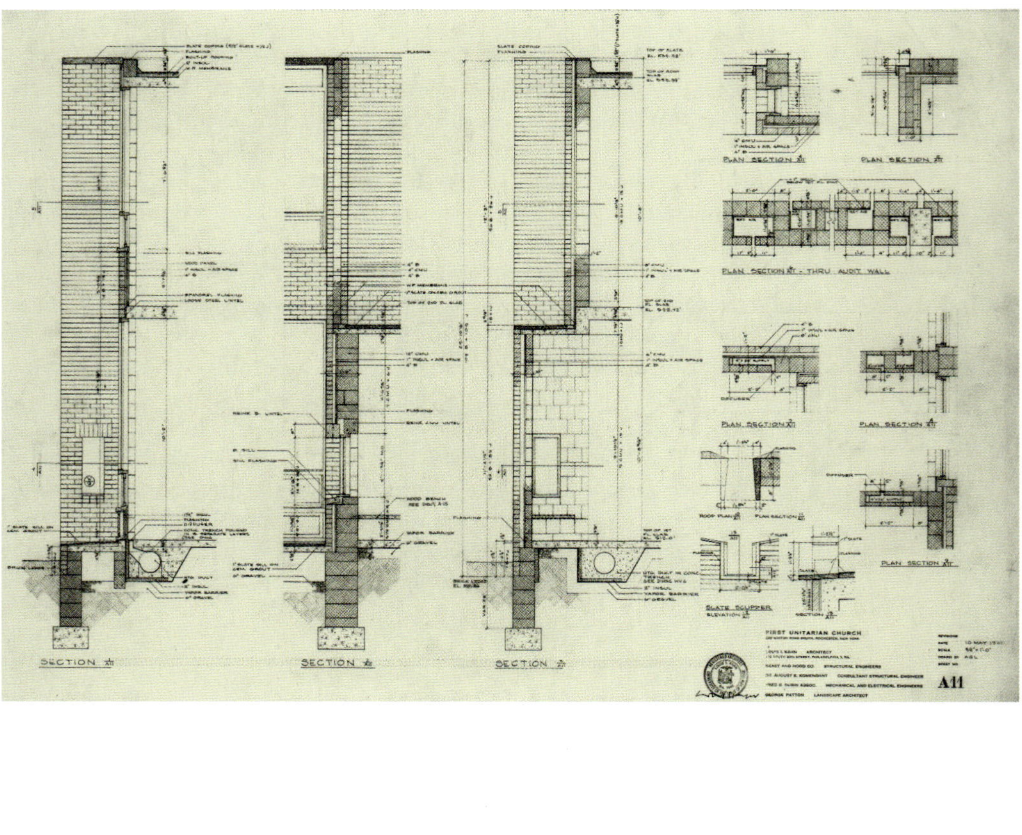

289-291. First Unitarian Church and School, Rochester, New York, 1959-69.

289. Sections and plans (LIK, 030.I.C.525.1)

290. Construction site photograph (LIK, 030.IV.A.525.3.25).

291. Detail of the concrete surfaces of the basement (photograph by the author, 2010).

fact that he alternates vertical and horizontal assembly of the boards to bring out the meaning of the walls. Therefore, in the specifications he indicates that the boards should be assembled vertically for the walls of the auditorium, or in a "vertical pattern," and horizontally for the walls of the staircases, to give them different characters. In this perspective, the "vertical pattern" cannot help but appear as a way of indicating a metaphorical orientation of the gaze of the congregation that runs over the surfaces of the *béton brut*, guided precisely by the imprints of the boards towards the light arriving from above, as in the case of the gaze that follows the ascending ribs of the cluster pillars in a Gothic cathedral, up to where they meet the ribbing of the vaults and the light.

The panels of boards are held in position with spreaders of the Cone Fast Coil Ties type, with "threaded Coil Cones." Unlike what was done in the Yale Art Gallery, Kahn forbids the use of chamfers, relying on the solidity of the formwork system for the perfection of the corners ("Forms for outside corners shall be reasonably tight to prevent forming of rough fins").[7] The plywood panels for "all exposed concrete," like those of the ceilings, must be "full size sheets 5/8" or 3/4"," "plastic-surfaced plywood" of the G.P.X. type produced by the Georgia-Pacific Plywood Company, one of the leading American suppliers of panels for formwork.

The treatment of the surfaces of the formwork varies depending on the exposure of the concrete in main or secondary spaces: for the exposed concrete sectors the surfaces have to be covered with plastic film of the Formfilm type of the A.C. Horn Company; in other cases they are treated with special formwork oil.

The composition of the conglomerate also varies, depending on whether the concrete is exposed or unexposed. Above the basement level

[5] "Specifications," 12 May 1961, LIK, 030.II.A.14.11.
[6] "Specifications," 12 May 1961, LIK, 030.II.A.14.11.
[7] "Specifications," 12 May 1961, LIK, 030.II.A.14.11.

292-293. First Unitarian Church and School, Rochester, New York, 1959-69.
Construction site photographs (LIK, 030.V.A.23.12; LIK, 030.IV.A.525.3.64).

Kahn wants a concrete that is "as light in color as possible," obtained with Portland Cement and additives like Plastiment or Pozzolith. Each layer of pouring has to have a height of 12 inches and the operation must be accompanied by "hand-spading, rodding and tamping" (vibration of the formwork is expressly forbidden). "Care shall be exercised to eliminate honeycombing of exposed concrete surfaces," we read in the specifications, "but excessive vibration of exposed concrete work shall be avoided and more care shall be taken with hand-spading and rodding."[8]

The "construction joints" for columns and walls have to be located at the meeting point with beams or top slabs, while the treatments to apply between one phase of pouring and the next should avoid defects in the exposed surfaces: "When placing new concrete at a construction joint, the older concrete shall first be thoroughly cleaned and then covered with a slush coat of 1:1 Portland Cement and sand grout immediately before depositing the new concrete."[9]

Demolition is ordered for all the concrete parts that do not comply with the specifications, displaying "poor workmanship or carelessness, or containing foreign substances or materials."[10]

For the floorings of the auditorium, corridor or lobby, Kahn calls for a surface in sheets of "grained Vermont" stone (18 x 18 inches), though in the construction they are replaced by concrete. Nevertheless, even in the phase documented by the specifications the concrete of the First Unitarian

Church is not only that of the blocks, walls and slabs. The gray compound also pervades the surfaces of certain floorings. To avoid any signs of manual working whose effects cannot be controlled on the worksite, Kahn specifies the final way of working the floorings: the poured concrete has to be smoothed "to produce a dense, smooth surface free from trowel marks." To increase the strength of these surfaces, a coat of Lapidolith or similar products is to be applied. So while all the surfaces of the walls and slabs are textured by lines in different patterns, those of the floorings are free of any joints, forming a neutral backdrop consistent with the constructive process, because that concrete is the only one not generated by molds, but by the repeated movement of the trowel.

The masonry of the perimeter walls is composed of clay bricks that form the external layer and "brick and/or concrete masonry units" (there are reinforcements).

For the concrete block walls in the main parts, an element with a square section is chosen (8 x 8 inches); instead, below the first floor the blocks are the usual 8 x 16 inches or 18 inches size. "All concrete masonry units used in visible or exposed locations shall be uniform in color, of a light shade and selected for fine uniform size of aggregate and must have sharp unspalled arises."[11] The concrete of the prefabricated architraves to insert in the block walls is also used to create chromatic tunings: "Face of lintels to match face of concrete masonry unit. Precast lintels shall be of same materials as the concrete masonry units."[12]

Kahn's pursuit of truth in exposed materials extends to the indication of three types of mortar of different strength and different colors, almost as if the joints were like painted lines to weave into the surface of the walls.

The mortar of the joints of bricks and concrete blocks has to be shaped with the trowel to underline the horizontal pattern of the masonry (the indication is changed for the laying of the concrete blocks, adding the same kind of work to the vertical joints as well). The specifications also indicate how to treat the joints between the sectors in exposed reinforced concrete and the block masonry: "the vertical joint between masonry and concrete shall be raked out to a depth of 1", and finished smoothly."[13]

With concrete poured in formwork made of wooden boards, in keeping with the aesthetic of *béton brut*, Kahn shapes the structure of the most important part of the First Unitarian Church, that of the auditorium, with the roof of inclined planes arranged in a cross and four colossal skylights.

[8] "Specifications," 12 May 1961, LIK, 030.II.A.14.11.
[9] *Ibid.*
[10] *Ibid.*
[11] *Ibid.*
[12] *Ibid.*
[13] The drawings of the "First Floor Plan" and "Second Floor Plan" dated 10 May 1961 contain indications of all the materials and their finishes (LIK, 030.IV.C.525.001). Concrete is present in a number of forms: "Concrete Masonry Unit" (CMU); "Concrete with integral steel trowel finish" (C1); "Concrete with architectural finish from pattern of small boards in forms" (C2); "Concrete with smooth finish from plywood forms with approved joint pattern" (C3). The repertory is completed by "T&G Wood Boards" (W), "Ceramic Mosaic Tile" (CT), "Asphalt Tile" (AT), "Acoustical Tile" (AC), "Slate" (S). Kahn assigns concrete blocks the role of determining the character of the main classrooms. The table makes it clear that C3 is only for the ceilings, with the significant exception of the auditorium, where C2 is used for the ceiling. C1 is only for floors. C2 appears as fragments in different rooms and corridors. While brick is not mentioned, in the drawing conventions four types are indicated, for "hollow core CMU," "solid core CMU," "brick" and "concrete." All the masonry is covered with a layer of bricks.

Worksite chronicles: experiments in formwork

The ceremony for the laying of the first stone takes place on 26 June 1961. William Lyman Porter and Des A. Putterman of Kahn's office, and Stanwood Hyde, Project Inspector representing the committee of the First Unitarian Church, are appointed to supervise the construction. Starting in July, work is done on the foundations, beginning from the South Wall, with excavation, assembly of the reinforcements and formwork, and pouring of concrete.

During a worksite meeting on 19 July the construction sequence of the concrete structure of the auditorium is discussed, to decide whether to first erect the walls and slabs in reinforced concrete and then lay the concrete blocks of the walls, and to define the phases of pouring and deployment of the formwork of the walls. "Concrete – formwork and pouring: Discussed placement of narrow board forms in the pouring of the Auditorium wall. Two pours may be necessary from the 2nd floor ceiling up. Below the

294-296. First Unitarian Church and School, Rochester, New York, 1959-69. Construction site photographs (LIK, 030.V.A.525.3.75; LIK, 030.V.A.23.20; LIK, 030.IV.A.525.3.57).

boards will run full height from floor to ceiling. Board at joint were conical ties cocur [sic] will be replaced when forms are moved. Contractor may purchase prelacquered plywood form boards for location where narrow board forms are not required."[14]

Kahn selects the materials to make the "experiments in formwork" and the "sample wall panels" with which to check on the assembly of bricks, blocks and formwork, the pouring, the connections between the parts, as well as the safety of the scaffolding.[15]

To check on the quality of the clay bricks and concrete blocks, Kahn has samples sent from the companies of Rochester.[16] In this phase he considers the possibility of using a brick cladding color among those produced by O.W. Ketcham Incorporated of Philadelphia, of the tone of the bricks laid in the Richards Laboratories, and a darker color.[17] With bricks in these two shades of color two trial masonry sectors are made, laid with vertical joints that are thinner than the horizontal joints, and with two types of working of the horizontal joints, one with the trowel as in the specifications, and the other with a specific grapevine jointer.[18] Kahn examines the "color and texture" of the concrete blocks, and recommends elimination of chipped or discolored blocks, saying "beware of the line caused by splashing mortar at each level where the scaffold is set."[19] Then a trial wall is prepared in only blocks "with both faces exposed."[20] Known as "block manholes," the hollow blocks laid in a course form a series of small bottomless coffers to fill with concrete. The stacking of the courses can also be seen as a system of permanent formwork. Where switches and sockets are planned, metal tubes are passed through the cavities. This is another form of the "hollow stones" Kahn wrote about regarding the tetrahedra.

During a worksite inspection on 8 September, Kahn is able to examine the various trial masonry sectors, those with the formwork for the auditorium and those in bricks and blocks. In the case of the walls of the auditorium, to avoid the formation of marks during pouring, the possibility is discussed of inserting "rubber gaskets" in the formwork. To obtain the proper "quality of concrete" the compound is made without additives (Pozzolith, Plastiment or other).[21] Pozzolith is added only later. Between the two types of brick of different colors, Kahn chooses the darker one, probably because

[14] William Lyman Porter, "Job Meeting," 19 July 1961, LIK, 030.II.A.16.16. After assembly of the first plywood formwork of the base sectors, Kahn's office recommends greater care in the "alignment of forms" (Stanwood T. Hyde, "Daily Report," 31 July 1961, LIK, 030.II.A.15.23).
[15] Stanwood T. Hyde, "Daily Report," 31 July 1961, cit.; see also William Lyman Porter, letter to Robert F. Hyland & Sons, Inc., 23 August 1961, LIK, 030.II.A.15.31.
[16] William Lyman Porter, letter to Stanwood T. Hyde, 31 July 1961 LIK, 030.II.A.15.37.
[17] Jacob Mandel, letter to Harold Pundt, 21 August 1961, LIK, 030.II.A.15.31.
[18] William Lyman Porter, letter to Robert F. Hyland & Sons Inc., 23 August 1961, cit.
[19] *Ibid.*
[20] *Ibid.* For the trials, the mortar compound is different from the one indicated in the specifications (dark Nazareth cement with lampblack is used), and is the same for the blocks and bricks.
[21] Stanwood T. Hyde, "Daily Report," 16 August 1961, LIK, 030.II.A.15.23.

297-299. First Unitarian Church and School, Rochester, New York, 1959-69.

297. Construction site photograph (LIK, 030.V.A.23.15).

the tone tends to establish a better dialogue with the shadings of the various types of concrete. For the joints, he prefers the effect obtained with the grapevine jointer, but advises using a smaller tip to get thinner grooves. Finally, he decides to make the color of the mortar of the joints a bit less pale.[22]

The concrete walls of the underground spaces are made with an ordinary system of plywood panel formwork, held in position by metal brackets with washers, left in place after completion of the work.

The desire to achieve a continuous texture of lines and tones prompts Kahn to modify certain details, to obtain the graphic effect developed in the drawing of the elevations. After having examined the wall samples and having chosen the color shade and the design of the joints, he decides to replace the slate sheets specified for the roof of the protruding brick masonry sectors with sheets of terracotta covered with a layer of concrete. All the textures, from those of the formwork imprints to those of the various types of masonry, have to create a homogenous surface pattern. The contents of a document indicate the reasons for the elimination of the slate sheets: "The homogeneity of the wall surface and lack of distraction for the eye will allow the wall to read clearly in all its dimensions. By using terra cotta in place of slate, the character of the building will be more in keeping with the character of the renderings in the April Progressive Architecture."[23]

In the light of this statement, the question becomes clear of the perception of the texture of a surface that forms the basis of Kahn's choices for the architraves in the same shade as the contiguous blocks, and even of construction details like that of the architraves reduced to slender bands in the block walls that wrap the auditorium. This is the same question Kahn approaches in the design of the concrete surfaces.

In January 1962 the positive effects of the use of "rounded plastic chamfer strips" on the pouring are noted, thanks to which the beams appear massive: "No weeping at joint, no marred surface caused by pulling form. Texture is uniform around two surfaces. The chamfer leaves two hardly distinguish-

[22] "Report of Inspection," 8 September 1961, LIK, 030.II.A.16.6.

[23] Porter, letter to Sergeant W. Wise, 3 October 1961, LIK, 030.II.A.15.34.

298-299. Construction site photographs (LIK, 030.V.A.26.7; 8).

300-303. First Unitarian Church and School, Rochester, New York, 1959-69.

300-301. Construction site photographs (LIK, 030.V.A.26.4; LIK, 030.IV.A.525.3.1).

302. View (photograph by the author, 2010).

303. View of a classroom (photograph by the author, 2010).

able lines where material laps to the form, and this accentuates the corner. The beams in appearance are quite massive, and the corners or chamfers do not seem to detract from the delineation of the beam."[24]

The walls in exposed concrete of the auditorium are marked by the imprints of vertical boards, 2,5 inches wide, and by the holes of the ties positioned at intervals of every ten boards (the hole has a diameter of 1,25 inches and occurs along the joint between two contiguous boards; the metal tie remains visible). The positioning of the rare vertical joints between the boards is studied to be at the same height, but spread out in the pattern of imprints to become imperceptible, without creating a pattern of its own. This confirms the fact that the laws of artistic perception are transferred to

the assembly of the formwork, and that with the boards Kahn wants to design only vertical imprints. Unlike what is indicated in the specifications, the boards have a surface that has not been smoothed, leaving clear imprints of the grain on the concrete.

A deep horizontal groove separates the assembly of the formwork and the pouring of the concrete in the walls of the skylights. The groove is placed to coincide with a significant point of the composition of the structure: the contact of the inclined planes of the roof with the walls. That groove also serves to make the imperfection of the assembly of the formwork in the two overlaid sectors less visible – the imprints of the boards and holes in the two sectors are not lined up. The assembly of the formwork in the beams is also studied to obtain a design. In the cantilevered beam under the slab of the gallery for the organ, along the intrados, at the center, two strips of plywood are joined with cut corners, leaving a V-shaped line in relief on the pouring. This type of joint, still destined in Rochester to be just a marginal detail in the making of the formwork, would become – at the start of the 1960s – one of the distinctive signs of Kahn's concrete.

Kahn sets out to create an order of pillars in *béton brut* with the design of the formwork, playing with the width of the boards depending on whether the side of the pillars in question faces towards the auditorium or towards the corridor. But the design is not always rigorous, and variations occur in the assembly of the pieces of the formwork. In general, the boards or strips of plywood on the visible sides are assembled to obtain symmetrical imprints and a design diversified according to the position of the trunk of the pillar with respect to the wall of blocks – clamped, in a niche, or free. In the four median pillars Kahn studies a connection with the beams and inclined planes so as to create a sculptural piece marked by an opening corresponding to the groove between the inclined planes, and with the usual imprints of the construction process, almost like a capital in the structural system of *béton brut*.

The scaffolding and formwork of the roof of the auditorium are developed with Komendant and Keast & Hood in March 1962.[25] When the time comes to assemble the formwork of the inclined planes of the roof, there is discussion of the type of artificial lighting to install in the auditorium. Again in this case, the issue is the perception of the *béton brut* surfaces. In the initial phases of the project Kahn had envisioned spotlights along the gallery,

[24] Stanwood T. Hyde, "Daily Report," 3 January 1962, LIK, 030.II.A.15.23. "Cleaning and painting 10" wall forms," is written in another report (Stanwood T. Hyde, "Daily Report," 11 January 1962, LIK, 030.II.A.15.23).

[25] Stanwood T. Hyde, "Daily Report," 2 February 1962, LIK, 030.II.A.15.23.

304-310. First Unitarian Church and School, Rochester, New York, 1959-69.

304. View of the auditorium (photograph by the author, 2010).

305-306. View and detail of the auditorium roof (photographs by the author, 2010).

307. Detail of the structure supporting the roof (photograph by the author, 2010).

308-309. View and concrete wall detail of the skylight in the auditorium (LIK, 030.IV.A.525.1.36; photograph by the author, 2010).

310. Detail of the concrete blocks (photograph by the author, 2010).

311-315. First Unitarian Church and School, Rochester, New York, 1959-69.

311-312. View and detail of the corridor leaning the church (photographs by the author, 2010).

313. Detail of the concrete beam with V-joint, in the gallery for the organ (photograph by the author, 2010).

314. Service spaces (photograph by the author, 2010).

315. Lintel of the door of a service space (photograph by the author, 2010).

beyond the concrete block walls; then he had designed a series of cavities in the concrete inclined planes of the roof in which to install the light ducts. The decision to be made in June has to do with the replacement of the light ducts in the cavities with other hanging fixtures. What Kahn, in any case, wants to avoid is going back to the original solution, because with spotlights pointed towards the inclined planes "the light might have a ghostly effect, and would show up imperfections in the concrete unnecessarily."[26]

The boards of the formwork of the inclined planes of the cruciform roof of the auditorium are 2,5 inches wide, like those of the formwork of the walls, but they are planed so as not to leave visible grain imprints. The boards are assembled to form a perfect geometric fit at the meeting point at the center of the cross. Strips placed in the formwork create grooves separating the inclined planes. The strips are inserted next to the boards, so as not to create imprints of bands of different width. The artificial lighting is composed of prismatic lamps that emerge from holes left in the inclined planes, screened by metal sides to project the light downward, without directly lighting the surfaces in *béton brut*.

The concrete block envelope is inserted in the structural system in *béton brut* of pillars, walls and inclined planes. For the masonry, Kahn develops an arrangement to obtain a pattern of square faces (8 x 8 inches circa).

The concrete blocks made by Domine Builders Supply are not always of the quality desired by Kahn. When the blocks of the auditorium are laid, at the start of December, the company's representative makes a visit to the worksite "for examination of concrete block color variation."[27] Batches of blocks are not infrequently rejected, due to their "color and texture."[28] Halfway through May 1962 "all blocks walls except those for auditorium and corridor are completely erected."[29] The progress of the worksite, with

[26] Duncan Buell, letter to Maurice Van Horn, 26 June 1962, LIK, 030.II.A.15.3.
[27] Stanwood T. Hyde, "Daily Report," 5 December 1961, LIK, 030.II.A.15.23.
[28] Stanwood T. Hyde, "Daily Report," 31 May 1961, LIK, 030.II.A.15.7.
[29] Stanwood T. Hyde, "Daily Report," 16 May 1961, LIK, 030.II.A.15.7.
[30] "Different batches of block, delivered to the site at different times, created unwanted construction patterns in the walls […] zippered patterns caused by

the long interruption to permit installation of the physical plant systems inside the block walls of the auditorium, causes damages to the masonry.[30] Towards the end of November, Harold Pundt washes the "walls in the high towers, and this has not yet unified the color of the concrete."[31] However, a similar treatment has not yet been applied to the "interior masonry." The blocks of the auditorium remain with different shades of gray that create either irregular motifs or specific chromatically defined sectors.

In the manner of Le Corbusier, who accepts the *malfaçons* produced in the concrete, Kahn too defines the unexpected variations in the fabrication as "natural 'mysteries' of concrete": "Mr. Kahn hopes, of course, that people will become adjusted to the natural 'mysteries' of concrete. He did state however that if there is a persistent wish to unify the surface color, that a thin cement wash would be the most appropriate method."[32]

An important choice for the character of the main halls, including the auditorium, has to do with the elimination of the stone for the floorings, called for in the specifications. Now the range of grays of the auditorium is completed by the continuous concrete surface of the flooring, polished to a glow, and without seams. The chancel is identified by a large platform, raised by two steps; the pulpit is its appendix, at the same level. The material selected by Kahn for the chancel and the pulpit is wood (oak, in particular, for the pulpit). The fact that the boards have a width of 2,5 inches, like those of the formwork, confirms the attempt to match the various textures. The presence of large wood surfaces is seen as a means of improving the acoustics of an auditorium enclosed only by concrete, though the possibility of building even the platform and the pulpit in concrete is also considered.[33]

When the *béton brut* walls of the auditorium penetrate into the volume of the building, becoming part of the closures of the rooms and spaces

the wall's having been kept partially open for months until the mechanical ducts were installed, then closed up with non-matching blocks" (William S. Huff, *Louis Kahn: 'Sorted Recollections and Lapses in Familiarities.' Interview with Jason Aronoff*, in, "Little Journal," vol. V, September 1981, No. 1, now in, Latour, ed., *Louis I. Kahn: l'uomo, il maestro* cit., (pp. 405-31), p. 423).

[31] Duncan Buell, letter to Maurice Van Horn, 7 December 1962, LIK, 030.II.A.15.34.

[32] *Ibid.*

[33] Duncan Buell, letter to Maurice Van Horn, 16 August 1962, LIK, 030.II.A.15.34.

316. First Unitarian Church and School, Rochester, New York, 1959-69. "Auditorium Fabrics," 14 June 1962 (LIK, 030.I.C.525.1).

around the auditorium, Kahn introduces a different arrangement of groups of formwork panels, also assembled horizontally, never with the decorative concerns of Le Corbusier, but simply to indicate the different role of the wall sectors: the horizontal boards are for the wall sectors suspended over the spaces created for the doors or large openings (that of the kitchen).

Outside the auditorium, Kahn uses a precious wood facing for the more prestigious ceilings of the First Unitarian Church, especially in the Library, where he creates special grille panels with parallel slats paced by gaps, inserted between the powerful exposed concrete beams. The effect of wood and *béton brut* is part of the tradition of 20th-century exposed concrete architecture, represented by works known to Kahn, like the apartment by Perret on Rue Raynouard and the Musée des Travaux Publics in Paris.[34]

Hanging tapestries: the "idea of the spectrum"

The photographs taken as soon as the construction in Rochester was complete and the church was opened show the concrete blocks of the auditorium without any decorative treatment. But this is simply an intermediate phase of the work, because Kahn intends to create special fabric panels to hang on the wall sectors identified by openings, "hanging tapestries" as they are called in the documents. Again in this case, Kahn gets inspiration from the work of Le Corbusier, who is making use of tapestries to create chromatic contrasts with *béton brut* and to improve the acoustics of rooms. But while for Le Corbusier the tapestry is a grand, easily transported fresco, or *muralnomade*, for Kahn it is a system of colors that react to light, steeped in the logic of the wall on which the panels are hung, and calculated in its shadings to create a horizontal movement capable of reasserting, when seen from a distance, that continuity challenged by the openings for the ventilation and the pillars.

From the first studies, Kahn's panels are rectangular strips of fabric, with different heights and widths, to hang on the vertical wall sectors, taking the measurements generated by the crevices in the block walls into account, as well as the presence of the pairs of doors at the center of the two walls. The colors chosen for each panel are primary (blue, red, yellow), though shaded in the passage from one panel to the next. The form of this succession of colors varies in the different studies, using slender lines or thick bands. In general, all of Kahn's drawings for the panels are without patterns, aiming only at the effects of gradual changes of color. This is why the line of color becomes the main motif. This is also what Kahn would start to pursue for concrete: the absence of any pattern shaped by the formwork.

For the decoration of the panels Anni Albers, the wife of Joseph, called in by Kahn, proposes a pattern of fretted lines that is simply the transposition in fabric of the order of the square concrete blocks in the masonry sectors on which the panels are to be hung. With this choice, Anni Albers replicates Kahn's idea of decorating the masonry sector of the Bath House with a mural painted on the basis of the pattern of the concrete blocks. "Each square of my paper to be 4" so that the whole is related to your 8" x 8" cinderblocks," Anni Albers explains to Kahn (though she later bows out of the project).[35]

The documents bear witness to how Kahn's principles of vision, his perception of the value of lines and colors and their resonance in space, optical and illusionary effects, are based on knowledge of the works and teachings of Joseph Albers at Yale University. The same geometric patterns of lines or bands, the rhythm of the successions, the calculation of their

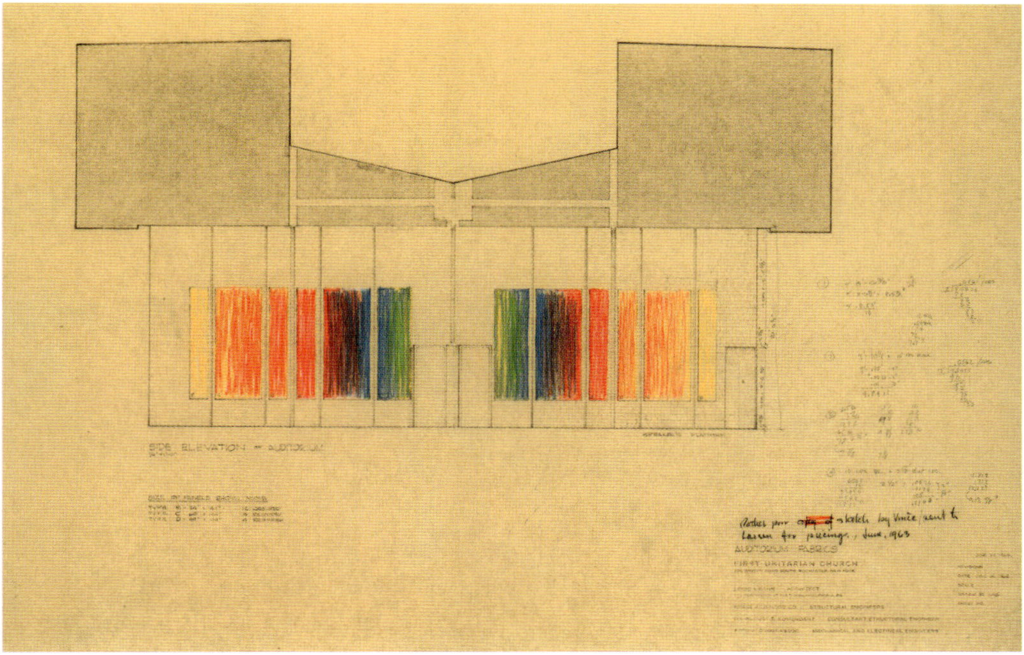

thicknesses, which Kahn and his staff develop in a series of drawings using colored pastels, coincide with the studies of Albers and the exercises he assigns to his students at Yale. It is against the backdrop of Albers' studies on vision that we can interpret certain expressions that circulate in Kahn's office regarding the fabric panels. To seek, as Kahn puts it, the "idea of the spectrum" means identifying that sequence of lines that is capable of competing with that "breathing line" theorized by Albers.[36] "The design lends itself to either high or low intensity and the space environment of the fabric must dictate the choice. If the initial idea of the spectrum is good, then we know what to do and how to do it is the next step."[37]

The correspondence between the Larsen Design Corporation of New York, hired to produce the panels, and Kahn's office regarding the choice of the weave and color of the fabric, shows that Kahn wants to avoid any weave that does not correspond the overall effect he is seeking, and that he is selecting the range of colors from yellow to red, blue to green, and the combinations of blue-green and red-gold, which he feels would be the color "right for the space,"[38] while the type of weave is "brocade" or "double cloth."[39] At the same time, he is considering the exact width of the various panels, also taking the weaving technique into account.[40] The fact that the

[34] For Kahn's knowledge of the works of Perret see Louis I. Kahn, letter to Marion Tournon-Branly, 12 November 1963, LIK, 030.II.A.55.37, and Marion Tournon-Branly, letter to Louis I. Kahn, 23 March 1962, LIK, 030.II.A.55.37.
[35] Anni Albers, letter to Louis I. Kahn with attached sketch, 27 June 1962, LIK, 030.II.A.15.44. See also Anni Albers, letter to Duncan Buell, 20 June 1962, LIK, 030.II.A.15.44.
[36] Horowitz, Danilowitz, *op. cit.*, p. 162.
[37] Duncan Buell, letter to Maurice Van Horn, 16 July 1963, LIK, 030.II.A.15.34.
[38] Duncan Buell, letter to Maurice Van Horn, 5 September 1962, LIK, 030.II.A.15.34.
[39] *Ibid.*
[40] Jack Lenor Larsen, letter to Duncan Buell, 30 August 1962, LIK, 030.II.A.15.34. See also Win Anderson, letter to Duncan Buell, 20 September 1962, LIK, 030.II.A.14.7.

auditorium is almost without echo prompts Kahn and his staff to focus on panels of thicker fabric (also for the wall behind the pulpit), reducing the distance of the panels from the walls and attaching them directly "against the concrete block."[41] Kahn's preference for a thicker fabric is dictated by the desire to eliminate any possible wrinkles that might create an unexpected motif on the panel surface. It is the same concern that guides the development of the formwork system, or the profile of the mortar joints between concrete blocks.[42] The fact that Kahn is looking for a gentler spectrum in the passage of colors is borne out by various phrases found in the correspondence: "On the Yellow to Red, I think this would also be better with the addition of Orange which would help the value change. Now the Red seems so much darker than the Yellow."[43]

The series of panels forms a sequence of colors coordinated according to an almost musical rhythm, in the manner of the *pan de verre ondulatoire*. In the design of the fabric, Kahn seeks an agreement of sizes and colors between the textile panels and the vertical wall sectors of the auditorium. The sizes of the panels are developed on the basis of those of the masonry sectors, so the panels will be proportional to the latter. The application is always done by placing them near the narrow vertical openings of the air intakes. While in the Bath House the mural took the pattern of the cinder blocks into account, here the fabric panels are shaped not in terms of the blocks, but in terms of the masonry sectors. The openings in the walls become the generating lines of the panels, imposing their measurement and pace, but at the same time the color shadings guide the eye in a horizontal movement, from one panel to the next, spanning the openings to restore the continuity of the wall.

317. First Unitarian Church and School, Rochester, New York, 1959-69. Detail of the tapestry hanging on the concrete blocks walls (photograph by the author, 2010).

The thickness and layout of the colored lines and the size of the yarn of the fabric are studied to achieve a particular effect in the perception of the wall surfaces. The eyes of the congregation should register only the chromatic variations, in a continuous succession of shadings against the homogenous, vibrant backdrop of the concrete blocks. In this way, the surfaces are colored in a spectrum of light, as if the space were flooded with light filtering through multicolored glazings. Thus the textile panels become petrified light in the moment of the intense coloring of the rays on the construction, like solidified concrete; they evoke the sight of the space of certain cathedrals. "Mr. Kahn stated he did not want the pattern to be apparent from across the room, but rather to become apparent as one approaches closer to fabric."[44]

If we consider the sequence of vertical textures that extends from the textile panels to the masonry sectors, the symbolic reasoning behind the vertical arrangement of the boards in the formwork of the skylight sectors becomes more comprehensible. The imprints of those boards simply amplify those patterns that express an ascending, mystical movement, or a movement of penetration of light.

The worksite of the First Unitarian Church, when the construction of the auditorium is coming to a close, receives a visit on 8 June 1962 from a representative of the contractor George A. Fuller, Allen Katz, responsible for the construction of the Salk Institute, "to inspect concrete work for Salk Project now under building in California." The first version of the concrete works for the retaining walls of the Salk Institute would be based on the walls of the auditorium in Rochester. But at this point, with the construction of the Salk Institute, we can see a definitively surpassing of the phase of coming to terms with the technical and formal principles of *béton brut*, principles that are still at work in the First Unitarian Church. The Salk Institute marks the beginning of a new chapter in Kahn's research on the expressive qualities of the process of the fabrication of concrete and on the nature of its compound, a chapter that will take him towards what will become the degree zero of his concrete.

[41] Duncan Buell, letter to Maurice Van Horn, 10 October 1962, LIK, 030.II.A.15.41.
[42] Win Anderson, letter to Duncan Buell, 2 July 1963, LIK, 030.II.A.14.9.
[43] Barbara Wallance, letter to Duncan Buell, 23 September 1963, LIK, 030.II.A.14.7.
[44] Duncan Buell, letter to Jack Lenor Larsen, 16 July 1962, LIK, 030.II.A.15.28.

Acknowledgements

The books *Exposed Concrete and Hollow Stones, 1949-1959*, by Roberto Gargiani, and *Towards the Zero Degree of Concrete, 1960-1974*, by Anna Rosellini, are the result of direct research into all of Kahn's works in concrete from the 1940s onwards and all of the main archival materials. Without the support, the availability, and the welcome of the Architectural Archives of the University of Pennsylvania in Philadelphia it wouldn't have been possible to write these books. We thank all the members of the archives in Philadelphia and in particular the collection manager William Whitaker who always supported each initiative that has contributed to the publication of these books, and the archivist Nancy Thorne who helped us consult the archival documents. Our thanks also goes to Nathaniel Kahn, Sue Ann Kahn, Samantha Sizemore of the Kimbell Museum, Kathleen Mylen-Coulombe of the Yale Art Gallery for having put at our disposition the material conserved in their respective archives, and to Patricia Cummings Loud, the Kimbell's curator of architecture, who accompanied Anna during the visit to the museum and facilitated the consultations of the documents conserved in the museum archives.

Our understanding of some important concrete construction processes in Kahn's work come from the fundamental discussions Anna had with Nicholas L. Gianopulos and Frederick Langford – to whom we are indebted for their contributions and willingness to help.

There are many people who have facilitated our visit to Kahn's architectures. We would like to thank in particular the personnel of the Swiss Embassy in Dhaka and the architect Nurur Rahman Khan, and Garry Van Gerpen and Bob Lizarraga who accompanied us during our visit to the Salk Institute and who have made some archival materials available.

The fabrication of the books, the translation of texts and the page layout were opportunities to collaborate with the Presses polytechniques et universitaires romandes (PPUR) of the Ecole polytechnique fédérale de Lausanne (EPFL). We thank in particular Frederick Fenter and Nicholas Tappin and our most sincere acknowledgement goes to the director Olivier Babel.

An affectionate thanks goes also to our friend Paola Ranzini Pallavicini for taking care of all graphic design of the book.

In particular our thanks goes to deans of the Environnement Naturel, Architectural et Construit faculty (ENAC) of EPFL, for the support and the impetus that on several occasions we have wished to honour.

Finally, we are grateful to the European Research Council for a grant that has supported our research on the topic of exposed concrete.

Name index

Figure Credits

The Architectural Archives, University of Pennsylvania, August Komendant Collection
Fig.#: 193, 194

The Architectural Archives, University of Pennsylvania, George Kreier Collection
Fig.#: 261, 263

The Architectural Archives, University of Pennsylvania, G. Robert LeRicolais Collection
Fig.#: 100

The Architectural Archives, University of Pennsylvania, Marshall Meyers Collection
Fig.#: 128, 134, chapter three frontispiece, 151, 186, 189, 191, 195

The Architectural Archives, University of Pennsylvania, Anne Griswold Tyng Collection
Fig.#: 41, 85, 102-105, 109, 121-124, 147

The Architectural Archives, University of Pennsylvania by the gift of Joseph Esherick
Fig.#: 250

The Architectural Archives, University of Pennsylvania by the gift of Richard Saul Wurman
Fig.#: 143

Louis I. Kahn Collection, University of Pennsylvania and the Pennsylvania Historical and Museum Commission. Housed in the Harvey and Irwin Kroiz Gallery, the resources of the Kahn Collection are used by the permission of the Architectural Archives, University of Pennsylvania (hereafter cited as Kahn Collection).
Fig.#: chapter one frontispiece, 4, 5, 7-12, 17-20, 23-27, 31, 36, 40, 45-46, 51-52, 54-60, 62-66, 70, 83, chapter two frontispiece, 84, 86, 88-93, 95, 97-98, 106-108, 110-120, 125-126, 129-133, 138, 140-142, 144-145, 148, 150, 153-170, 173-181, 188, 192, 196-200, 206, 212, 214-231, 240-242, 46-249, chapter five frontispiece, 262, 264-266, 268-269, 287-290, 292-301, 316

Kahn Collection, photo by Ronald Binks
Fig.#: 182, 183, 185

Kahn Collection, photo by Robert Damora
Fig.#: 146

Kahn Collection, photo by John Ebstel / © Keith DeLillis Gallery, New York
Fig.#: 28, 127, 135, 152, 267, 308

Kahn Collection, photo by Joseph Molitor
Fig.#: 209

Collection Sue Ann Kahn
Fig.#: 32, 34, 35, 47, 48, 136

Collection Harriet Pattison and Nathaniel Kahn
Fig.#: 33